SUSTAINABLE MINING PRACTICES

# Sustainable mining practices – a global perspective

Edited by

## Vasudevan Rajaram, Ph.D., J.D.
*Complete Environmental Solutions, Oak Brook, IL, USA*

## Subijoy Dutta, P.E.
*S&M Engineering, Crofton, MD, USA*

## Krishna Parameswaran, Ph.D., P.E.
*ASARCO LLC, Phoenix, AZ, USA*

A.A. BALKEMA PUBLISHERS   LEIDEN / LONDON / NEW YORK / PHILADELPHIA / SINGAPORE

Published by: A.A. Balkema Publishers Leiden, The Netherlands, a member of Taylor & Francis Group plc
www.balkema.nl, www.tandf.co.uk, www.crcpress.com

Library of Congress Cataloging-in-Publication Data

British Library Cataloguing in Publication Data

ISBN 90-5809-689-0

Printed in Great Britain

# Contents

This book is dedicated to:

Mrs. Usha Rajaram (Wife of Dr. Raj Rajaram)
Mrs. Santwana Dutta (late mother of Mr. Subijoy Dutta) and
Mrs. Krishnamma and Mr. Krishnaier Parameswaran
(late parents of Dr. Krishna Parameswaran)

# About the editors

Dr. Rajaram is a licensed professional engineer, a certified professional geologist, and an attorney with over 30 years of experience in mining and environmental fields. He obtained his B.E. in mining engineering from Osmania University, Hyderabad, in 1970. Subsequently, he obtained his Master of Science degree from South Dakota School of Mines, and Ph.D. from the University of Wisconsin, Madison. He has a Juris Doctor degree from Illinois Institute of Technology, Chicago, Illinois, U.S.

Dr. Rajaram is currently at Black & Veatch in Chicago, Illinois. He has completed several mining, environmental and tunnelling projects in the United States, Canada and India. He has been involved in the design of large open pit and underground mining projects in the western United States, and in the design and construction of waste management facilities in the United States, Canada and India.

Mr. Subijoy Dutta is a registered professional engineer (P.E.). He has recently authored a book, "Environmental Treatment Technologies for Hazardous and Medical Wastes – Remedial Scope and Efficacy", published by Tata McGraw Hill Company. Mr. Dutta has over 15 years of experience in Remedial Investigations, Feasibility Studies (RI/FS), Remedial Design, and Remedial Actions (RD/RA) pertaining to the RCRA and CERCLA regulations. He also provides expertise in the treatment, storage and disposal aspects of medical waste. He has designed modified cover systems for Landfills and has supervised installation of Landfill covers for several hazardous waste sites. He has prepared reclamation plans for several abandoned mine sites in Eastern Oklahoma. Most recently he is involved with the watershed management for a very fast growing County in Virginia.

Mr. Dutta has more than 30 publications in major technical journals. He also has two equipment patents pending. He has receiving the 1991 National Award for Individual Excellence in Environmental Restoration from the U.S. Air Force chief of Staff General Merrill A. McPeak in Washington D.C. He has also received the Unsung Hero Award in 1997 from the EPA Administrator, Carol Browner for his voluntary work towards the cleanup of the highly polluted Yamuna river in New Delhi, India.

Dr. Krishna Parameswaran received a Bachelor of Technology degree in metallurgical engineering from the Indian Institute of Technology, Bombay, in 1968, a Master of Science degree in metallurgical engineering from the University of Missouri-Rolla in 1971 and a Doctor of Philosophy in metallurgy from Pennsylvania State University in 1974. He is a licensed professional engineer in Pennsylvania.

Dr. Parameswaran joined ASARCO Incorporated (Asarco) at South Plainfield, New Jersey, in the Central Research Department as a senior research engineer in 1981. He has held various positions in the research and regulatory compliance areas and in 2001, was transferred to the Phoenix office as Director of State and Federal Regulatory Affairs. He is currently Director of Environmental Services – Operations and Compliance Assurance in Asarco's Environmental Affairs Department. His responsibilities include oversight of environmental operations at Asarco's operating facilities; coordinating environmental activities among the facilities; providing technical support to environmental programs within the operations; developing environmental policy, goals and objectives for the corporation and each individual operation and ensuring internal consistency within ASARCO on such issues. In addition he coordinates the company's Environmental Management System (EMS) and manages Asarco's compliance and system audit programs.

His interests include sustainable development as it applies to mining, recycling of metals, energy use in metals production and life cycle energy use and life cycle assessments of products. He presented a keynote lecture at the International Symposium on Environmental Management of Mining and Metallurgical Industries in Bhubaneshwar, India (August 1998) entitled, "The 21st Century: Challenges and Opportunities for the Mining Industry." He chaired a panel on sustainable development at an international environmental conference "Minerals, Metals and the Environment," sponsored by the Institution for Mining and Metallurgy in Manchester, U.K. (February 1992) and represented the company in a United Nations Workshop on Sustainable Development in New York (October 1998). He participated as an industry representative on the Sustainable Mineral Roundtable co-sponsored by the U.S. Forest Service and U.S. Geological Survey to develop indicators of sustainable development for minerals/materials and energy systems.

# Contributors

Bonnie Robinson
Geologist,
Yamuna Foundation for Blue Water Inc.,
Crofton, MD, USA

S.L. Chakravorty
Ex. Hony Secy. General, NISM,
Kolkata,
India

Jim Altham
Curtin University,
Western Australia,
Australia

Michael Dunbavan
Horizon Environmental P/L,
Sydney,
Australia

Frederick T. Graybeal
ASARCO LLC, (formerly ASARCO Inc.)
Chatham,
NJ, USA

Harold L. Owens, P.E.
U.S. Department of Labor,
Pittsburgh,
PA, USA

Ian Douglas
School of Geography,
University of Manchester,
Manchester, UK

K.K. Jain
Director, Ministry of Environment & Forests
New Delhi,
India

John W. Fredland, P.E
Mt. Hope, WV,
USA

Kelvin K. Wu, PhD, P.E.
Mine Safety and Health Administration,
Pittsburgh, PA,
USA

R. Knol
Former HSE Manager,
Geita Gold Mine,
Tanzania

Krishna Parameswaran, PhD, P.E.
Director
Environmental Services and Compliance Assurance
ASARCO LLC, (formerly ASARCO Inc.)
Phoenix,
AZ, USA

Nigel Lawson
School of Geography, University of Manchester
Manchester,
UK

Paul Worsey
University of Missouri-Rolla,
Rolla,
MO, USA

Raj Rajaram
President, Complete Environmental
Solutions Private Limited,
Oak Brook,
IL, USA

Ravi Bhargava
Ecomen Laboratories,
Lucknow,
India

D. Renner
Managing Director,
Iduapriem,
Gold Mine,
Ghana

Robert E. Melchers
Centre for Infrastructure Performance and Reliability,
The University of Newcastle,
Australia

N.C. Saxena
Center for Mine Environment,
Indian School of Mines,
Dhanbad,
India

M.M. Singh
Engineers International, Inc.,
Chandler,
Arizona,
USA

Stuart A. Bengson
ASARCO LLC, (formerly ASARCO Inc.)
Tucson,
AZ, USA

Subhash Bhagwat
Illinois Department of Natural Resources,
State Geological Survey Division,
Urbana/Champaign, IL, USA

Subijoy Dutta
S&M Engineering Services,
Crofton, MD, USA

Turlough Guerin
Shell Distribution Oceania,
Rosehill,
NSW, Australia

Dr. D. Watson-Jones
Clinical Lecturer,
London School of Hygiene & Tropical Medicine,
London,
England, UK

William Langer
U.S. Geological Survey, Denver,
Colorado,
USA

# Foreword

The current debate on sustainable development (SD) has included discussions, initiatives, reports and recommendations on a number of issues, including:

- Defining Sustainable Development and the Economic, Environmental and Social Spheres of Sustainability
- Presenting the Case of Mining, Minerals and Sustainable Development (i.e., The MMSD Initiative)
- Developing Processes for Multi-Stakeholder Input and Community Engagement
- Building the Business Case of Sustainable Development
- Developing Global Reporting and Sustainable Development Criteria
- Presenting Guidelines for Transparency, Assurance and Verification
- Engaging in Multilateral Agreements and Protocols on a Variety of Issues.

Individual companies, trade organizations, national and global institutions, local, state and federal governments, academic institutions, professional and scientific societies and NGOs all have been actively involved in the sustainable development process.

Significant technical developments have contributed to the recovery of the mining industry in the past 20 years. These new technologies have been responsible for the outstanding productivity, performance and the excellent health, safety and environmental record of the industry. Technological development remains a vital component to the sustainability of the minerals industry. The Toronto Declaration of May 2002, which concluded The Global Sustainable Development Initiative (or MMSD), states:

Research and technology are vital to the SD initiative, if the mining industry is to keep or be granted a "social license" to operate.

In addition to R&D, we must also recognize the new opportunities offered to the educational community in adopting SD concepts and values as the new beginning for revitalizing, if not reinventing, the old paradigm of minerals education and educational partnerships. Some steps towards achieving this goal may include:

- Integrate SD concepts within the undergraduate curriculum
- Introduce SD elective and "core curriculum" courses
- Promote interdisciplinary, post-graduate programs focusing on SD
- Engage in the discussions to establish an R&D agenda based on SD.

Mining education and research programs today require interdisciplinary approaches and talents and are, or should be, driven by the enabling technologies that are fueling the high technology revolution. Institutions involved in SD should also consider forming global partnerships and understand the need to collaborate, share resources and develop cohort groups, irrespective of location. Such educational partnerships, focusing on SD, should:

- Encourage faculty mobility within participating institutions
- Develop joint undergraduate and post-graduate programs
- Promote student exchanges
- Organize, teach and disseminate training courses and materials.

The educational processes should emphasize the need to include Sustainable Engineering in the curriculum in order to create a culture that can assist in "operationalizing" SD practices at the engineering/mine level.

Last year (2002/2003) I had the honor of serving as President of two societies, the Society for Mining, Metallurgy and Exploration (SME), a premier global society in the minerals field and the Society of Mining Professors (Societaet der Bergbaukunde), a group of senior mining engineering educators. Both these entities contributed to the notion and development of a declaration that was subsequently signed and endorsed by the major global scientific/professional minerals-societies and presented at a series of conferences held on the Greek Island of Milos in May 2003. Known as the *Milos Statement*, this declaration presents the vision of the minerals scientific community, comprised of engineers, scientists, technical experts, and academics, regarding its contribution to Sustainable Development.

The declaration states that Society's transition towards a sustainable future cannot be achieved without the application of the professional principles, scientific knowledge, technical skills, educational and research capabilities practiced by our community.

The Declaration also makes important statements on the professional responsibilities of the community and on its values, goals and aspirations pertaining to Education, Training and Development and Communication. Updates of the Milos Statement can be followed at the SME website (www.smenet.org).

The natural resources sector has the potential to contribute to wealth creation, quality of life and sustainable development. But, for this to happen, new technologies for natural resources and environmental management must continually be developed and innovative educational programs and processes must be instituted. Industry, academia and government must work together to take advantage of the new opportunities afforded by the SD debate.

This textbook, Sustainable Mining Practices – A Global Perspective, is a major step towards developing the tools and concepts necessary to introduce and apply sustainable development practices in the minerals sector. Whether in education or in mining practice, the goal is to introduce "Sustainable Engineering," to promote an understanding of the environmental, ethical, social and business aspects of the discipline and their application to design methodologies and practices. I would like to congratulate all those who contributed to the concept and development of this book, editors and authors, for providing a much-needed reference in the area of sustainable mining engineering.

Dr. Michael Karmis

President 2002, Society for Mining, Metallurgy and Exploration, Inc. (SME)

Stonie Barker Professor, Department of Mining and Minerals Engineering & Director,
Virginia Center for Coal and Energy Research
Virginia Tech, Blacksburg, Virginia, USA

December 2003

# Preface

Mining has evolved from a focus on resource development at a minimum cost to a more holistic approach based on economic, environmental and social aspects of a given project at the present time. With the inception of our consciousness towards the environment, the regulations of the 1970s and public activism for the environment in many parts of the world have made progressive steps towards improvement of the environmental quality and led to sustainable mining practices. This book documents several sustainable mining case histories from the Americas, Australia, Africa, Asia, and Europe. The main differences observed in these case studies were lack of training amongst mining and regulatory personnel in Asia and Africa as compared to the other continents. This is likely due to the dearth of financial resources in Asia and Africa.

The MILOS Statement described in the Foreword defines a cooperative program amongst educators and mining industry professionals. This cooperation should be practiced in all parts of the world to ensure that educators are producing employees who will implement sustainable mining practices in industry. In addition, vocational training for the various trades associated with mining should also be emphasized. With our sheer advancement in technologies and increasingly more cross-functional programs, there is a need for a multidisciplinary approach where the core mining courses are supplemented with relevant environmental, economics and risk management issues and challenges in all colleges that offer mining and allied programs.

It is hoped that the sustainable mining practices presented in this book from Asia, Africa, Australia, Europe, and the Americas will assist mining companies to implement such practices in their operations. It is our duty and obligation to leave the world a better place than we found it, and in mining, which is an essential industry, it is imperative that we learn from the experiences in various continents to constantly improve the mining and reclamation practices and strive to make that sustainable. When we make sustainable mining practices as Standard Operating Practices (SOPs), it will be easier for the mining companies to obtain permits from regulatory agencies and retain the support of the population that benefits from the products made available by the mining industry.

It is the sincere hope of the authors and editors of this book that mining companies and operators in various parts of the world will learn from the case histories and other relevant information on sustainable mining and reclamation practices presented in this book, and make mining a truly sustainable enterprise.

# Acknowledgements

Dr. Rajaram would like to express his sincere thanks to Turlough Guerin for helping in several chapters of the book, and in giving valuable suggestions for the book structure. The numerous weekends sacrificed by my wife Usha Rajaram made this book possible. Special thanks go to Sumit Dutta who worked tirelessly during his summer break from school to edit the chapters of the book and place the figures at appropriate locations within the book.

Subijoy Dutta expresses his sincere appreciation to Bryan Watts, geologist, for providing numerous sources of literature and introducing an expert who has contributed a unique paper involving the blasting area in this book. He greatly appreciates the invaluable support and field information provided by Dale Dalton and Daryl Jackson of Coilwell, Inc., Oklahoma. His sincere thanks and appreciation goes to his wife, Urmi Dutta, for providing strong and sustained domestic support during the long hours that he had to dedicate for completing this book.

Dr. Parameswaran would like to express his appreciation to ASARCO LLC (Asarco) management for permission to contribute to this book and to Juan Rebolledo of Grupo Mexico de S.A. de C.V (Grupo Mexico) for his encouragement in the endeavour. He also expresses his sincere thanks to Ray Krauss, formerly with Homestake Mining and now a consultant to Barrick, for information, photographs and review of the case study of closure at Barrick's (formerly Homestake) McLaughlin mine. He greatly appreciates the effort of Fred Graybeal and Stu Bengson who co-authored portions of certain chapters and Don Robbins, Tom Aldrich, Doug McAllister, Brian Munson and Kathy Arnold and other colleagues at Asarco who provided ideas and inspiration and also assisted with the review of the book. He expresses his thanks and appreciation to his wife, Padma Parameswaran, and his father, Krishnaier Parameswaran, for their support and encouragement in writing this book.

We thank Michael Karmis, a leader in sustainable mining in the United States, who took time from his busy schedule to review the book and write a Foreword.

# CHAPTER 1

## Introduction

Raj Rajaram[1] and Krishna Parameswaran[2]
[1]*Complete Environmental Solutions, Oak Brook, IL, USA*
[2]*ASARCO LLC, Phoenix, AZ, USA*

The World Commission on Environment and Development published a report in 1987 entitled "Our Common Future". The report, commonly referred to as the Brundtland Report, defined Sustainable Development as "development that meets the needs of the present without compromising the ability of future generations to meet their own needs" (WCED [1])[1]. The report influenced our thinking about development. Since then the United Nations and governments have started putting systems in place for Sustainable Development and some governments have started developing national indicators of sustainable development. The United Nations has conducted International Conferences in 1992 (Rio de Janeiro Summit) and in 2002 (Johannesburg Summit) to monitor the progress toward sustainable development. The objective of the authors in this book is to highlight the progress that we have made in sustainable mining practices and to present case histories from the Americas, Europe, Australia, Africa, and Asia.

---

[1] References for Chapters 1–1.3 are listed at the end of Section 1.3.

# 1.1 What is sustainable mining?

Raj Rajaram[1] and Krishna Parameswaran[2]
[1]Complete Environmental Solutions, Oak Brook, IL, USA
[2]ASARCO LLC, Phoenix, AZ, USA

Although many attempts have been made to clearly define sustainable mining, there is no simple answer to this question. The concept of sustainability is more difficult to comprehend when applied to a resource that is considered non-renewable. The following subsections provide an analysis of various aspects of sustainability in mining. Several different attempts to characterize sustainable mining are also presented there. This should provide a comprehensive and working definition of sustainable mining practices.

## 1.1.1 IS MINING REALLY SUSTAINABLE?

This section is based largely on the discussion of sustainability as it relates to metal mining (Muth and Parameswaran [2]). The same principles of sustainability would be applicable for coal and other minerals.

At first blush it seems that mining and "sustainable development" are not compatible. Many years ago, a commentator in a financial journal noted that industrial development was fully compatible with sustainability "except, of course for mining." Others have considered the concept applied to mining an oxymoron because we are dealing with a finite and non-renewable resource.

Mining is vital for the economic and industrial development of many countries. Every item of modern commerce must be produced from something that is either mined or grown and that the idea of sustainable development, or indeed any development, *not* based to some degree on mining is impossible. In many quarters, there is the simplistic belief that somehow one can create a future in which mining, and in particular, the mining of metals, will be unnecessary. There is a school of thought that extracting minerals from the ground at a rate greater than can be replaced by geological processes violates the principles of sustainable development.

These beliefs spring from three commonly misunderstood observations: First, that no single mine can operate for ever, and, in that sense mining is not sustainable; second that mineral resources are finite and nonrenewable; and third, certain historic mining practices have had negative impacts on the environment, that are no longer acceptable.

There are responses to all these points. It is true that no single mine can last forever but this is really not relevant. This is because sustainability "encompasses many more values than the continuing availability of the resource being developed. Indeed it is the very fact that mineral development will end some day that makes the integration of those other sustainability considerations into the mining process highly appropriate" (Pring [3]).

In that regard, the newest facility for the manufacture of advanced computer chips will, for that matter, not last forever. This does not mean that advanced chip making technology is not "sustainable." Indeed, individual mines have much longer lives than commonly believed. In the United States the Bingham Canyon copper mine, 25 miles southwest of Salt Lake City, Utah, operated by Kennecott Utah Copper Corp. has been producing copper for over 100 years. Over this period, the mine has produced over 17 million tons of copper, 23 million ounces of gold, 190 million ounces of silver and 890 million pounds of molybdenum and currently is estimated to have twelve years of additional life. It is unlikely that any single manufacturing facility of any product or commodity can rival that.

Experience tells us that even as demand for metals grows year after year so do the proven and probable resources. This is due, in part, to discovery of new resources through continued exploration even as mining proceeds and, in part, due to development of new technologies that allows us to process ores or even wastes that were previously uneconomic to extract. Examples include development of froth flotation for processing non-ferrous metal sulfide ores, the Solvent Extraction – Electowinning (SX-EW) process for leaching copper ores and cyanide heap leaching process for gold ore, each of which resulted in major additions to the inventory of economically mineable reserves. Thus, continuing exploration by mining companies and adoption of new technologies can make mining sustainable.

Another aspect contributing to the sustainability of metal mining is the recyclability of metals. Many metals, once produced, are capable of being recycled, smelted time and again to their original elemental form and refined to the most demanding specifications. In a large part due to recycling, the majority of all copper mined is still in use, as is at least 99% of all the gold ever produced and upward of 60% of all the silver mined is still in existence as bullion, coins or fabricated products. Metals are, through recycling, infinitely renewable and, therefore, useful to future generations. This is indeed the essence of sustainability.

If mining as a worldwide industrial undertaking is "sustainable" what about the community, cultural, human health and environmental impacts of a particular mining operation?

A prosperous, sustainable society is built on several forms of capital. There is, of course, financial or economic capital and natural capital that include, besides clean water and air, the mineral deposits being extracted. Mining operations must be conducted in a manner that is protective of the human health and the environment by minimizing impacts. To do otherwise limits the opportunities for development available to future generations. Then, there is human capital, by which is meant trained and educated people who can play a constructive role in both the production of wealth and the development of social institutions that are important to civil society. The social institutions that are developed are themselves a form of capital. They are what we may call social capital, contributing to a stable social order and community well-being. A detailed economic analysis of sustainability is provided in Chapter 3.

Mining can contribute to sustainable development of a particular community in a number of ways: by creating financial capital that is in turn usefully employed in more diverse enterprises; building human capital by increasing the levels of education and technical skills of local employees and others who provide goods and services to the operations; and generating social capital when a share of the wealth generated finds its way back into the community for the development of socially useful infrastructure and institutions The World Bank states that "mining is compatible with sustainable development" emphasizing "not only efficient use of economic capital, or the maintenance of ecosystem integrity and natural resource productivity, but also social equity and mobility, participation and empowerment." (Akermann [4]).

Ideally it is an important function of government to promote community development, building on the base created by economic development, but should governments fail, then mining companies should be prepared to shoulder the responsibility (West [5]).

Mining is sustainable when it is conducted in a manner that balances economic, environmental and social considerations, often referred to as the "triple bottom line," Sustainable mining practices are those that promote this balance.

## 1.1.2   WHAT DOES SUSTAINABLE DEVELOPMENT MEAN FOR THE MINING INDUSTRY?

In the 2003 Jackling Lecture (Yearley [6]), Mr. Douglas Yearley, Chairman Emeritus of Phelps Dodge Corporation, notes that, after years of debate, there is consensus that three things that are required for sustainable mining:

1. Integrated approaches to decision making on a full, life-cycle basis that satisfy obligations to shareholders and that are balanced and supported by sound science and social, environmental and economic analysis within a framework of good governance.
2. Consideration of the needs of current and future generations.

3. Establishment of meaningful relationships with key constituencies based on mutual trust and a desire for mutually beneficial outcomes, including those inevitable situations that require informed trade-offs.

There is also consensus that sustainable development is a journey rather than a destination, in that the concept has to adapt to ever changing social values, priorities and needs and that the industry needs to adopt and create a new way of doing business.

This includes research and development leading towards better methods to harness our natural resources with minimal impacts on the environment. Communication tools and transparent mechanisms need to be developed to address the social consequences of mining. Mining companies are improving the way in which they manage their social responsibilities. For example, in Africa, there are examples provided later in this book where mining companies are proactive in dealing with the HIV/AIDS virus, and are taking steps to help their workforce and the neighboring community to prevent and treat HIV/AIDS.

In the Jackling lecture, Mr. Yearly discusses the three major activities of the Global Mining Initiative (GMI), undertaken by major mining companies, as follows:

1. Establishment of the Mining, Minerals and Sustainable Development Project (MMSD), an effort to improve the performance and enhance the mining sector's contribution to sustainable development. The MMSD report was launched in September 2002 at the World Summit on Sustainable Development in Johannesburg.
2. Conducting a Global Policy Conference entitled "Resourcing the Future" which took place in Toronto in May 2002. The MMSD report findings were discussed with leaders of government, international organizations and non-governmental organizations (NGOs).
3. Formation of the International Council on Mining and Metals (ICMM) to represent and lead the industry in meeting the challenges of sustainable development.

Mr. Yearley makes the business case for sustainable development citing the following benefits:

– lower labor costs – providing clean, safe working conditions can improve productivity, result in fewer labor disputes, and increase retention rates;
– lower health care costs – healthy communities and workers will increase productivity, and reduce costs for health care;
– lower production costs – waste reduction, recycling and energy efficiency can significantly lower production costs;
– lower regulatory burden – a company working beyond compliance will have the trust of regulators and have a smoother permitting process;
– lower closure costs – terminal liabilities can be more accurately predicted, controlled and managed;
– lower cost of borrowing – managing risks and running an efficient operation will result in lower interest rates for borrowing;
– lower insurance costs – insurance companies reduce rates for safe and efficiently run companies; and
– improved investor relations – investors are attracted to companies that practice sustainable development and demonstrate the above mentioned benefits.

He notes that by practicing sustainable development principles, companies can improve their access to land and markets and thus derive a competitive advantage. Their improved reputation as a result of commitment to the principles of sustainability ensures that they can keep the social license to operate and thereby make the company a better investment prospect.

Despite these benefits, many companies remain cautious when it comes to stating publicly their commitment to sustainability. Although business leaders can cite a catalog of environmental and social initiatives that their companies are implementing, few feel comfortable in labeling these as "sustainability programs." There are a number of reasons for this hesitancy. Some businesses find the concept of sustainable business to be too large and vague to break down into manageable pieces so that they can understand the impact on their business. Others are concerned that it may be an impossible goal because their business appears to be or is characterized by their detractors as inherently "not sustainable." Others fear that their commitment to sustainability may hinder their achieving their business objectives. Finally, many worry that by publicly stating their commitment, they will only increase the public's expectations and expose themselves to attacks by non-governmental organizations (NGOs) and pressure groups. The evidence gained from companies who have successfully implemented such programs is that these concerns need not deter those who wish to move towards a more sustainable business model. Companies making well planned, business focused and strategic decisions can successfully articulate their own pathway towards the goal of sustainable development that will bring long lasting value to their business and to their stakeholders (Obbagy and Doyle [7]).

### 1.1.3   SUSTAINABLE DEVELOPMENT FROM THE PERSPECTIVE OF THE MINERAL PROFESSIONAL COMMUNITY

The business benefits mentioned above cannot be realized without the involvement of a mineral professional community consisting of engineers, scientists, technical experts, and teachers. These professionals met in May 2003 at the International Conference on Sustainable Development Indicators in the Mineral Industries on the island of Milos, Greece. The findings of this conference have been summarized in the Milos Statement (Karmis [8]). The goals stated in the Milos Statement are summarized below:

- Professional Responsibility:
  - Employ science, engineering, and technology as resources to people, catalysts for learning, providers of increased quality of life, and protectors of the environment, human health and safety.
  - Encourage the development, transfer and application of technologies that support sustainable actions throughout the product and mine life cycles.
  - Give high priority to identifying solutions for environmental and development challenges as related to sustainable development.
  - Address social equity, poverty reduction and other societal needs as issues that are integral to minerals and mining related endeavors.
  - Participate in the global dialogue on sustainable development.
  - Engage in all stages of the decision making process, not only the project execution phase.
- Education, Training, and Development:
  - Attract the best people to the fields of mining and minerals by encouraging, facilitating and rewarding excellence.
  - Build and maintain a critical mass of engineering, technical, scientific, and academic capacity through improved education and training.
  - Promote the teaching of sustainability principles in all engineering programs at all academic levels.
  - Support and commit funding to the infrastructure that enables nations to provide mineral education, professional training, information and research.
  - Prevent the loss of core competencies.
  - Encourage a global exchange in academic training, as well as apprenticeship and internship programs.
- Communications:
  - Support professional growth and interaction through books, articles, symposia, short courses and conferences on minerals and mining in sustainable development.
  - Share and disseminate to the public sound information, knowledge and technology, including information on every aspect of minerals and mining, through print, electronic and other appropriate media.

- Disseminate technical information on sustainable development and the role of minerals, metals and fuels in sustainable development. This includes information on the role of minerals in maintaining a high quality of life.
- Promote the achievements and capabilities of mineral community professionals to managers and executives, policy makers and the general public.

The following groups and organizations have endorsed the Milos statement (Karmis 2003 [8]):

*The American Society of Mining and Reclamation*
*The Australasian Institute of Mining and Metallurgy*
*The Canadian Institute of Mining, Metallurgy and Petroleum*
*The European Federation of Geologists*
*The Iberoamerican Association of Mining Education (The Asociación Iberoamericana de Enseñanza Superior de la Minería)*
*The Institute of Geologists of Ireland*
*The Peruvian Institute of Mining Engineers (Peruvian El Instituto de Ingenieros de Minas del Peru)*
*The Society for Mining, Metallurgy and Exploration*
*The Society for Mining, Metallurgy, Resource and Environmental Technology (Gesellschaft fuer Bergbau, Metallurgie, Rohstoff- und Umwelttechnik e.V.)*
*The Society of Mining Professors (Societaet der Bergbaukunde)*
*The South African Institute of Mining and Metallurgy*
*The Spanish Association of Mining Engineers (Consejo Superior de Colegios de Ingenieros de Minas)*

It is heartening to note that the industry executives, the academic community and technical professionals are thinking alike in bringing about sustainable practices in the minerals industry. With the emphasis on global dialogue and communication with policy makers at the national and international levels, the best mining practices that are practiced in one part of the world can be disseminated to other parts of the world. Once these are codified and become the industry norm, sustainable mining can truly improve economic, social and environmental conditions worldwide.

# 1.2 International sustainability reporting

Michael Dunbavan
*Horizon Environmental Pty Ltd., Sydney, Australia*

## 1.2.1 INTRODUCTION

Sustainable mining practices have been evolving over the last two decades, and are an offshoot of the environmental movement started worldwide in the 1970s. The need to manage our mineral resources in harmony with the natural environment, and the needs of the community in which mining plays a significant role have led to many innovations in sustainable mining practices.

## 1.2.2 SUSTAINABLE DEVELOPMENT AND ECO-EFFICIENCY

Sustainable development recognizes that development, meaning change in and improvement of quality of life, will occur within the present generation, but that this development should not diminish the opportunity for continued development in the future. An overall balance must be maintained amongst economic, social and environmental factors, with no single factor dominating any other.

The principle of sustainable development recognizes that the quality of human life, and the opportunity for individuals and societies to improve that quality, varies greatly around the world. Those people with higher standard of living generally have more power to improve their own lives than do people who are impoverished. Thus, the principle of sustainable development is applicable at a global scale, and not just at a national or market scale. The challenge is in the application of such a broad principle to a particular business sector, and individual businesses within that sector.

## 1.2.3 SUSTAINABILITY REPORTING AND THE TRIPLE BOTTOM LINE

The Global Reporting Initiative (GRI) was launched in 1997 as a joint initiative of the US non-government organization CERES and the UN Environment Program with the goal of enhancing the quality, rigor and utility of sustainability reporting (GRI [10]). Sustainability reporting is essentially an approach to detailed reporting of financial, environmental and social performance indicators (referred to as triple bottom line reporting) which was formalized initially in the GRI 2000 Sustainability Reporting Guidelines. These guidelines have been revised and were released as the GRI 2002 Sustainability Reporting Guidelines at the World Council for Sustainable Development in Johannesburg in August, 2002.

"Eco-efficiency is achieved by the delivery of competitively priced goods and services that satisfy human needs and bring quality of life, while progressively reducing ecological impacts and resource intensity throughout the life-cycle to a level at least in line with the earth's estimated carrying capacity." (World Business Council for Sustainable Development).

A comparison of sustainable development, sustainability reporting and eco-efficiency shows that the latter two focus on current performance of economic, social and environmental elements. Sustainability reporting is silent on providing for future needs, and eco-efficiency expresses the ability of future generations to meet their "needs and aspirations" in terms of the earth's estimated carrying capacity. If sustainability reporting or eco-efficiency is promoted as an indicator of sustainable development, then the tacit assumption in this approach is that the needs and aspirations of future generations are the same as those of the present generation. A simple reflection on the needs and aspirations of communities from two generations before the present clearly demonstrates that this assumption of identifying future needs may not be "sustainable". That is, consideration of the past shows that any reliable prediction of the expectations of future generations is very unlikely.

Sustainable development can only be measured in retrospect. The second half of the twentieth century showed exponential growth of the human population, consumption of materials and destruction of habitat to the point where the demands of the human race exceeded the carrying capacity of the earth. Until that combination of circumstances was realized, any development could be classed as sustainable.

Fundamental changes in values related to maintaining quality of human life on earth may be required in the future. Thus, selection and measurement of indicators of sustainable development, as defined, for the purpose of forward projection is inherently uncertain and this uncertainty is substantial. The presence of such uncertainty should not be taken as an excuse to discard the promotion of sustainable development. This uncertainty needs to be acknowledged and that, with this uncertainty, the predicted outcomes of actions today may not always be realized. A form of the precautionary principle applies, in that progress toward more efficient and thoughtful use of natural resources should not be deferred because of the uncertainty surrounding future needs.

## 1.2.4   MEASUREMENT TOOLS FOR SUSTAINABILITY

The GRI 2002 Sustainability Reporting Guidelines provide a detailed standard for measurement of economic, environmental and social performance. Because this standard was developed for application by all business sectors, it offers a means for intercompany comparison within the same business sector and among different business sectors. The drawback with these Guidelines is their apparent complexity indicated by measurement of 5 economic aspects, 10 environmental aspects and 21 social aspects. Each aspect can be complex within itself; for example, the aspect of biodiversity which is one of the 10 aspects included in the environmental category.

The Dow Jones Sustainability Index (DJSI) is a product from cooperation of Dow Jones Indexes, STOXX Ltd. and SAM Group. DJSI offers a sustainability indexing service to companies on the basis that investors will consider those companies listed on the DJSI as more reliable vehicles for long term investment. DJSI has available a Guide to corporate sustainability assessment (www.sustainability-index.com), however, the details of the assessment process are commercially confidential. The end result is similar to that intended by the GRI, in that a company which has committed to assessment for the DJSI can be compared with other companies within the same business sector and among different business sectors.

The World Business Council for Sustainable Development (WBCSD) has produced a guideline "Measuring Eco-Efficiency – a guide to reporting company performance". This document is detailed; however it does not appear as proscriptive as the GRI Guideline. Although there is little difference in the indicators for measurement of sustainability and eco-efficiency, the fundamental difference arises in reporting. Eco-efficiency measurement is based on the normalization of indicator data to a unit of product or service value. For example, for primary steel production, eco-efficiency could be reported per ton of steel produced, or per ton of steel sold, or per dollar net value of each ton of steel made.

Such flexibility has advantages, in that an interested industry group can agree on a set of indicators which are particularly relevant; and disadvantages such as inter-company comparisons within a business sector are less certain and may not be possible among different business sectors.

# 1.3   Scope of the book

Raj Rajaram[1] and Subijoy Dutta[2]
[1]*Complete Environmental Solutions, Oak Brook, IL, USA*
[2]*S&M Engineering Services, Crofton, MD, USA*

This book is intended to demonstrate that mining, an essential industry for maintaining our standard of living, can be conducted in a manner that is compatible with environmental and socio-economic needs of the society. Case histories that describe the mining practices in five different continents are contained in this book, and through these case histories, sustainable approaches and practices adopted in various mining operations have been highlighted. Coal and metal mining are focused, and some case histories of industrial minerals and aggregates are included. This book encompasses a wide array of information pertaining to sustainability. Some of the unique highlights amongst them include the following:

- International sustainable mining activities,
- Status of current mining practices in Americas, Asia, and Europe,
- Economics of sustainability,
- Mine closure planning,
- Biodiversity,
- Energy use management,
- Impacts of blasting,
- Surface and groundwater impacts,
- Mine subsidence,
- Environmental indicators for mining,
- Emerging mining technologies for sustainability,
- Reclamation of mining legacies,
- Small scale mining in developing countries,
- Managing maintenance wastes,
- Best Practices for Sustainable Mining, and
- Diverse case histories from Africa, Americas, and Asia on reclamation and closure.

Mining has had a history of developing mineral resources with primary emphasis on cost without sufficient regard to the environmental, socio-economic, or post-mining land use impacts. In the past 30 years, most countries have passed formal environmental legislation, detailing the acceptable standards of human impact to air, water, and land. As these environmental laws are being enforced by governmental agencies, mining companies are striving to change their mining, reclamation, and waste disposal practices to conform to these laws. In this process, mining is becoming a sustainable enterprise and more thought is being paid to coexistence with other land uses in the community. There is a progressive trend for mining companies to employ environmental and land use planning professionals, and to pay special attention to solve the tough issues related to environmental and socio-economic impacts of mining.

The issues affecting sustainable mining have been discussed in depth in Chapter 3 of the book. The problem of large volumes of waste generated during open pit mining and their management within a mine need to be carefully managed as accentuated in Section 3.1. It requires the careful planning of materials movement within the mine, and examples illustrating this aspect are provided in Section 3.2 (case study of the McLaughlin Mine in California, USA) and Section 3.5 (case studies from many mining operations). Proper mine closure planning which will eliminate the problems associated with mining legacy sites has also been discussed at depth in Section 3.2. Economic issues must be considered for sustainable mining as addressed in Section 3.3, and the importance of internalizing the life cycle costs, especially the closure and post-closure costs are detailed. Managing environmental impacts and ensuring biodiversity at mining sites have also been emphasized in this book under Section 3.4.

Land use is a critical part of the mine planning process, and the increased pressure on urban land is discussed in detail in Section 3.5. The importance of recycling and reuse, and case studies of land use planning in various mining situations are provided. The mining industry uses a large amount of energy in mining and processing of the ores to produce a final product. Energy efficiency improvements in the mining industry and energy use management are detailed in Section 3.6.

Sustainable mining systems start with incorporation of Cleaner Production (CP). CP results in improving profitability, improvements to the environment, and acceptability of the project to the local, State and Central governments. Section 4.1 is an exhaustive compilation of case studies of CP in various Australian mining operations, from small scale improvements to large improvements in water management and energy efficiency. Cleaner Production is made possible through the incorporation of International Standards Organization (ISO) 14001 in all aspects of a company's operation. These case studies provide a starting point for a mining company to plan their own CP program.

The importance of planning and implementing a proper blasting program and avoiding liabilities associated with blasting are dealt with in Section 4.2. Airblast, ground vibrations, flyrock, fumes, and dust impacts of blasting are discussed in the Chapter, and suggest various ways of minimizing the environmental impacts of blasting. Minimizing surface water and ground water impacts are discussed in Sections 4.3 and 4.4, respectively, and the importance of these aspects in obtaining a mining permit are detailed. Ground water recharge using a soil adsorption system is described in Section 4.4.

Mining subsidence has created several mining legacy sites in the United States and impacts the use of the land overlying the underground mine. A detailed discussion of the mechanism of subsidence and its control is provided in Section 4.5. The importance of comprehensive planning and coordination of surface and underground development is also provided in this chapter. Environmental indicators provide a useful and inexpensive method of monitoring the mining impacts at a site. Section 4.6 provides several indicators that can provide early warning systems to prevent environmental damage from mining operations.

There are several emerging mining technologies that allow the mining of resources while minimizing environmental impacts. Two such technologies are described in Section 4.7. As environmental regulations become more stringent, and public pressure on surface land increases, hydroevacuation mining method for coal, coal bed methane, and other mineral extraction is expected to be more utilized. This emerging technology has been designed to cause minimal surface disturbance and is also suitable to recover coal and other minerals from steeply dipping reserves which could not be otherwise mined by standard mining methods. Time Domain Reflectometry is another technology which is emerging to aid in determining ground movement in tailings, dams, surface mines, and subsidence resulting from underground mines. Use of these emerging technologies should provide the mining industry with better options for a sustainable operation.

Mineland reclamation has made rapid strides in the last few decades. Techniques of surface mine reclamation in coal and non-coal mines, and reclamation of abandoned mines are covered in great detail in Chapter 5. Two case studies of reclamation of abandoned mines in Montana, USA, illustrate the complicated process involved in completing the reclamation of abandoned mines and returning them to beneficial use. The case studies also reflect the new methodologies adopted by the Office of Surface Mining in the United States to reduce environmental hazards associated with abandoned surface and underground mines. Lessons learned in the United States should provide valuable guidance and directions for other countries that are beginning to undertake such tasks.

A major issue in the permitting, operation and closure of surface and underground mines is proper management of the waste generated. Several tailings dam failures in the United States and other countries have led to the use of improved design and risk management techniques as addressed in Section 6.2. Case studies of tailings dam design, operation, and emergency response to failures of tailings dam are provided in Section 6.2.3. Waste rock and soil handling part of waste management have been addressed in Section 6.3. Maintenance wastes from mining operations are detailed in Section 6.4. The importance of recycle, reuse, and treatment of maintenance wastes are described using the survey results from many mining operations in Australia and Asia.

Use of best management practices is a critical aspect of sustainable mining. Best Mining practices in small scale mining, tailings pond design, and construction and maintenance, are detailed in Chapter 7. The importance of small scale in India's mining sector, and the issues related to sustainability in small scale mining have been addressed in great detail under Section 7.1. The United States Mine Safety and Health Administration (MSHA) is a leader in many aspects of mine safety and health in the world. Their role and the technical issues they consider in permitting a tailings management facility are elaborated in Section 7.2. Best management practices (BMPs) to minimize cross-media transfer of waste during mining operations have been covered in Section 7.3.

Improvements in the mining industry have been made by learning from case histories of mining operations that introduced novel ways of dealing with complex problems. A variety of case histories of mining from the Americas, Asia (India), and Africa is covered at great depth in Chapter 8. Case histories from the Americas include an innovative way of dealing with conflicting land uses in an aggregate quarry located within an urban setting, sustainable exploration techniques from a project in French Guyana, and innovative reclamation practices developed by ASARCO LLC in the southwestern United States. Case histories from India include several mines operated by the private sector in India, and in particular, mines operated by Tata Iron and Steel Company. Human Immunodeficiency Virus/Acquired Immune Deficiency Syndrome (HIV/AIDS) is devastating the mining industry in Africa, and a case history of the efforts being made by mining companies to prevent and treat this disease is presented in Section 8.3. In addition, mine closure issues in Tanzania and Ghana are presented in this chapter.

Some in the developing world may look at industrial development in the West and conclude that it is necessary to "dirty up first, clean up later". This happened primarily because countries focused on economic development first and regulatory programs evolved later. These old practices from the industrialized countries should be avoided and not pursued by the developing countries. Modern technology and management practices allow the production of goods with far less economic impact today than what was possible before. Pollution prevention is generally much more economic and effective than remedial action and cleanup. The cost of protection of environmental and other natural resources from the very onset of mining operation is small compared to remediation after contamination has occurred (Muth and Parameswaran [2]).

By implementing the sustainable mining practices, the land use conflicts that often tend to occur when a mining development is proposed to a community, is expected to be resolved with much less time and resources by mining companies and regulatory agencies. Michael Karmis, the 2002 President of the U.S. Society of Mining Engineers (SME), said in his interview with Mining Engineering magazine (Karmis [11]) that "companies must recognize that environmental and social responsibility and an open information policy can translate into competitive advantage and economic benefits". This book addresses the environmental and socio-economic issues faced by the mining industry, and suggests through examples how mining companies can adopt policies that clearly demonstrate to the public and the regulatory agencies that they are adequately considering the environmental and social impacts of their activity.

## REFERENCES

1. World Commission on Environment and Development, (WCED 1987), Our Common Future, Oxford University Press, Oxford.
2. Muth, R.J. and Parameswaran, K., 1998, The 21st Century: Challenges and Opportunities for the Mining Industry, International Symposium on Environmental Management of Mining and Mineral Industries (EMOMAMI-'98), Bhubaneshwar, India, pp. 3–7, 13.
3. Pring, G., 1998, Sustainable Development: Historical Perspectives and Challenges for the 21st Century, United Nations Development Programme (UNDP) and United Nations Revolving Fund for Natural

Resource Exploration (UNRFNRE) Workshop for the Sustainable Development of Non-Renewable Resources Toward the 21st Century, New York, October 15–16, 1998.

4. Akermann, R., 1998, Is Mining Compatible with Sustainable Development? A World Bank Perspective, International Council on Metals and the Environment (ICME) Newsletter, Volume 6, No. 2.
5. West, M., 1998, Report in the May 8, 1998 edition of the Mining Journal.
6. Yearly, D.C., 2003, Sustainable Development for the Global Mining and Metals Industry, Mining Engineering, Society of Mining Engineers (SME), Littleton, Colorado, USA, August, pp. 45–48.
7. Obbagy, J.E. and Doyle, C., Making Sustainable Choices: Defining a Pathway for Sustainable Development, ICF Consulting Viewpoint, pp. 1–4.
8. Karmis, M., 2003, Sustainable Development Conference Held in Greece, Mining Engineering, SME, Littleton, CO, USA, August, pp. 31–32.
9. International Institute for Environment and Development (MMSD), 2002, Mining, Minerals and Sustainable Development Report, London, England.
10. Global Reporting Initiative, 2002, Sustainability Reporting Guidelines, Prepublication Release, August 8.
11. Karmis, M., 2002, Interview with the Editor, Mining Engineering, SME, Littleton, CO, USA, May.

# CHAPTER 2

## Current status of mining practice

Raj Rajaram[1] and Krishna Parameswaran[2]
[1] Complete Environmental Solutions, Oak Brook, IL, USA
[2] ASARCO LLC, Phoenix, Arizona, USA

The important contributions that coal, ferrous and non-ferrous metals make to the U.S. economy have been repeatedly demonstrated by mineral economists and groups representing the mining industry. As environmental issues have come to the forefront of public attention, it has become increasingly difficult to obtain the permits necessary to operate a mine, smelter or refinery. However, this has made mining companies evolve innovative solutions to minimize the impact on the environment, both during active mining and subsequent reclamation/closure phases of a mining operation and in downstream processing activities and metallurgical operations. With the information flow through the Internet, environmental improvements made in mine planning or reclamation in the United States and Europe are available in a relatively short time to others around the world.

The member countries of the International Copper Study Group, International Lead and Zinc Study Group, the International Nickel Study Group, and the members of the International Council on Metals and the Environment have made significant progress since 1999 in integrating economic, environmental and social considerations in planning, operating, and reclaiming a mine site. The importance of several key issues in improving mining practice has been recognized and significant studies have been completed. These include:

1. stewardship programs to demonstrate responsible management of processes and products throughout the life cycle of a mine, including exit strategies for the mine site;
2. community consultation and involvement in the decision making process;
3. promotion of recycling, both in developed and developing countries;
4. research and development to improve mining, processing, and reclamation methods, and dissemination of the material to facilitate decision making by governments, industry, and others;
5. open and transparent mechanisms to improve communication, consultation, and cooperation between all stakeholders; and
6. information development and dissemination relating to measuring performance and reporting on economic, environmental, and social factors at all stages of the mining cycle, from exploration to mine reclamation (IIED [1]).

The current status of mining varies significantly from continent to continent, with the most advanced practices in North America, Australia, Europe and South Africa, and the least advanced practices in Asia and other parts of Africa. Even within each continent, there is significant variability, as seen in the practices in South Africa and other parts of Africa. The importance assigned to the environment in planning and operating a mine also varies from continent to continent, and is the primary reason for wide variability in mining practices. The availability of trained labor, and the availability of capital to introduce mechanization and computerization in mine operations also vary from continent to continent. This chapter presents the status of current mining practices and regulations relating to mining operations in the Americas focusing on U.S.A., Asia and Europe to demonstrate the wide variation in the current status of mining practice.

# 2.1 Americas

Krishna Parameswaran
*ASARCO LLC, Phoenix, Arizona, USA*

The Americas are richly endowed with mineral resources comprising ferrous and non-ferrous metals, industrial minerals and fuel minerals. For many countries in the region, the development of these resources plays an important role in sustaining their economies by contributing to export revenues and foreign exchange reserves and providing business opportunities. The Americas produced 59% of the world's copper; 57% of silver; 39% of zinc; 34% each of aluminum, gold and lead; 33% of iron ore and 32% of nickel, based on 2000 statistics. Amongst industrial minerals, the region produced 42% of the world's gypsum, 41% of sulfur, 39% of salt and 33% of phosphate rock. The region also produced 38% of the world output of natural gas, 27% each of crude oil and coal; and 33% of refined petroleum products (USGS [2]). The climate in the region is variable from temperate to tropical and approach to regulations differs from country to country. Information on natural resources, environmental regulatory framework, national minerals policy and permitting processes for the United States are furnished in the following subsections.

## 2.1.1 NATURAL RESOURCES (NMA [3])

The United States is rich in mineral resources, both fuel and non-fuel minerals. Mining has an economic impact of nearly half-trillion dollars, or nearly 5% of the country's Gross Domestic Product (GDP).

The United States' metallic mineral resources include copper, lead, gold, zinc, and iron ore. Other domestically produced metallic minerals include magnesium and molybdenum. The United States is also an important coal producer, with coal and uranium providing over three quarters of the nation's electricity supply. All 50 states have some form of mining activity and the U.S. produces 78 minerals and major commodities. Mining provided direct employment for over 267,000 people in 2001. Metal mining employed 33,269; non-metal mining 26,342; coal 79,771; sand and gravel 44,987 and stone 82,675.

## 2.1.2 ENVIRONMENTAL REGULATORY FRAMEWORK (NRC [4], Holland and Hart [5], Mining Environmental Handbook [6])

Hardrock mining operations in the United States are subject to a complex set of federal and state laws and regulations designed to protect the environment. The scope and degree of regulations depends on the size of the mining operation; the type of land, water and biological resources impacted; the state in which the operation is located; the organization of state and local permitting agencies and how they implement and enforce regulations.

Federal laws and regulations impacting mining include:

1. *General Mining Law of 1872 and Federal Land Policy and Management Act (FLPMA) of 1976*
   The General Mining Law is the basic statute governing hardrock mining on federal lands. Private entities obtain rights to minerals under the location system. The Mining Law originally applied to all mineral deposits on federal lands except coal. Congress removed various minerals from the Mining Law of 1872, making them leasable or saleable. Locatable minerals include gold, silver, mercury, lead, tin, copper and other hard rock mineral deposits. Prospectors who discover a valuable mineral deposit can stake a mining claim or claims, in accordance with state and federal requirements. Owners of valid claims may be entitled to a patent (fee title) to

the deposit and the lands containing it. However, valid mining claims need not be patented to mine and market the extracted minerals, as long as the claims are properly maintained.

Leasing of mineral deposits applies to coal, oil, gas, shale tar sands, sodium, phosphate, potassium and sulfur and is governed by the 1920 Mine Leasing Act. Direct sale applies to clay, stone, gravel minerals and is governed by the Materials Act of 1947 and Surface Resource Act of 1955.

The bulk of federal public lands are managed by two agencies: the Bureau of Land Management (BLM) in the Department of Interior, and the U.S. Forest Service (USFS) in the Department of Agriculture. Other major land management agencies include: National Park Service, the U.S. Fish and Wildlife Service and the Bureau of Reclamation.

The Federal Land Protection and Management Act (FLPMA) provides policy guidance on federal land management. It requires BLM to conduct comprehensive land use planning, manage lands based on the principles of multiple use and sustained yield, to balance the need for environmental protection and the need for domestic sources of minerals, food, timber and fiber, while giving priority to protecting areas of critical environmental concerns.

Proposed mining projects trigger the application of BLM regulations (CFR [7]) to comply with the "unnecessary and undue degradation of public lands" principle. These regulations provide for three levels of surface-disturbing activities:

- Casual use involving "negligible" surface disturbance – does not require agency notification or approval.
- Activities not qualifying for casual use but disturbing five acres or less per year (including access) – requiring at least 15 days prior written notice to the BLM and may require consultation on access, and
- Activities disturbing more than five acres per year or any activity in certain protected areas such as wilderness areas or areas of critical environmental concern require notification and approval of plan of operations.

Approval of a plan of operations may require preparation of an Environmental Impact Statement (EIS) by BLM. Activities conducted under the notice provision must comply with specific performance and reclamation standards. This may include limiting access routes, performing reclamation at the earliest possible time and measures to preserve top soil and control erosion. All operations including casual use must comply with applicable state and federal laws. Most mining projects need to post a financial bond to ensure compliance with reclamation activities.

Environmental protection is a major factor in land use planning that is to be considered by the land management agencies. The land use planning process determines which lands are available for exploration/discovery as well as restrictions that may affect development once minerals are found. Once mineral rights are obtained, numerous environmental permits and approval are required before mineral development can commence.

The 1891 Organic Act, Forest and Rangeland Renewable Resources Planning Act of 1974, as amended by the National Forest Management Act of 1976, provides the U.S. Forest Service (USFS) direction on management of forest lands. USFS's planning process is similar to that of the BLM. It manages surface-disturbing activities related to mining claims under the 1891 Organic Act. USFS regulations (CFR [8]) require the filing of a "notice of intent" for any activities that might cause disturbance of surface resources. If the District Ranger determines that such operations will cause "significant disturbance" of surface resources, the operator must submit a proposed plan of operations for approval. This may require the preparation of an EIS.

2. *The National Environmental Policy Act (NEPA) of 1969*
NEPA integrates decision making by BLM and USFS with the assessment of other environmental concerns on proposed mining projects. For operations on federal lands that are expected to have a significant impact on the environment, an EIS needs to be prepared. The EIS process requires public "scoping" of the issues, identifying alternatives to be evaluated and results in a record of decision that determines the content of the plan of operations and mitigation requirements. For smaller projects with less potential impact, an environmental assessment (EA) is done, which allows the land management agency to determine if impacts are significant.

3. *The Clean Air Act (CAA) of 1970 and 1990 Amendments*

CAA's primary focus is the setting, attainment and maintenance of National Ambient Air Quality Standards (NAAQS) for criteria pollutants, i.e., particulate matter (PM), sulfur oxides, nitrogen oxides, ozone, carbon monoxide and lead. The United States Environmental Protection Agency (EPA) is required to set primary and secondary NAAQS for criteria pollutants and to review the standards every five years. Primary NAAQS need to be set at levels that are "requisite to protect public health" with an adequate margin of safety. The CAA places most of the responsibility on states to implement and enforce its goals and requirements. This is usually achieved through the establishment of "State Implementation Plans," (SIPs) which outlines how NAAQS will be achieved. Secondary NAAQS are to be set at levels "requisite to protect the public welfare." Welfare effects include: effects on soils, water, crops, vegetation, manmade materials, animals, wildlife, weather, visibility, and climate, damage to and deterioration of property, and hazards to transportation, as well as effects on economic values and on personal comfort and well being.

In addition, major new sources of air emissions and major modifications to existing major sources are required to undergo permitting before beginning construction under one of two NAAQS related programs. In attainment areas (where NAAQS are being met), such sources are subject to the Prevention of Significant Deterioration (PSD) program, meant to prevent air quality from deteriorating to or below NAAQS. This program works by placing a limit on allowable deterioration or increment and by requiring the use of "best demonstrated available control technology" (BACT). In non-attainment areas the requirements are more stringent. Major new or modified sources must not only install controls reflecting "lowest allowable emission rate" (LAER) but also demonstrate an offset representing one-to-one emissions reduction from another source.

The CAA establishes New Source Performance Standards (NSPS) for air emission sources falling under one of over 60 industrial source categories, including mineral processing and smelting. These standards require new or modified sources within a listed source category to meet best demonstrated technology for that source.

The 1990 amendments to the CAA mandated that each state institute an operating permit program for major sources. The operating permit program requires a renewable permit for each stationary air emissions source subject to any of the CAA's numerous provisions. EPA has concurrent authority, along with the states to enforce CAA and any issued permit, as well as any state requirement contained in an approved SIP. These amendments also require establishment of an "enhanced monitoring program," with the goal of demonstrating continuous compliance. This for some sources can translate into installation of continuous emissions monitoring systems. Further, these amendments required stationary sources to prepare risk management plans to deal with accidental releases of regulated substances that are present at a facility in quantities exceeding established thresholds.

The 1990 amendments also broadened the focus from criteria pollutants to the regulation of a wide array of specified hazardous air pollutants (HAPs) (including eleven metallic HAPs) for many categories of stationary sources, including primary copper and primary lead smelters. These sources are required to install "maximum available control technology" (MACT) for major sources of HAPs. A major source is one that has a potential-to-emit 10 tons per year of any HAP or 25 tons per year all HAPs. The amendments also require EPA to study the residual risk after implementation of the MACT and, if warranted, require additional controls.

4. *The Federal Water Pollution Control Act of 1972 or Clean Water Act (CWA)*

The CWA goals were to restore and maintain the integrity of the nation's waters by controlling discharge of pollutants into "navigable waters" and to achieve and protect fish, wildlife and recreational uses often referred to as the "fishable, swimmable" goal. The Act made EPA responsible for setting pollutant discharge standards for each industry. EPA was also given permitting authority for discharge of pollutants along with oversight responsibilities over state permit programs through the National Pollutant Discharge Elimination System (NPDES). Most states have been delegated permitting authority and consequently enforce the Act's permit program. States continue to have the primary responsibility for setting in stream water quality standards, which EPA must approve. The NPDES or state permits set limits on pollutant discharges and require self-monitoring of compliance. These limits are derived from the more stringent of technology-based effluent limitations promulgated by EPA for various industry sectors (e.g., ore mining and dressing), or water quality standards for streams and other water bodies adopted by states to protect designated uses, such as public water supply, aquatic life, recreation or agriculture.

The U.S. Army Corps has been given permitting authority for dredge and fill permits, although EPA has a role in reviewing and enforcing these permits. Over the years, "navigable waters" have been construed broadly to include, besides surface waters, wetlands and ephemeral streams and "discharge of dredge and fill material" is defined to encompass a wide range of excavation and fill activities in surface waters and wetlands.

Storm water runoff is regulated through general permits for construction of industrial (which includes mining) and large and medium urban sites.

5. *The Endangered Species Act (ESA) of 1973*
ESA prohibits the "taking" of a wildlife species classified as endangered or threatened. Before a federal agency can issue a permit for an exploration or mining activity, it must determine if endangered or threatened species are present in the area where these activities will occur. If they are present, the agency must consider what impact these activities might have on the species. If the agency determines that proposed activities might affect the species, it must consult with the United States Fish and Wildlife Service or the United States Marine Fisheries Service to ensure that the action is not likely to jeopardize the continued existence of endangered or threatened species of wildlife or vegetation or adversely affect the species critical habitat. ESA and other wildlife and vegetation protection laws can restrict and even preclude mining activities in some locations. Compliance with these requirements may mean project design changes such as changing access routes or require relocation of project facilities to avoid sensitive areas such as breeding and calving grounds, migration routes "critical" to protected species, or endangered or threatened species' habitats.

6. *Resource Conservation and Recovery Act (RCRA) of 1976 and Hazardous and Solid Waste Amendments (HSWA) of 1984*
This statute requires EPA to regulate the management of hazardous wastes from generation to disposal, i.e., cradle to grave. It also sets forth a framework for the management of non-hazardous wastes, underground storage tanks and medical wastes. RCRA focuses only on active and future facilities and does not address abandoned or historical sites. As with other environmental laws, RCRA allows EPA to delegate permitting and enforcement authority to the states.

With a few exceptions (waste in domestic sewage, irrigation return flows, discharges regulated by CWA permits and radioactive materials regulated under the Atomic Energy Act) RCRA defines all "discarded" materials as solid waste regardless of whether a material is solid, liquid or gas. The scope of discarded materials has been subject to debate and litigation. EPA's regulatory definition of solid waste includes materials that are "abandoned," "recycled," or "inherently waste-like," all of which are defined in regulations. EPA's definition of "hazardous wastes," with certain exceptions includes solid wastes that:

- Are on a list of hazardous wastes: nonspecific source wastes, wastes from specific sources and commercial chemical products;
- Exhibit characteristics of ignitability, corrosivity, reactivity or toxicity;
- Constitute a mixture of a listed hazardous waste and solid waste or are derived from a listed hazardous waste (with some exceptions)
- Are a mixture of a listed hazardous waste and solid waste or are derived from a listed hazardous waste (with some exemptions)

One of the exemptions to EPA's definition of hazardous wastes arose from a statutory provision called the Bevill amendment, which required EPA to study extraction, beneficiation and mineral processing wastes and make a regulatory determination on how these wastes should be regulated. EPA determined that all extraction and beneficiation wastes and twenty large volume mineral processing wastes (including primary copper, lead smelting slags and copper slag tailings) would be regulated as non-hazardous.

EPA's hazardous waste regulations impose requirements for generators, transporters, and operators of treatment, storage, and disposal (TSD) facilities. Generator requirements include obtaining an EPA ID number, determining whether their wastes are hazardous, preparing their waste for transportation (according to Department of Transportation (DOT) packaging and placarding requirements), preparing the Uniform Hazardous Waste Manifest (manifest) and complying with storage, training, planning, record keeping and reporting requirements. The manifest is the key ingredient of RCRA's cradle to grave regulatory scheme. Each time a waste is transferred (from generator to transporter, from one transporter to another or from transporter to the designated facility) the manifest must be signed to document receipt. The designated facility on receipt of the waste sends a copy of the signed manifest back to the generator. There are three categories of generators: large quantity generators, small quantity generators and conditionally exempt small quantity generators. A large quantity generator must obtain a permit to store hazardous wastes for more than 90 days.

Transporters of hazardous wastes must obtain an EPA ID number, comply with manifesting requirements, meet DOT transportation requirements, must properly deal with accidental spills and accidents, and report serious accidents and spills to the National Response Center and DOT.

TSD facilities are subject to RCRA's Subtitle C permitting requirements and have to meet standards for TSD units, e.g., containers, tanks, surface impoundments, waste piles, land treatment, landfills and incinerators. TSD facilities must comply with siting requirements. They must analyze incoming wastes, meet security requirements, inspect their facilities, and comply with employee training and record keeping requirements. Other requirements may include: ground water monitoring and reporting, control of air emissions, closure and post-closure and financial assurance. In addition, as part of the issuance of a RCRA permit, EPA and authorized states require corrective action to remediate past releases of hazardous constituents or wastes.

HSWA, the 1984 amendments to RCRA, require phasing out land disposal of hazardous waste. Hazardous wastes have to be treated to meet land disposal restrictions levels, thus increasing the cost of disposal. Some of the other mandates of HSWA include increased enforcement authority for EPA, and a comprehensive underground storage tank program.

7. *Comprehensive Environmental Response, Compensation and Liability Act (CERCLA) or Superfund*
   CERCLA established a "Superfund," funded through a tax on chemicals and petroleum products to clean up past disposal of hazardous substances, pollutant and contaminants. It authorizes EPA to seek out those parties responsible for such releases. Potentially responsible parties (PRPs) include: (1) current owner of a facility from which there is a release (2) former owner or operator of a facility from which there is a release (3) generators who arrange for disposal or treatment or arranged with a transporter for disposal or treatment at a facility from which there is a release and (4) transporter who accepted any hazardous substance for transport to disposal or treatment facility from which there is a release. Liability under CERCLA is "strict," meaning liability without proof of fault and "joint and several," meaning that an individual PRP can be held liable for the entire cost of cleanup and it is up to that party to recover costs from other PRPs.

   EPA maintains a computer database (called the Comprehensive Environmental Response, Compensation and Liability Information System (CERCLIS)) of sites where releases of hazardous substances are suspected. The National Priorities List (NPL) is the list of sites EPA believes pose the greatest threat to human health and the environment. Sites are put on the NPL when they achieve a high score on the Hazard Ranking System (HRS). Only sites on the NPL qualify for Superfund financing.

   EPA cleans up orphan sites when potentially responsible parties cannot be identified or located, or when they fail to act. Through various enforcement tools, EPA obtains private party cleanup through orders, consent decrees, and other small party settlements. EPA also recovers costs from financially viable individuals and companies once a response action has been completed.

   EPA is authorized to implement the Act in all fifty states and U.S. territories. Superfund site identification, monitoring, and response activities in states are coordinated through the state environmental protection or waste management agencies.

CERCLA authorizes two kinds of response actions:

- Short-term removals, where actions may be taken to address releases or threatened releases requiring prompt response.
- Long-term remedial response actions, that permanently and significantly reduce the dangers associated with releases or threats of releases of hazardous substances that are serious, but not immediately life threatening. These actions can be conducted only at sites listed on EPA's NPL.

CERCLA also enabled the revision of the National Contingency Plan (NCP). The NCP provided the guidelines and procedures needed to respond to releases and threatened releases of hazardous substances, pollutants, or contaminants. The NCP established the NPL.

CERCLA was amended in 1986 by the Superfund Amendments and Reauthorization Act (SARA) making several important changes to the program. SARA:

- stressed the importance of permanent remedies and innovative treatment technologies in cleaning up hazardous waste sites;
- required Superfund actions to consider the standards and requirements found in other State and Federal environmental laws and regulations;
- provided new enforcement authorities and settlement tools;
- increased State involvement in every phase of the Superfund program;
- increased the focus on human health problems posed by hazardous waste sites;
- encouraged greater citizen participation in making decisions on how sites should be cleaned up; and
- increased the size of the trust fund from $1.6 to $8.5 billion.

The Emergency Planning and Community Right-to-Know Act (EPCRA or Title III of SARA) is intended to inform the public regarding the presence of certain chemicals and the release of those chemicals to the environment. EPCRA is also intended to assist planning at the local level to respond to potential releases of specified chemicals. Information on releases of toxic chemicals is reported to EPA and maintained in a database called the Toxics Release Inventory (TRI). Manufacturing operations have been subject to these requirements since 1989. TRI reporting was extended to coal and metal mining in 1998.

8. *Toxic Substances Control Act (TSCA) of 1976*
   TSCA gives EPA the authority to regulate chemical during the manufacture, distribution and use of chemicals. As various chemicals are used by the mining industry, its operations could be subject to TSCA. The goal of TSCA is to protect human health and environment from chemicals that pose an unreasonable risk. For the most part, TSCA programs do not appear to regulate mining activities. However, one TSCA program that has impacted the mining industry is the regulation of polychlorinated biphenyls (PCBs). These regulations (CFR [9]) establish prohibitions of, and requirements for the manufacture, processing, distribution in commerce, use, storage, disposal and marking of PCBs and PCB items including electrical equipment such as transformers and capacitors.

9. *The Antiquities Act of 1906, the Archaeological Resources Protection Act (ARPA) of 1979 and the National Historic Preservation Act (NHPA) of 1966*
   Under the Antiquities Act, the President is authorized to designate national monuments on federal lands in order to protect structures of historic, prehistoric or scientific interest. ARPA requires a permit authorized by the federal land manager with jurisdiction over the land for excavation and removal of archaeological resources. NHPA requires the identification and protection of properties that qualify for listing under the National Register of Historic Places.

10. *The Migratory Bird Treaty Act*
    This Act implements provisions in various international treaties for the protection of migratory birds. There are more than 1,000 protected bird species including common birds such as duck and geese. It is a criminal offense to kill a protected bird. Liability is strict which means it attaches even if death is accidental and measures were taken to avoid the killings.

11. *The Surface Mining Control and Reclamation Act (SMCRA)*

    SMCRA regulates surface coal mining, including the surface effects of underground coal mining. It establishes minimum standards for permitting and reclamation of surface coal mining operations and allows states primacy for regulation of such operations upon submittal and approval of state programs. In addition to securing a permit, SMCRA requirements include monitoring baseline conditions, establishing vegetation reference areas, designing and building sedimentation facilities, preparation of reclamation plans, meeting minimum requirements for post-mining land use and calculation and posting of performance bonds. NHPA and related acts require performance of transect surveys, records review, and impose conditions during operations with respect to identification and notification. SMCRA also has provisions to designate lands as unsuitable. SMCRA contemplates cooperative regulation of surface coal mining by the Office of Surface Mining Reclamation and Enforcement (OSM) and the states.

States have laws and regulations applicable to metal mining operations. These include:

12. *Reclamation laws*

    Laws requiring reclamation exist in all western states and generally include:

    • Thresholds for regulation of exploration activities and mining operations;
    • Application content, review and approval procedures, public participation requirements;
    • Requirements pertaining to characterization of overburden and ores, prediction of acid drainage and management of acid generating materials;
    • Requirements relating to management of reagents in beneficiation (acids, cyanides);
    • Requirements for stabilization and reclamation of the site;
    • Requirements for closure of tailings disposal areas and spent ore areas;
    • Requirements for revegetation;
    • Requirements for financial assurance;
    • Reporting requirements; and
    • Monitoring requirements.

13. *Groundwater laws*

    These include laws ranging from a detailed permitting program requiring aquifer protection permits to Colorado's law requiring each regulatory agency to integrate groundwater protection in its permitting and regulatory decisions. Under Arizona's Aquifer Protection Permit (APP) Program, a facility that discharges to groundwater needs to obtain an APP. Certain operations are regarded as discharging facilities, unless determined by the Arizona Department of Environmental Quality (ADEQ) that the facility is designed, constructed and operated so that there is no migration of pollutants directly to the aquifer or the vadose zone. Discharging facilities include: solid waste disposal facilities, mine tailings ponds, and leaching operations. The permit requires the facility to be designed, constructed and operated to ensure the greatest degree of discharge reduction through the application of Best Available Demonstrated Control Technology (BADCT). The permit also requires that discharges do not cause or contribute to a violation of the aquifer water quality standard at the point of compliance. The permits may also have requirements pertaining to monitoring, record keeping and reporting, discharge limitations, contingency planning and a compliance schedule.

2.1.3   MINING & ENVIRONMENTAL CONCERNS (Mining Environmental Handbook [6])

Hardrock mining occurs where minerals are concentrated in economically viable deposits. Less than 1% of the earth's crust contains mineral deposits that are economically viable. Therefore, there is little discretion where mines can be located.

Hardrock mining has the potential to cause environmental impacts from exploration to closure and post closure. These include disturbance to the land surface, impacts on groundwater, surface water, aquatic biota, aquatic and terrestrial vegetation, wildlife, soils, air and cultural resources. The complex array of federal and state regulations in place in the United States considerably mitigates these impacts, allowing mineral development to occur while protecting human health and the environment.

Mining is a transitory land use that invariably requires disturbance and alteration of the land surface in order to access and extract the minerals essential for modern society. The extent of disturbance will depend on the type of mining-open pit mining of metals or area or strip mining of coal and uranium, mountain top mining of coal in steep areas, placer mining or alluvial mining. The nature of the land surface is also changed through the removal and placement of overburden, waste rock and tailings. The amount of surface land disturbance associated with underground mining and in-situ mining is considerably less but includes development of surface facilities in support of operations, and impacts to the surface resulting from planned and unplanned subsidence.

Water quality issues include metals, cyanide, acid drainage, pit lakes and placer mining, which can have short-term and long-term impacts on water chemistry, aquatic biota and aquatic habitat. Hardrock mining has the potential to release to the environment metals, metalloids (arsenic, antimony and selenium), sulfate, nitrate, suspended solids and cyanide. These constituents are present in the ore and remain in waste rock, tailings, pit walls, pregnant and barren solution ponds, or result from reagents used in mining.

Acid mine drainage (AMD) is produced by the exposure of sulfide minerals, such as pyrite, to air and water resulting in the oxidation of sulfur and the production of acid and elevated concentrations of iron, metals and metalloids (copper, lead, zinc, arsenic, cobalt, mercury, nickel, molybdenum and antimony) and sulfates. Sulfide minerals occur in coal overburden, sulfide ore and associated waste rock and processing wastes. Although the factors that create acid drainage and that may mitigate impacts are well understood, there is little long-term data to predict the extent of acid drainage at a particular site. Carbonate-bearing waste rock or tailings can produce drainage that is alkaline to neutral pH with elevated concentrations of cadmium, zinc, manganese, arsenic, molybdenum and selenium.

Pit lakes have the potential to impact pre-mining water quality and quantity. Contaminated water in a pit lake can impact down-gradient groundwater or surface water quality. Even if the water does not flow downstream, the concentration of metals and other contaminants and salinity in the pit through evaporation may impact migratory birds and terrestrial wildlife.

Groundwater withdrawn to dewater pits is often discharged into local streams. These waters usually have higher dissolved metals than the receiving streams. Although some dilution is likely to occur, contaminants can concentrate in stream segments potentially impacting aquatic biota. More serious problems arise if the streams end in an evaporative sink, allowing accumulation of the contaminants in the water column and sediments, which provide habitat and food for migratory birds. In contrast, the presence of an evaporative sink will operate to mitigate any discharge to groundwater (i.e., groundwater is in fact moving into the pit).

Road construction during exploration and mining can result in soil erosion that can be washed into local streams by stormwater. Also, railroad beds in mining areas may be constructed using mine wastes. Storm events could leach contaminants into adjacent streams or drainages. Also, occasional spills from railcars can contaminate surface waters.

Groundwater withdrawal for mineral processing and to prevent filling of open pits and underground workings can affect groundwater quantities and levels. In the former case, most of the water is kept on site while in the latter excess water may be stored in impoundments and eventually discharged.

## 2.1.4   NATIONAL MINERALS POLICY

The Mining and Mineral Policy Act of 1970 (predecessor to FLPMA) states:

> The Congress recognizes that it is the continuing policy of the federal government to foster and encourage private enterprise in (1) the development of economically sound and stable domestic mining, minerals, metal and mineral reclamation industries, (2) the orderly and

economic development of domestic mineral resources, reserves and reclamation of metals and minerals to help assure satisfaction of industrial security and environmental needs...

This policy statement and the principle of multiple use i.e., "a combination of balanced and diverse resource use that takes into account the long-term needs of future generations." in FLPMA provides the basis for fostering sustainable mineral development. However, the complex regulatory scheme has made it difficult to permit mines and is leading to mineral exploration occurring outside the U.S.

### 2.1.5 PERMITTING PROCESS

Hardrock mining is subject to many federal state permitting, operating and reclamation requirements. Recent compilation of hardrock mining regulatory requirements and tabulations of permit requirements from Environmental Impact Statements are summarized in a study for Congress (NRC [10]). Similarly, coal and industrial mineral mining operations are subject to federal, state and local approval processes. The permit process is designed to make mining sustainable by protecting the public health and environment, and reclaiming the land for future generations.

Mining permits require the use of best available control technologies, and thus ensure that the mine operator is using currently available control technologies to minimize impacts to the community and the environment, while producing a return on the investment to shareholders. The public participation process has evolved significantly in a manner that the mine operator has to develop and make transparent the mining and reclamation plans. After the permits are issued, continuous monitoring of impacts to the environment is required, and ensures sustainability of the mining operation.

### REFERENCES

1. International Institute of Environment and Development (IIED), 2002, Mining, Minerals and Sustainable Development Project, London, England.
2. U.S. Geological Survey Minerals Yearbook (USGS), 2000, The Mineral Industries of Latin America and Canada.
3. National Mining Association (NMA), 2002, Facts about Coal and Minerals, pp. 2–56.
4. National Research Council (NRC), 1999, Hardrock Mining on Federal Lands, National Academy Press, Washington D.C., pp. 37–57.
5. Holland & Hart L.L.P., 1996, Environmental Regulation of the Mining Industry, from the Second Edition of the American Law, Mathew Bender & Company Inc., New York, N.Y.
6. Mining Environmental Handbook, 1997, Ed. J.J. Marcus, Imperial College Press, Chapters 3, 4 and 5, pp. 38–98, and 132–189.
7. 43 CFR 3809, 2004, Code of Federal Regulations (CFR), The Office of Federal Register, National Archives and Records Administrations, Washington D.C. (43 refers to Title 43: Public Lands: Interior and 3809 refers to subpart 3809: Surface Management). Can be accessed electronically at http://www.gpoaccess.gov/help/index.html.
8. 36 CFR 228, 2004, Code of Federal Regulations (CFR), The Office of Federal Register, National Archives and Records Administration, Washington D.C. (36 refers of Title 36: Parks, Forests and Public Property and 228 refers to part 228: Minerals). Can be accessed electronically at http://www.gpoaccess.gov/help/index.html
9. 40 CFR 761, 2004, Code of Federal Regulations (CFR), The Office of Federal Register, National Archives and Records Administration, Washington D.C. (40 refers of Title 40: Protection of Environment and 761 refers to part 761: Polychlorinated Biphenyls (PCBs) Manufacturing processing, Distribution in Commerce, and use prohibitions). Can be accessed electronically at http://www.gpoaccess.gov/help/index.html
10. National Research Council (NRC), 1999, Hardrock Mining on Federal Lands, National Academy Press, Washington D.C., pp. 169–196.

# 2.2   Asia

Ravi Bhargava[1], K.K. Jain[2] and N.C. Saxena[3]
[1]Ecomen Laboratories, Lucknow, India
[2]Director, Ministry of Environment & Forests, New Delhi, India
[3]Center for Mine Environment, Indian School of Mines, Dhanbad, India

## 2.2.1   INTRODUCTION

Mining is a vital sector of the economy in several Asian countries. Some of the largest coal and metal mines are in China, India, and Indonesia. The climate in these countries ranges from tropical to temperate, and the regulations relating to sustainable mining practices vary from country to country. The current status of regulations and mining practice in India is given as an example of the balance between mining and the environment in Asia.

## 2.2.2   INDIA – CURRENT STATE OF MINING PRACTICE

India is the biggest democracy in the world having the seventh largest geographical area and a steadily growing nation of over one billion people. There is much diversity in geographical features; Himalayas in the North; Thar desert, Aravalli Hills, and semi-arid plains in the West; Vindhyachal mountains in the Centre; Deccan plateau in the South; Western and Eastern ghats with vast coastal plains to the East and the West; and North-Eastern region with significant biodiversity. The country's coastline is 7,500 km long. There are 14 major rivers besides numerous smaller perennial water bodies and a great diversity of natural eco-systems. A general map of India with its major mineral resources is depicted on Figure 2.1.

### 2.2.2.1   *Natural resources and mining history*

India is endowed with significant mineral resources, which include fossil fuels, ferrous and non-ferrous ores and industrial minerals. Even the coastal tracts contain beach sand rich in rare earth and heavy minerals. These non-renewable resources are finite. Many of these resources are located in environmentally sensitive areas. Our best iron ore and bauxite deposits in eastern, central and southern India are located in forests.

History of Indian mining industry indicates that mining of coal dates back to 1774 when shallow mines were developed in Raniganj Coalfield along the west bank of river Damodar.

Iron ore mines were opened in 1878, gold mining from Kolar Goldfields was started in an organized way in 1880, the first mechanized iron ore mine was commissioned in 1958 at Noamundi by a private sector company, and main shaft in Jaduguda uranium mine was commissioned in 1968. The mining industry has achieved significant development in the country during the last five and half decades since 1947. As a result of the concerted and planned exploration activities carried out through the various Five Year Plans, today the country is comfortably placed in respect of many minerals essential for core sector industries like iron & steel, cement, aluminum, refractory, and ferro-alloys.

### 2.2.2.2   *Global challenge and mechanization trend*

Today, it is difficult to compete in the global market if cost of production is not competitive and timely delivery of quality products is not ensured. To meet this challenge, there is a need to optimize the

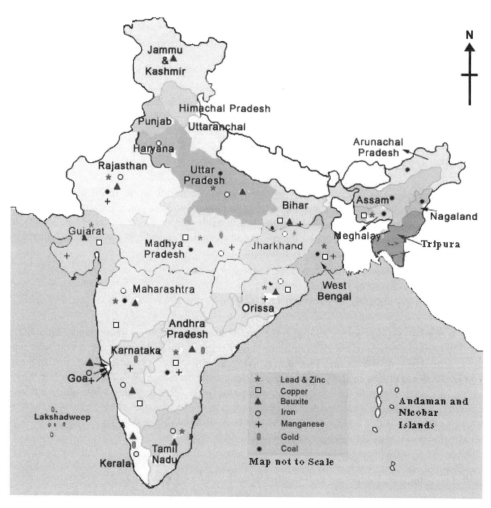

Figure 2.1.   A general map of India showing its major mineral resources. Courtesy: www.mapsofindia.com.

size of indigenous mining operations which at present are scattered and environmentally incompatible. Optimization measures include: amalgamate small mineral concessions; develop matching infrastructure; upgrade technology adopting best mining practices; and promote research and development in exploration, exploitation and process performance. For environmentally sustainable development, mineral resources are not to be exploited recklessly as in the past. The concept of mine closure planning has also to be an integral part of mining plan, which at present is in an embryonic stage in India.

Limestone and iron ore along with coal account for a major share in deployment of mining machinery. There are around 350 opencast mechanized mines in the country of which two thirds produce limestone and iron ore. There is a progressive increase in average size of mine due to adoption of heavy earth moving machinery. As blasting operations are not allowed in the proximity of settlements or eco-sensitive areas, high capacity continuous mining systems are becoming more popular. In some iron ore mines, bulldozers with ripper attachment are being increasingly used. Backhoes are also finding more and more use.

## 2.2.3   ENVIRONMENTAL MOVEMENT IN INDIA

The constitutional provisions to protect and improve the environment and to safeguard the forests and wildlife of the country date back to 1950. Issues relating to protection of environment and sustainable use of natural resources first received attention in the planning process in the early seventies. Even before India's independence in 1947, several environmental laws existed but the real momentum for bringing about a well-developed framework was made only after the 1972 Stockholm UN Conference on Human Environment. This was the first attempt to address relationships between environment and development at the global level. The conference created an important impetus in recognizing and addressing emerging environmental problems. Under the influence of this conference, the National Committee on Environmental Planning and Coordination (NCEPC) within the then Department of Science & Technology was set up in 1972. This was the turning point in national policies to integrate environmental issues into economic and developmental objectives. There was focus on improving air and water quality, control of chemicals and conservation of forests and wildlife. A number of important acts like the Wildlife (Protection) Act, the Water (Prevention & Control of Pollution) Act, the Air (Prevention & Control of Pollution) Act and the Forest (Conservation) Act were enacted in 1972, 1974, 1981 and 1980 respectively. NCEPC later evolved into a full fledged Ministry of Environment & Forests (MoEF) in 1985, which is the nodal agency in the administrative structure in the Central Government in the country for planning, promotion, regulating and ensuring environmental protection.

Power to regulate mines and minerals in minerals development is shared by the Central and State Governments. The Ministry of Coal and Mines is the primary agency in the Central Government for formulating and implementing policies for exploration, development, operations in respect of all mines and minerals other than natural gas and petroleum. Environmental clearance is required for all major mining projects having a mining lease area of more than 5 ha and forestry clearance is required for all mining proposals involving forest land, irrespective of the size of mining lease granted by MoEF.

### 2.2.3.1   *Environmental regulatory framework*

It is estimated that there are nearly 200 laws in the country with provisions relating to environment protection. Most countries in the world, both developed and developing, already have comprehensive legislation and monitoring systems to regulate mining activities from the environmental perspective.

Mining of minerals other than coal, lignite, natural gas and petroleum is regulated under the Mines and Minerals (Development & Regulation) Act, 1957, amended in 1994; Mineral Concession Rules, 1960; and the Mineral Concession and Development Rules, 1988. These acts and rules have provisions for environment preservation and protection while carrying out mining operations. In addition, there are following five main environmental acts which cover the mining industry:

- The Water (Prevention and Control of Pollution) Act, 1974 (MoEF, 1974 [1]) (amended in 1988)
- The Air (Prevention and Control of Pollution) Act, 1981 (MoEF, 1981 [2]) (amended in 1988)
- The Environment (Protection) Act, 1986 (MoEF, 1986 [3]) (amended in 1989)
- The Forest (Conservation) Act, 1980 (amended in 1988)
- The Wildlife (Protection) Act, 1972 (amended in 1991).

The important features of the above Acts and Rules are as follows:

1. *The Mines and Minerals (Development & Regulations) Act, 1957 (MOM, 1957 [4]) (amended in 1999)*
   This act provides for general restriction on undertaking prospecting and mining operations; procedure for obtaining prospecting licenses or mining leases; and conservation and systematic development of minerals.

2. *The Mineral Concession Rules (MCR), 1960 (MOM, 1960 [5]) (amended in 2000)*
   These rules framed under the MMDR Act, 1957 and subsequent amendments stipulate that a "Mining Plan" shall incorporate, amongst others, a plan of the area indicating water sources,

limits of forest areas, density of tress, impact of mining activity on forest, land surface and environment including air and water pollution; scheme for restoration of the area by afforestation, adoption of pollution control device and such measures as may be directed by the concerned Central and State Government agencies. An environmental management plan forms a part of the mining plan.

3. *The Mineral Conservation & Development Rules (MCDR), 1988 (MOM, 1988 [6]) (amended in 2000)*
   These rules contain a chapter devoted to environment. There are 11 provisions in this chapter pertaining to storage and utilization of top soil, storage of overburden, waste rock, reclamation and rehabilitation of land, measures against ground vibrations, control of surface subsidence, measures against air and noise pollution, discharge of toxic liquids, and restoration of flora.

4. *The Water (Prevention and Control of Pollution) Act, 1974 amended in 1988 and The Air (Prevention and Control of Pollution) Act, 1981 amended in 1988*
   These two acts provide for the control of water and air pollution and to tackle environmental problems in an integrated way by the establishment of one Central Board and individual State Boards. The acts prohibited waste emissions without written consent and payment of fees to the State Pollution Control Board.

5. *The Environment (Protection) Act, 1986*
   This act has precedence over the previous pollution control acts and provides for an overall protection and improvement of the environment. The Government of India is empowered by EPA to take all measures deemed necessary for protection and improving the quality of the environment, and preventing, controlling and abating environmental pollution including authority to direct closure of any industry or operation. The powers of Government of India under the act provide for laying down of standards for emissions and discharge of pollutants from all sources.
      The Ministry of Environment and Forests has notified the Hazardous Waste (Management & Handling) Rules in 1989 under EPA. These rules provide for control on generation, collection, treatment, transport, import, storage and disposal of hazardous wastes. The rules clearly state that import of waste for dumping is completely prohibited. The implementation of these rules is through the State Pollution Control Boards and the State Departments of Environment.

6. *The Forest (Conservation) Act, 1980*
   This act provides for the protection of two classes of forest; reserved and protected forests. Prior approval of Government of India is required for any change in status of reserved forest or non-forest use of protected forest land. Reserved forest has the highest conservation status and the area so classified cannot be used for any non-forest purpose. Surface and underground mining are deemed non-forest activities and, therefore, MoEF approval is required for mineral concessions in any forest area.
      The act stipulates that, for any area of forest lost due to development, the developers have to pay for purchase of an equivalent area of non-forest land as near as possible to the site of diversion, or twice the degraded forest area, for transfer to the State Forest Department with sufficient funds for compensatory afforestation, which is then declared as protected forest.
      The Safety zone for mining operations cannot form part of the replacement forest area. The developers have to provide funds to the State Department for one and half times the forested area of the safety zone. The act has now been modified with respect to underground mining so that only the area actually damaged by subsidence need be replaced by duly afforested non-forest land.

7. *The Wildlife (Protection) Act, 1972*
   This act provides power to the authorities for regulating hunting of wild animals, declaration of any area to be a sanctuary, national park or closed area, protection of specified plants, sanctuaries, national parks and closed areas and miscellaneous matters.

### 2.2.3.2   *Liberalization measures and the National Mineral Policy*

The Government of India issued the Mineral Development Policy statement in 1993 to achieve business excellence in the mineral sector, opening the exploration and mining for minerals to foreign companies. This was followed by further liberalization process through amendments in the mining law and the Foreign Direct Investment Policy. Today, the State Governments are fully empowered to grant mineral concessions for most non-fuel and non-atomic minerals and can also review, transfer or amalgamate minerals concessions without reference to the Central Government. Time frames have been fixed for conveying decision on mineral concessions applications and proposals for approval of mining plans. Foreign Direct Investment is possible in all sectors of non-fuel and non-atomic minerals through the automatic route. While the cap for automatic approval is 74% for diamond and other precious stones, even 100% approval is available for all other non-fuel and non-atomic minerals including gold and silver. Reconnaissance surveys have been identified as a distinct activity prior to prospecting and provisions has been made in the law for granting of reconnaissance permit.

The National Mineral Policy for non-atomic and non-fuel minerals prohibits mining operations in identified ecologically fragile and biologically rich areas and strip mining in forest areas. Opencast mining could be permitted only when accompanied by a comprehensive time bound reclamation. It states further that the environmental management plan should have adequate measures for minimizing environmental damage, restoration of mined out areas and plantation as per prescribed norms. As far as possible, reclamation and afforestation have to proceed concurrently with mineral extraction.

### 2.2.3.3   *Mining and environmental concerns*

Mining industry is one core economic activity, which has literally a "deep" relationship with the Mother Earth. More than 0.8 million hectares of lands is under mining, a substantial portion of which lies in forest areas. Important coal, bauxite, iron ore and chromite deposits in India are found in forests. Good limestone deposits are also available near the wildlife sanctuaries, national parks and coastal areas. Mineral production is often not in consonance with conservation of forests since at many places commercial reserves exist below thick forests.

Initially, the development projects from the public sector undertakings of the Central Government requiring approval of the Public Investment Board of the Ministry of Finance were normally considered for environmental clearances. For example, the multi-purpose river valley projects were cleared by the Planning Commission, Government of India in consultation with NCEPC in regard to environmental aspects. Environmental impact assessment based environmental clearance procedure was adopted as an administrative measure in late seventies for the river valley projects. The procedure was later extended to cover other sectors like industry, thermal power, nuclear power, and mining. India has over 24 years of experience in conducting environmental impact assessment of development projects. On 27th January, 1994 the Ministry of Environment & Forests had issued the Environmental Impact Assessment (EIA) Notification under EPA, 1986, imposing certain restrictions on undertaking new development projects or expansion and modernization of existing ones, unless prior environmental clearance has been obtained from the Ministry. The different processes involved in mining and processing of minerals along with the associated environmental impacts is detailed in Figure 2.2.

The EIA notification provides for two-stage clearance for mining projects. Site clearance is also mandatory for proposals for prospecting and exploration of major minerals when area is more than 500 ha. However, for carrying out test drilling on a scale not exceeding 10 bore holes per 100 sq km and for prospecting and exploration, no site clearance is required. The site clearance is given in the first stage and environmental clearance in the second stage. Besides the environmental clearance, forestry clearance under the provisions of the Forest (Conservation) Act, 1980 is also mandatory for all the mining proposals involving diversion of forest land for non-forest purpose. The forestry clearance is again accorded in two-stages. In the first stage, the proposal is agreed "in principle" subject to certain conditions. The second stage approval is given after receipt of compliance report from the concerned State Government regarding compliance with the conditions stated in the first stage.

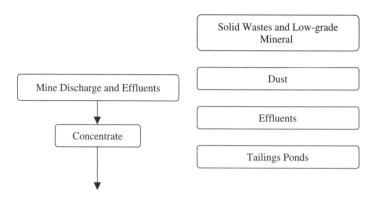

Figure 2.2.   Mining processes, wastes, and environmental issues.

### 2.2.3.4   *Stakeholder participation*

The opening of mines besides impacting the physical environmental in general, has social and economic consequences at the local level including displacement of people from the project site in some cases affecting their environment, health and culture. Learning from the experience of other counties, public hearing procedure was introduced in the decision-making process in India from 10th April, 1997. Public hearing has been made mandatory for all the development projects attracting the provisions of the EIA Notification. This has helped in the consideration of concerns of the affected local communities. The large mines address the communities concerns in much better manner than small mining company owners. However, a few small mining companies have also responded better with regard to environment and their relationship with the local people, setting a working model for others to follow.

The project proponent is invited to give presentation at the public hearing on salient features of the project, associated environmental issues, environmental protection measures, social welfare program for the local community and provide clarifications/answer to queries. Commitments to comply with certain suggestions are made by the project authorities. Thereafter, the State Pollution Control Board sends detailed report of the public hearing panel to MoEF. A time frame of 60 days for completion of public hearing has been set. If public hearing has already been held at the site clearance stage of the project then hearing is not required for environmental clearance. The public hearing procedure is not applicable to prospecting and exploration proposals.

### 2.2.3.5   *Environmental appraisals*

The following documents are required for environmental appraisal of the projects:

1. prospecting & exploration proposal – site clearance
   - information as per format for mineral prospecting
2. new mining proposal – site clearance
   - information as per site clearance format
   - public hearing report
   - No Objection Certificate (NOC) from the concerned State Pollution Control Board (desirable)
3. new mining proposal – environmental clearance
   - application as per schedule – II of the EIA notification
   - information as per format for environmental appraisal
   - rapid environmental impact assessment[*] and environmental management plan (EIA/EMP) report ([*]based on one full season baseline environmental data except for monsoon period)
   - details of public hearing[**] ([**]not to be repeated if public hearing has already been held at the site clearance stage)

- "Consent to Establish" from the concerned state pollution control board
- a copy of approved mining plan
- a comprehensive rehabilitation plan if more than 1,000 people are to be displaced from the project site, otherwise a summary plan
- commitment regarding availability of water and power from the concerned state authorities when supply is to be met from the public source. in case the project site falls in an area notified by the Central Ground Water Authority, a copy of their approval to extract ground water is to be furnished
- detailed hydro-geological report if ground water pumping including mine water discharge is heavy or mine site falls in a water scare/drought prone region
- a copy of forestry clearance in case the mining lease includes forest land
- in case of strategic minerals, NOC from the Atomic Energy Department
4. mining (operating mine) – environmental clearance
    - expansion or modernization of the existing operating mine requires environmental clearance. documents required are the same as for the new proposals. in addition, the project proponent has to submit the following:
        - a detailed report on compliance with the conditions of the clearance letter issued by the ministry of environment and forests, if any
        - compliance status report in respect of the conditions of the "Consent to Operate" issued by the State Pollution Control Board
        - last year's environmental statement
        - a report on implementation of resettlement & rehabilitation of project affected people, if any
    - the factors that cause delay in decision-making can be grouped in the following heads:
        - incomplete baseline data and information
        - old baseline data
        - inadequate maps
        - inconsistency in data
        - inadequately covered issues.

### 2.2.3.6 *Improvements proposed in the environmental appraisal process*

The following steps are being taken by the government to improve the environmental appraisal process in India:

1. *Scoping EIA*

    Currently, screening and scoping are not part of the EIA studies for mining projects. Consequently, some of the environmental issues which need to be discussed in detail, are not properly covered in the EIA/ EMP reports. The purpose of the scoping EIA is to define the key issues and resources to be analyzed in the EIA for determining potential impacts.

2. *EIA/EMP process*

    The MoEF has improved the process as follows:

    - provided guidelines and timetable for a formal process to determine the scope of EIA/EMP for each project
    - provided guidelines for collection and documentation of environmental baseline data for the key resources like surface water, groundwater, flora & fauna, land use, meteorology, air quality, and socio-economics
    - provided guidelines for inclusion of full operation, mitigation, reclamation, mine closure and monitoring in the EIA/EMP.

3. *Size of mining lease*

    In India, the State Governments grant mining leases as a small as 0.5 ha for mining of certain minerals. In such a situation, scientific working and compliance with environmental protection measures and standards is not feasible. A two-fold approach is being adopted to tackle the problem. First, for the existing mines, an amalgamation of small leaseholds wherever feasible or permission to prepare a collective EIA/EMP for a cluster of mines, has been adopted. Second, it has been decided to discourage such small-scale mining operations from the environmental management point of view. A committee has been constituted to look into the issues and make recommendations about the minimum size of mining lease that should be followed in the interest of systematic and scientific mining.

4. *Environmental Standards*

The existing environmental standards applicable to the mining industry are a composite of general environmental standards except for coal. These standards are inadequate in certain respects for effective environmental protection and management of minerals sector when compared to international norms. The Department of Mines has prepared a mineral-wise air quality and effluent standards for nine minerals. These standards are for chromite, copper, iron and manganese ore, lead and zinc, bauxite, limestone and dolomite mines.

5. *Demonstration of Best Management Practices*

Two demonstration sites having different topographic features and environmental management issues have been selected in Goa and Himachal Pradesh to demonstrate best management practices. The purpose is to demonstrate actual implementation of specific mitigation measures in a time bound and cost effective manner for replication at other mine sites. This would offer advantages such as:

- develop feasible mitigation measures and techniques for mining projects
- train mine operators in principles of mining, designing, implementing and monitoring the mitigation measures
- the projects will generate information that could be used for policy making and standards setting processes
- development of technical and management guidelines for different minerals mining projects.

6. *R&D Efforts*

The MoEF has identified a total of 25 centers throughout the country under its Environmental Information System (ENVIS) network for the collection, storage, retrieval and dissemination of information on various environmental topics. The Center for Mining Environment (CME) at the Indian School of Mines has funded 14 research projects in the following priority areas to develop indigenous know-how:

- management of solid wastes
- optimization of tailing pond /settling pond designs
- application of models to predict pollution levels
- utilization of industrial wastes such as fly ash
- stabilization and control of overburden dumps
- study of air borne respirable dust concentrations at work places
- digital modeling of groundwater contamination
- development of fugitive dust dispersion model for mining areas
- determination of emission rates of dust and validation of air quality models
- methodology for cumulative impact assessment in non-coal mining projects.

7. *Data Bank and Networking*

There are many organizations in the country, which could effectively serve as a resource in select areas of mining environment issues. However, because of inefficient networking capabilities, coordination among them often does not take place. An effective database management system and networking among the concerned institutions and the selected Central and State Government Departments and Pollution Control Boards is being established for facilitating dissemination and sharing of mining and environmental data. The MoEF has initiated a project to establish and operate pilot Environmental Information Centres (EIC) in three States

(Andhra Pradesh, Gujarat and Maharashtra) to act as a repository of validated environmental data which can be accessed and used by the stakeholders in the environmental clearance process, to evolve a suitable mechanism for net working with data providers/generators, and to identify constraints in developing a national EIC for mining.

8. *Institutional Strengthening*
   Institutional capacity building of Government Departments who are responsible for policy making, legislating, standard setting and enforcing compliance is being enhanced to ensure effective monitoring of EMPs and enforcement of the prescribed environmental standards by mine operators. Institutional strengthening will also focus on developing expertise in planning, designing and implementing environmentally friendly projects in the institutions working in the field of mining and the environment.

9. *Training – Resource Group Development*
   Environmental management training courses based on the best practices from around the world have been taken up to upgrade the knowledge and know-how of selected personnel from mining industry, research institutions, Pollution Control Boards, MoEF and other Central and State Departments in planning, designing and implementation of environmentally friendly mining projects.

10. *Baseline Data Generation*
    It is also conceived that baseline information on micro-meteorological and environmental (air, surface and groundwater quality, soil physico-chemical characteristics; flora and fauna) data should be generated by all agencies during mineral exploration activities.

### 2.2.3.7   *Environmental performance monitoring*

The mine operators are responsible for sending periodical reports (six bi-monthly and one annual report) on compliance with conditions of clearance letters to the MoEF and its regional offices. Site inspection is carried out by the regional offices to oversee enforcement of environmental safeguards. Interactions are held with the project representatives and the concerned State Pollution Control Boards. Besides this, Special Monitoring Committee(s) are constituted for sensitive projects to monitor compliance with the environmental protection measures and give advice to the company in the matter. Midterm corrective measures, if required, are also stipulated for environmental protection. However, non-compliance, deficiencies and violations, if any, are brought to the notice of the company. Time is given to them to ensure compliance with the stipulations and standards. In case of serious violation, action is taken under the provisions of the relevant Acts. Analysis of the past compliance reports indicates that the following are the main areas where compliance is not satisfactory:

- management of check-dams and drains
- plantation on inactive waste dumps
- fugitive dust control
- rehabilitation of project affected families
- particulate matter higher than ambient air quality standards
- stabilization of dump slopes
- treatment of water before discharge
- green belt development.

Even achievements in terms of reclamation of the forest area are not very encouraging. This calls for much greater attention to environmental matters on the part of the mine management. Most of the remediation measures are well within the competence and capabilities of the mine operators and all that needs to be ensured is a definite commitment to compliance of the measures set out in the EMP.

## 2.2.4   CASE STUDY

The Gujarat Ambuja Limestone mine is owned by Gujarat Ambuja Cements Limited. It is located in Junagadh district of Gujarat in western India. The mine occupies an area of 366 ha, and annually produces about 3 million tons of limestone and one million tons of marl as raw material for cement production. The limestone is soft to medium hard in nature. Initially, drilling and blasting of the limestone in conjunction with rippers was used. Since 1995, in an effort to reduce the environmental impacts, five surface miners excavate the limestone and marl, and convey it to the cement plant through a series of closed conveyor belts.

Environmental concerns involved in Figure 2.2 include:

- abandoned mines – fire, subsidence, air and water pollution
- reclamation and rehabilitation
- mine closure
- slopes stability
- reclamation of inactive dumps
- safety and health aspects.

Environmental issues involved in Figure 2.1 include:

- change in natural drainage pattern
- disruption of water regime/lowering of water table
- contamination of water bodies
- land degradation/localized subsidence
- ecosystem disruption
- air pollution
- displacement of people
- disruption of wildlife and loss of vegetation
- chemical leakage
- slope failures
- tailings ponds structural stability
- noise pollution
- radioactivity
- occupational and community health and safety.

In the lease area, the completely mined out area of over 63 ha has been concurrently reclaimed or rehabilitated. Mined out areas have been rehabilitated for use as cement plants, sewage water reclamation plant, and creating an artificial reservoir. Other areas are developed as orchards and forested land. The reservoir recharges the groundwater and increases the well yields in the surrounding areas. The reservoir attracts a number of migratory birds, and fish are also stocked in the reservoir. Over 58 different species of plants, totaling over 100,000 plants, are grown in the reclaimed areas to improve the biodiversity in the area. Some of the land that is not as productive is converted to pasture land.

An environmental management division has been setup by the company, and it carries out environmental monitoring using a van equipped with a satellite. Air borne dust surveys are carried out using gravimetric and personal dust samplers. Ambient noise levels near mining machines are regularly measured. The domestic waste is segregated into biodegradable and non-biodegradable fractions, and the biodegradable waste is composted. Mobile vacuum cleaners are used to keep the work areas dust free. An oil water separator is used in the plant to remove oils from wastewater before disposal. The company has obtained the ISO 14001 certification.

The rural development division known as Ambuja Cement Foundation undertakes development activities in the fields of education, community development, soil moisture conservation, water harvesting, agriculture and horticulture, and promotion of renewable energy sources. In summary, Ambuja cement operates its limestone mine in a sustainable manner.

REFERENCES

1. Ministry of Environment and Forests (MOEF) 1974, Water Prevention and Control of Pollution Act, Gazette of India, New Delhi, India, Amended 1988.
2. Ministry of Environment and Forests (MOEF) 1981, Air Prevention and Control of Pollution Act, Gazette of India, New Delhi, India, Amended 1988.
3. Ministry of Environment and Forests (MOEF) 1986, Environmental Protection Act, Gazette of India, New Delhi, India, Amended 1989.
4. Ministry of Mines (MOM) 1957, Mines and Minerals Development and Regulation Act, Gazette of India, New Delhi, India, Amended 1999.
5. Ministry of Mines (MOM) 1960, Mineral Concession Rules, Gazette of India, New Delhi, India, Amended 2000.
6. Ministry of Mines (MOM) 1988, Mineral Conservation and Development Rules, Gazette of India, New Delhi, India, Amended 2000.

# 2.2.5 Small scale mining in India

Subijoy Dutta[1] and S.L. Chakravorty[2]

[1] S&M Engineering Services, Crofton, MD, USA
[2] Ex. Hony Secy. General, NISM, Kolkata, India

## 2.2.5.1 INTRODUCTION

Although the focus and attention of the mining sector is placed on large scale mining, a considerable fraction of all mining activities in most of the developing countries fall under artisanal and small scale mining. In small scale mining people usually work in the informal sector with simple tools and equipment, outside the legal and regulatory framework. The vast majority are generally very poor, exploiting marginal deposits in harsh and often dangerous conditions – and with considerable impact on the environment (IIED [1]).

The relative contribution of Artisanal and Small Mining (ASM) to sustainable development depends on the priorities accorded to different objectives. In terms of meeting the world's need for minerals, large companies currently dominate the overall sector. For some minerals – such as emeralds and tungsten – virtually all production is from ASM. From an economic perspective, most resources can be mined far more efficiently and intensively using large-scale mining methods, and in terms of environmental damage, small-scale mining generally has a greater impact per unit of output. From a livelihoods perspective, ASM often provides the only means of obtaining income and is therefore important. Yet for many people it never provides more than a subsistence wage, so its actual contribution is often limited. For a detail global report on artisanal and small scale mining readers are referred to the MMSD [1] report.

To have an in-depth look at the sustainability of small scale mining, the ASM operations in India will be covered in this chapter. This should provide a realistic and relative perspective for small scale mining operations in terms of productions, socio-economic factors, minerals extracted, and the area and extent of ASM operations. It should provide a pragmatic parallel to compare and analyze the sustainability issues of another developing country.

## 2.2.5.2 ARTISANAL AND SMALL SCALE MINING

The liberalization and globalization of India's economy in the 1990s has provided a new vision of "world class" mining activities as it pertains to mining in India. And most of us think in term of multinational investment for mega-scale mining units. Quite often we forget that except for a few of our large-scale operations India is essentially a country of "small-scale mining" since as many as 95% operations can be defined as small-scale, and we can not ignore them. Although big and very big mines have big and strategic roles to play, we quite often fail to appreciate that the small/medium mines also play an important complementary role not only in terms of production but also in terms of resource discovery and utilization. We must remember that out of about 60/65 minerals extracted in India more than half are extractable only on small-scale because of geological constraints and techno-economic compulsions. However, they are an important adjunct of our industrial economy. Thus, if we try to objectively visualize the importance of small-scale mining, we have to accept and appreciate their positive role without any hesitation.

But unfortunately since 1947, except for occasional marginal statements nothing positive has been done in our national planning and executive actions to effectively place the small mining sector on a proper developmental track. Even in India's 1993 mining policy document there is only a cursory reference to small-scale mining. In India small-scale mining does not even have the status of "Small-Scale Industries" (SSI) to be entitled to different Government approved benefits.

The World Bank is formulating a positive policy for helping Artisanal/Small-Scale mining for which a Round Table Conference was organized in Washington D.C. on 17–19 May 1995, which one of the authors of this paper had the opportunity to attend. The Government seems to be serious about giving a major thrust to the mining industry as a whole including the small mining sector. Even for the underground coal mines which are basically small-scale operations, the nationalized coal giant Coal India Limited (CIL) is seriously considering making special efforts to increase both the production and the productivity.

## 2.2.5.3   MINERALS MINED ON SMALL/MEDIUM SCALE

Out of about 65 minerals mined in India, 38 came exclusively from small mines, 17 from small, medium and large mines, and 5 from comparatively larger mines confined to the non-ferrous sector. Broadly speaking, except for coal, base metals, iron ore and limestone, most of the other minerals are exploited in the small and medium mining sector, since the small deposits are predominant in number. And in the case of some minerals such small deposits are the only deposits. Even in the cases of iron ore, limestone, dolomite, bauxite, barite and the base metals there are large number of small deposits exploited by the small/medium mines. In the case of coal "we in India have been doing underground mining for more than a century now but our operations are still confined to small units of average size of 8,000–10,000 tons per month" (Kumar, 1996 [2]).

Some of the minerals exploited in Small/Medium and Medium/Large sectors are broadly indicated below:

1. Metallic (small/medium sector)
   - cassiterite
   - tungsten
   - nickel
   - manganese
   - iron ore
   - pyrophyllite
2. Non-metallic (small/medium sector)
   - agate
   - apatite
   - andalusite
   - ball clay
   - bentonite
   - barites
   - calc tufa
   - calcite
   - calcareous sand
   - china clay
   - chalk
   - corundum
   - dolomite
   - feldspar
   - fireclay
   - fuller's earth
   - fluorite, garnet
   - graphite, granite
   - gravel
   - gypsum
   - ilmenite
   - jasper
   - kyanite
   - lime kankar
   - lime shell
   - limestone
   - lepidolite
   - magnesite
   - mica
   - ordinary sand
   - moulding sand
   - silica sand
   - ochres
   - precious and semi-precious stones
   - pyrophyllite
   - rock phosphate
   - soapstone and talc
   - quartz
   - quartzite
   - building stone and road metals
   - staurolite
   - sillimanite
   - slate
   - wollastonite
   - vermiculite
3. Fuel minerals (small/medium sector)
   - coal (mainly underground mining)
4. Metallic (medium/large sector)
   - chromite
   - manganese

- iron ore
- magnetite
- copper
- lead and zinc
5. Non-metallic (medium/large sector)
  - bauxite
  - dolomite
  - limestone

- barite
- rock phosphate
- magnesite
- slate
- quartzite
6. Fuel minerals (medium/large sector)
  - coal (mainly open cast mining) and lignite

### 2.2.5.4   AN IMPORTANT ASPECT OF SMALL/MEDIUM-SCALE PRODUCTION

Quite often geological constraints prove to be the limiting factors allowing only small-scale operation. Apart from structural problems mineable reserves are quite often too small to allow large-scale operation. As indicated earlier, out of the 65 minerals produced in India about 38 or so could be produced on small-scale basis only because a large scale operation with large investment was not techno-economically viable.

If such geo-technically difficult deposits are not exploited by small-scale mining operations they would remain untapped making our economy totally dependent on import of many of such basic raw materials produced by small mines only. What is more significant is that the indigenous local industries, based on such minerals, would not develop fast enough if they have to fork out scarce foreign exchange for import of the raw material at the initial stage of development. In fact such small-scale mining gave impetus to the development of many of our indigenous mineral based industries (e.g. Ceramics, Granite, Glass, Fire Bricks, Ferro-Chrome etc.) creating employment in a dispersed manner. Such indigenous industries can be developed only under the initiative, drive and investment of the developing countries themselves because big investors from developed countries are not interested in such small deposits as we find in India today. But nevertheless, successful small-scale mining operations, planned and developed indigenously, may in a way lead to large-scale investigations and investment because small mines in many cases play the important role of catalytic agents. If non-exploitation of a viable resource is inimical to economic development, leaving such small and geo-technically complicated and difficult deposits unmined would be equally inimical to low-investment economic development.

Although large-scale mines can claim large production and high productivity at lower rates, high capital investment needs pose problems in setting up such operations.

### 2.2.5.5   EXTENT OF SMALL MINES IN INDIA

Since 1947 the number of operating mines (excluding "Minor Minerals") reporting to the Indian Bureau of Mines (IBM) increased from about 2,000 to over 4,800 in 1987. It stood at 4149 in 1990–1991 but came down again to about 3,750 or so in 1994–1995. There was however, a substantial change in the mineral wise distribution of mines both in terms of number and production (Indian Bureau of Mines, 1990–1991 [3]). The number of Mica mines has come down from about 1,100 in 1950 to 133 in 1990 and for Manganese Ore mines the drop has been from 604 in 1957 to 177 in 1990 primarily due to exhaustion of ore deposits and uneconomic workings. On the other hand for Limestone the number increased from 46 in 1947 to around 700 in late eighties; for Iron Ore the increase has been from 14 to 300 and for Bauxite Ore from 5 to 200 in 1990 due to increased demand of our developing economy.

Most of the increases have been in the small mining category (which can respond quickly enough to a developing economy) and more than 85–90% of the reporting mines (IBM Statistics) are probably in the category of tiny mines. Under the Mines Act, the Government of India reports only about half the number including also the Minor Mineral mines not considered in the IBM Statistics.

Table 2.1. Value of mineral production in India during 1988–1989 to 1995–1996 in Indian rupees in millions.

| Products | 1988–1989 (% of Prod.) | 1989–1990 (% of Prod.) | 1990–1991 (% of Prod.) | 1992–1993(P) (% of Prod.) | 1995–1996 (P) (% of Prod.) |
|---|---|---|---|---|---|
| All minerals | 148,050 (100) | 173,360 (100) | 190,210 (100) | 226,280 (100) | 332,120 (100) |
| Fuel minerals | 127,540 (86.1) | 147,620 (85.2) | 160,420 (84.3) | 188,280 (83.2) | 287,170 (86.47) |
| Metallic minerals | 9,070 (6.1) | 11,610 (6.7) | 12,970 (6.8) | 17,240 (7.6) | 21,420 (6.45) |
| Minerals | 5,510 (13.8) | 25,730 (14.9) | 29,780 (15.7) | 37,990 (16.7) | 43,990 (13.25) |

Source: Indian bureau of mines.

According to 1994–1995 figures maintained by the IBM, out of about 3,170 reporting mines (excluding fuel minerals) there are about 2,790 working mines (88%) in the Category B (small) producing on the average about 10,000 TPY. The secretary to the Govt. of India, Ministry of Mines, declared that "as much as 95% operations can be defined as small-scale" (Sen, 1996 [4]) and he undoubtedly considered only the formal figures reported to the IBM and other Govt. Departments. But if we consider also the informal sector, there are probably 8,000–9,000 mines.

Even in the nationalized coal mining sector (Coal India Ltd.) in 1995, most of the underground mines were in the category of small/medium mines with the average daily production of about 400 to 450 tons (Sengupta, 1996 [5]). The number of underground mines was 329 out of a total of 510 accounting for a production of 54.82 Mt (23%), the overall production of CIL being 237.5 Mt.

The values of mineral production in India during 1988–1989 to 1995–1996 are indicated in Table 2.1. It is evident from these figures that on the average roughly about 3/4th of the value of all non-fuel major minerals come from Metallic and Non-Metallic and about 1/4th from "Minor Minerals" groups.

The small mining sector (less than 100,000 TPY) was responsible in 1985 for about 35% (as against 16% of World average in 1982) of the total value of non-fuel mineral production in India. The calculation was based on the presumption that 70% value of Minor Minerals and 20% of other non-fuel minerals were from the small mines submitting statutory returns to the IBM and different State Governments. Out of this 35% about 3/5th (21%) came from Minor Minerals and 2/5 (14%) from all the other non-fuel minerals.

### 2.2.5.6 SOCIOECONOMIC ASPECTS

From the socioeconomic point of view small-scale mining has an important role to play in the developing countries. Since mineral resources are commonly distributed all over a country, small-scale operation makes direct contribution to the development of local economy in a dispersed manner. Apart from direct employment in mines, supply of many low-tech mining inputs and manufacture of many labor intensive mineral based products give rise to considerable local employment. Quite often shaping of stone and marble blocks is an important engagement for village people either in the artisanal or the organized sector.

Very often the tiny mines and also most of the small mines have to adjust their productive activities with the agricultural practices, particularly during monsoon when many such mines have either to close down temporarily or reduce production. Thus such small mines quite often supplement the agricultural income of the local people, effectively discouraging migration to the urban centers which is a very important aspect of our national planning.

At times, as it happened in Zimbabwe during 1991/1992, the small mines become an important source of sustenance during droughts and other emergencies (UN Guidelines, 1993 [6]). But there are some negative aspects, such as underpayment, health hazards and environmental degradation, which also need our attention.

It is more or less known that most of the small mines, particularly very small, with irregular and non-formal operators, under pay the labor below the Government determined wage rates. But in such cases the only redeeming point is that under village-level-economy the workers can manage better with lower wages than the urban workers. Moreover in many cases such low incomes from small mines are often taken as supplementary to agricultural and other income based on local resources. The national economy of India has not yet come to a stage where we can afford to stop all such underpayments because the alternative would be devastating to the local economy.

Health hazard is also a serious problem. There is hardly any provision in the widely dispersed small mines and crushers to protect the workers from dust pollution. Practically no medical facilities are available except for the first aid kit. As a result, a large number of workers suffer from pulmonary diseases. There are also cases of mercury poisoning in artisanal gold mining in many other countries, though not reportedly in India. Contamination of drinking water with traces of heavy metals and other substances of mining activities is possible, although extensive studies in this regard have not been made in India for small and medium mines.

Environmental damage is the other important factor to consider. Although such damages by individual and dispersed mines are not very significant, the contributions of small mines operating in clusters are considerable. Some times certain areas are completely devastated by a large number of small mines digging at random without any guidance and control. All such cases need special attention, guidance and assistance to motivate and persuade the miners to act in an eco-friendly manner. Complete stoppage of such mining operations is not considered to be a solution.

Thus small mines, apart from positive contribution to our economy, are also responsible for many negative impacts which need proper handling from the Government and different mining associations. Training and capacity building of the small mines is important for their sustainability.

REFERENCES

1. International Institute of Environment and Development (IIED), Mining, Minerals and Sustainable Development Project Report, London, England, 2002.
2. Mr. U. Kumar, President MGMI, MGMI NEWS LETTER, Vol. 21, No. 2, Jan/Feb, '96.
3. Indian Bureau of Mines (IBM), Indian Mineral Industry at a Glance, 1990–1991.
4. Ashoke C. Sen, Secretary to the Govt. of India, Ministry of Mines; T.H. Holland Memorial Lecture; The Mining Geological and Metallurgical Institute of India (MGMI) on 13th July 1996.
5. Mr. P.K. Sengupta, Chairman, Coal India Ltd., Valedictory address, International Seminar on High Production Technology in Underground Mining, MGMI, 13th July, 1996.
6. UN Guidelines for the Development of Small/Medium-Scale Mining – Selected Papers; UN the Interregional Seminar held in Harare, Zimbabwe; 15–19 February, 1993.

## 2.3 Europe

Subijoy Dutta[1] and Raj Rajaram[2]
[1] *S&M Engineering Services, Crofton, MD, USA*
[2] *Complete Environmental Solutions, Oak Brook, IL, USA*

### 2.3.1 INTRODUCTION

Europe accounts for only 2–3% of the global metal (non-ferrous) ore production, while it uses around 25% of the global total production of major non-ferrous metals. Together with the US, Western Europe is the largest user of metals. Further, the European mining and mineral sector plays an important role in the development of the economic activities of the European Union. Around 190,000 people are directly employed in the European minerals and metals mining industry, which generates a turnover in excess of 5 billion euros (MMSD [1]). The sector for construction minerals is by far the biggest employer, with direct employment of about 140,000 people. Many others are employed indirectly in associated industries such as equipment manufacture, exploration, processing, and manufacturing industries. In the European non-ferrous metals industry, for example, thousands of associated companies of different sizes employ more than 1 million people. Table 2.2 shows consumption of metals versus their population size for seven different countries representing a mix of industrially developed and developing nations. The data from this Table shows that Europe has a significant role to play in determining the patterns of metal use needed to make the transition to sustainable development.

With respect to the use of metals which has implications for practice and consumer trends in Europe the major perspectives of key stakeholder groups focuses on the following six initiatives.

- European Union (EU) Strategy for Sustainable Development,
- The Sixth Environmental Action Plan,
- The precautionary principle,
- Risk assessment strategies,
- The EU waste management and minimization strategy, and
- The revised EU Chemicals Policy.

All of these have implications for the mining, minerals, and metals sector. The issue raised by the European Union Sustainable Development Strategy and the use of the precautionary principle is succinctly presented here.

To get a clear reflection of the diversity in European climatic, geographic, and socio-economic environment, a recent study by MINEO, a European Research and Technological Development project was reviewed. MINEO brings together seven European Geological Surveys (France, Finland, Austria, United Kingdom, Denmark, Germany, and Portugal), members of EuroGeo-Surveys, the association of the geological surveys from the European Union, mining companies, an environmental institute (NERI) and the EC Joint Research Centre, Ispra. (Chevrel et al., 2002 [2]).

Six mining areas (test sites) in Europe have been chosen by MINEO to carry out the methodological developments. They have been flown at an altitude of 2,000–2,500 m above ground level during summer 2000, using the HyMap hyperspectral imaging spectroradiometer. Simultaneous ground radiometric measurements have been carried out for further radiometric calibration of the airborne data as well as for the generation of European-scale spectral libraries of contaminated areas. A map showing the Western European mine sites is shown on Figure 2.3.

Table 2.2.   Consumption of metals versus their share of world population for a few countries (MMSD [1]).

| Countries | % World population | Aluminum | | Copper | | Lead | | Steel | | Gold | |
|---|---|---|---|---|---|---|---|---|---|---|---|
| | | Kg/head | % Consumption | Kg/head | % Consumption | Kg/head | % Consumption | Kg/head | % Consumption | Grams/head | % Consumption |
| United States | 4.6 | 22.3 | 24.4 | 10.9 | 19.7 | 6.1 | 26.6 | 458.2 | 16.2 | 1.0 | 7.3 |
| Canada | 0.5 | 26.6 | 3.3 | 8.9 | 1.8 | 2.1 | 1.0 | 606.4 | 2.4 | 0.8 | 0.7 |
| Western Europe | 6.9 | 14.2 | 23.3 | 10.0 | 27.2 | 4.0 | 26.2 | 381.1 | 20.2 | 2.0 | 22.2 |
| Japan | 2.1 | 17.7 | 8.9 | 10.8 | 8.9 | 2.7 | 5.5 | 562.8 | 9.1 | 1.4 | 4.6 |
| Australia | 0.3 | 18.3 | 1.4 | 8.9 | 1.1 | 2.4 | 0.7 | 340.7 | 0.8 | 0.6 | 0.3 |
| Eastern Europe | 1.8 | 6.5 | 2.8 | 3.0 | 2.1 | 1.8 | 3.0 | 193.5 | 2.7 | 0.1 | 0.4 |
| China & India | 38.8 | 1.9 | 16.6 | 1.0 | 14.1 | 0.3 | 11.5 | 74.4 | 21.5 | 0.4 | 23.4 |

Figure 2.3. Location of the mining test sites (dark squares) (Source: MINEO [2]).

## 2.3.2 CURRENT STATUS AND SUSTAINABLE STRATEGY

In 2001, the European Commission proposed an EU Sustainable Development Strategy, later endorsed by the European Council, which was based on the need to integrate sustainable development into planning within the EU. From an industry perspective, the Sustainable Development Strategy raises several questions. One of the biggest concerns is to make sure that it does not limit industry's space to innovate; technological innovation must be placed at the heart of environmental strategy and not be impeded by over-regulation. Fabrizio d'Adda, Chairman of UNICEF's Environment Committee and CEO of the Italian company Enichem, states that "autonomous initiatives by companies are the main source of cost-effective progress in many environmental areas" as opposed to the "command and control" regulation of the European Commission.

Environmental organizations are also critical of the Sustainable Development Strategy and criticize the failure to tackle the international footprint of the EU's agriculture, fisheries, and trade activities. It is maintained that the strategy risks creating an "ecological Fortress Europe". Tony Long of the Brussels office of the World Wide Fund for Nature notes that the EU has for long been "strong on words and weak on action". The lack of clear targets and a timetable supports his observation. The European Environment Bureau, an NGO, regrets that the European Council did not adopt the phasing out of environmentally perverse subsidies, environmental tax reform, greening of public procurement, and strict environmental liability. Another key debating point is the feasibility of operation of the precautionary principle. The non-ferrous metals industry believes that the EU has given the impression that its use implies a search for zero risk. The principle should only be used after the completion of a risk assessment (that is, of exposure to risk) and when there is scientific uncertainty and reasonable grounds for concern as to the potentially dangerous effects of a substance. There is another school of thought that the definition of an "acceptable level of risk for society" has become politicized. Restrictions on the marketing and use of certain substances (subject to precaution) will hence be driven by a political agenda rather than scientific evaluation. Measures based on the principle should be proportionate and be preceded by a cost/benefit analysis that takes into account the impact of substitution of materials (MMSD [1]).

To the contrary of our most fervent wish of achieving a "zero-risk" environment, the reality echoes the impossibility of such wish. However, a sound environmental practice would zero in to the next best alternative of stimulating the prevention of harm. It is therefore in favor of an approach that includes:

- early action on the basis of reasonable suspicion of harm;
- the reversal of the burden of proof, because the traditional approach, which lies with legislators, may cause considerable delays before action is taken;
- the substitution principle: if safer alternatives are or may be available, they should be considered; and
- transparency and democratic decision-making to decide about the acceptability of technologies and activities and the ways to control them.

Some consider that environmental issues will increasingly come to the fore in the EU's policy agenda – this will affect the sector. It is widely recognized that the production and the use of target metals must be environmentally acceptable. A key debate is centered on precaution as a tool to manage hazard and risks from certain substances. This is an important debate for the non-ferrous metals sector since certain uses of some metals have the potential to present risks to public health and the environment. Examples of this are lead in gasoline and cadmium in batteries, if not properly recycled.

Several recommendations were put forth by the MMSD report. Some of the key recommendations are provided below:

- Development of a proactive stance (as opposed to a reactive one) towards legislation – The non-ferrous metals industry needs to share responsibility with authorities and civil society groups.
- Maintenance of an ongoing stakeholder dialogue, through which knowledge and information is shared – Different interest groups need to recognize their mutual dependency. The European Aluminum Association and Friends of the Earth Italy, for instance, cooperated on a study on the environmental performance of aluminum in road vehicles in the EU member states, plus Norway and Switzerland.
- Initiation of more meaningful dialogue at a national level between national associations, NGOs, governments of member states, and others – The United Kingdom Stakeholder Forum on Chemicals, for example, was set up by the UK Department of the Environment, Transport and the Regions in 2000 to promote a better understanding between different stakeholders (government, business, environment, and consumer groups) of people's concerns about chemicals in the environment.

- Increase in exchange of data and information between producers and downstream users.
- Development of more effective public policy and greater transparency to influence the growth and evolution of patterns of production and consumption.
- Development of better systems for determining the trans boundary impact of metals use within the community on other states and environments outside the EU.

## 2.3.3   CONCLUSIONS

There is a consensus among forecasters that consumption of aluminum and copper will continue to grow in the Western Europe at the historical rates of around 3%, at least over the next 5–10 years. The demand for lead is forecast to grow 1.1% annually in the next 5 years. Crude steel demand is expected to grow at between 1.8% and 2.1% per year. Consensus forecasts are invariably based on history and a "business as usual" approach to the future; they are often wrong. If transition and developing countries succeed in achieving a higher standard of living, barring some rather dramatic change such as development of alternatives to lead-based batteries, world consumption of lead could increase considerably. If 6 billion people in the world each consumed the 4.4 kg per capita that is today typical in industrial economies, world demand would be 26.4 Mt – over four times current world consumption. Renewable energy advocates have long suggested that countries without established power grids can electrify more effectively with decentralized generation based on wind or photovoltaics. This could create increased demand for batteries, which today are principally lead.

Recycling has an important role to play in the transitions towards sustainable development. In 2000, 15.6 Mt of aluminum scraps were recycled world-wide. The recycling rate is the percentage of material becoming available for recycling each year that is recycled. Recycling rates for building and transport applications range from 60% to 90% in various countries. The aluminum industry is working with automobile manufacturers to enable easier dismantling of aluminum components from cars in order to improve the sorting and recovery of aluminum. In 1997, over 4.4 Mt of scraps were used in the transport sector, and the use of aluminum in automobiles is increasing yearly (World Aluminum Institute [3]). The growth of packaging expected in South America, Europe, and Asia (especially China) may allow for growth in some parts of the scrap recycling industry. In the case of lead, 60–62% of refined lead production in the western world comes from recycled material. In the US, 90% of spent batteries are recycled. More than 50% of steel production in industrial nations comes from recycled materials. Despite the rapid growth rates and large volumes consumed in Asia, especially China, on a per capita basis, most consumption still occurs in the most industrialized countries (MMSD [1]).

## REFERENCES

1. MMSD, The Mining Mineral and Sustainable Development Project, a joint effort of a multidisciplinary team from 10 countries headquartered in London, 2001.
2. MINEO – Cheverell, S. Belocky, R. and Grösel, K., *Monitoring and assessing the environmental impact of mining in Europe using advanced Earth Observation Techniques* – MINEO First Results of the Alpine Test Site, http://www.brgm.fr/mineo, April 2002.
3. World Aluminum Institute at http://www.worldaluminum.org/production/recycling/index.html.

# CHAPTER 3

## Issues in sustainable mining practices

Raj Rajaram

*Complete Environmental Solutions, Oak Brook, IL, USA*

### 3.1   INTRODUCTION

Breaking New Ground, published as part of the Mining, Minerals and Sustainable Development (MMSD) project in 2002, devoted a chapter summarizing many issues under the topic Mining, Minerals and the Environment. Over 40 commercial and non-commercial sponsors of the project detailed the pressing issues affecting the minerals industry, and listed several issues critical to the sustainability of the industry as follows:

- large-volume waste
- mine closure planning
- environmental management
- land use planning
- energy use in the minerals sector
- threats to biological diversity.

These issues and their impact on the community that mining companies operate in will be discussed in this chapter. In addition, the economics of sustainable mining will be discussed in this chapter. Sustainable mining systems that address these issues will be addressed in the next chapter.

Since natural capital is depleted during mining, the conditions for sustainable development may still be met so long as other forms of capital, such as manufactured, human, social and financial capital increase (MMSD, 2002 [1]). If there is net gain in the total of these capital stocks over time, the development is sustainable. This view of sustainable development, termed as the "soft view", will be emphasized in this chapter, and throughout the book.

Although the above listed issues concentrate on the environmental impacts of mining, sustainable mining requires attention to the economic, social and governance issues. According to the deliberations that preceded publishing of Breaking Ground (MMSD, 2002), the issues that are being tackled by the mining industry in the economic, social and governance spheres include the following:

- Economy: Maintaining and enhancing the conditions for a viable industry, internalizing environmental and social costs, and maximizing human well-being.
- Social: Ensuring a fair distribution of costs and benefits, respecting and reinforcing the rights of all human beings, and ensuring that replacement of depleted natural resources with other forms of capital.
- Governance: Ensuring transparency through the transmission of relevant and accurate information to all stakeholders, ensuring that decisions are made at the appropriate level, and supporting participatory decision making and accountability for decisions and actions.

#### 3.1.1   *Large volume of waste generated*

Mining operations produce a large volume of generally low toxicity waste. This was recognized when the United States passed the Resource Conservation and Resource Recovery Act (US EPA, 1977 [2]) to manage hazardous wastes from industry. The Bevill Amendment to the Act specified different treatment for mining wastes. Mine wastes can be classified into four major categories (MMSD, 2002 [1]):

- overburden, the soil and rock that is removed to reach the orebody
- waste rock, the rock that is mined but is below the required grade for processing
- tailings, the waste resulting from the processing of ore to produce the desired concentrate
- heap leach spent ore, the rock remaining in heap leach pads (the low-cost method being used extensively for extracting valuables from lower grade ores).

These wastes are generally disposed on land; however, marine disposal has been successfully used in areas where either land disposal was not possible or environmental impacts could be managed adequately. Since marine disposal impacts are still being studied, this book will focus only on land disposal of mining wastes.

A major factor to consider in the disposal of mining wastes is acid mine drainage (AMD) since many ore bodies are in sulfide environments, and coal mining invariably results in AMD. The state of Wisconsin in the U.S. has passed legislation making waste management from mines in sulfide ore bodies so stringent that the government will not issue a permit until the mining company can satisfactorily prove (through case studies) that it will not pollute the surface water or groundwater from AMD or from the release of heavy metals (Wisconsin statute 293.50, 1997 [3]). Mining companies and governments have devised various methods to predict the AMD potential from mining operations, and methods to prevent or treat AMD (US Bureau of Mines, 1986 [4]). Methods to handle AMD, technically and administratively, are discussed in Chapter 6.

Overburden and waste rock can be backfilled in mined out areas, and is successfully done in many coal mining operations. In a large surface mine in Illinois, the backfilled areas are used for productive farming within a few years of completion of mining (personal communication, Arch Minerals). However, this is not possible in many metal mining situations where the mining ore from deep pits precludes backfilling. The increase in volume of waste when excavated (generally about 40%) means that the overburden and waste rock must be disposed of in nearby valleys. In underground mining, the waste rock can be left underground, and overburden is not removed. Waste disposal with out creating AMD impacts can be done without significant costs if the waste is characterized, and the location of the waste disposal site carefully selected. Chapter 6 discusses waste management of tailings and overburden/waste rock.

REFERENCES

1. International Institute for Environment and Development, 2002, Mining, Minerals and Sustainable Development (MMSD) Report, London, England.
2. U.S. Environmental Protection Agency, 1977, Resource Conservation and Recovery Act, Government Printing Office, Washington, DC.
3. State of Wisconsin, United States, 1997, Statute 293.50, Mining Sulfide Ore Bodies, Madison, Wisconsin.
4. U.S. Bureau of Mines, 1986, Pre-Mine Prediction of Acid Mine Drainage, Government Printing Office, Washington, DC.

## 3.2 Mine closure planning: reclamation, closure and post-closure issues

Krishna Parameswaran

*ASARCO LLC, Phoenix, AZ, USA*

### 3.2.1 INTRODUCTION

In its review of the adequacy of environmental protection provided by the existing regulations in the United States, the National Research Council (NRC [1]) discussed concerns relating to mine closure at active and inactive hardrock mining sites and identified some of the issues: backfilling of open-pit mines, reclamation planning for post-mining uses and post closure considerations. The NRC report was based on a request by the U.S. Congress to assess the adequacy of the regulatory framework for hardrock mining on federal lands. NRC appointed the Committee on Hardrock Mining on Federal Lands (Committee) to conduct the assessment. The NRC report recommended that land management agencies should ascertain that mine operators plan for and assure the reclamation and long-term post-closure management of mine sites on federal lands.

On the issue of the appropriateness of backfilling of open-pit mines, NRC acknowledged that it is a subject of considerable debate and indicated that it has no basis to contradict the conclusions reached by the National Research Council report by the Committee on Surface Mining and Reclamation (COSMAR) in 1979, which found:

> The [Surface Mining Control and Reclamation] Act requires that [coal-mined] land be restored to approximately its original contour. This provision is generally not technically feasible for non-coal minerals, or is of limited value because it is impractical, inappropriate or economically unsound.
>
> Further, to restore to the original contour where massive ore bodies have been mined by the open-pit method could incur costs roughly equal to the original costs of mining. Although technically possible, such backfilling of a large open-pit would be of uncertain environmental and social benefit, and it would be economically impractical to mine some deposits under the current cost structure.

Although the NRC Committee was of the opinion that backfilling may be environmentally and economically desirable in certain situations, such decisions must weigh the costs and benefits in a site-specific context.

Reclamation planning should consider all possible future uses. This could include showcasing historic mining features while promoting economic activity, such as tourism. Historic mining towns such as Cripple Creek and Leadville, Colorado, besides being an important part of the nation's cultural legacies continue to attract tourism. Pit lakes can provide aquatic resources for wildlife, high-walls of open pit mines can offer habitat for raptors and underground mines can offer habitat for bats. Historic mines also offer the opportunity for training earth science and engineering students, such as the Waldo mine in the Magdalena mining district by the New Mexico Institute of Mining and Technology, Socorro, New Mexico. In testimony to the Committee, the argument was made that the United States as the largest consumer of minerals in the world has a responsibility to preserve opportunities for mining on public lands and that experience has indicated that existing mining regions provide the most likely locations for future mineral development and extraction. While protection of the environment and safety considerations are of primary importance, adequate consideration must be given to future value of minerals that remain unextracted and uses of mine sites in reclamation planning.

One of the most common concerns expressed to the Committee was the post-closure condition of the site and impacts of post mining uses. Reclaimed mine sites require some degree of monitoring, inspection and maintenance over the long term. Such activities include maintenance of reclamation features such as soil covers, vegetation, closed impoundments, waste rock piles and water diversion structures. In addition, the quality of surface water and groundwater will have to be monitored.

Post-closure use of the land needs advance planning. Pit lakes could be used for boating and other water sports. Waste rock and leach piles may attract off-road vehicle enthusiasts and open-pit highwalls may attract weekend rock climbers. Some of these uses could adversely impact reclamation features and others could pose safety concerns. For example, extensive off-road vehicle use could damage soil and vegetative covers, resulting in water quality problems. These potential adverse environmental impacts and safety concerns must be addressed in reclamation planning.

Post-closure issues could also arise from unusual climatic or hydrological events. Floods or droughts in a desert environment could damage the reclaimed site resulting in additional costs to maintain reclaimed structures. Long-term monitoring to detect potential problems and trends is usually required. This generally involves periodic inspection by the site owner or regulatory personnel. Some sites may require more extensive and sophisticated post-closure monitoring programs. Monitoring can be used to determine if management options are properly carried out, to predict outcomes for closure plans and identify and characterize trends so that if problems arise, corrective action can be taken.

Laws requiring reclamation have been enacted in most states in the western United States where hardrock mining occurs on public lands, as noted by NRC [1]. They broadly resemble each other but differ in terms of specificity (prescriptive versus performance based standards) and financial assurance. They are generally administered by state agencies that regulate mining. Important features of state reclamation laws, as discussed in Section 2.1.1.2 in Chapter 2, include:

- thresholds for regulation of activities: exploration, small mining operations, large mining operations and designated operations utilizing certain chemical beneficiation techniques;
- application content, review and approval procedures, and requirements for public participation;
- requirements relating to characterization of overburden and ores, prediction of acid drainage and management of acid generating materials;
- requirements for management of beneficiation reagents (e.g., cyanides, acids, solvent extraction reagents);
- requirements for stabilization and reclamation of site;
- requirements for revegetation;
- requirements for closure of tailings ponds, waste rock piles and spent ores areas;
- financial assurance requirements: what assurance covers, amounts, forms of acceptable assurance (bonds, letters of credit, corporate guarantees, etc.); and
- reporting and monitoring requirements.

Smaller mining operations have to meet fewer obligations under state laws.

NRC noted that financial assurance for reclamation of mining sites is part of state regulatory requirements in all of the western states where mining is conducted on federal lands. In addition, all mines that are required to have Approved Plans of Operation under Bureau of Land Management (BLM) regulations and they are required to have some type of financial assurance for reclamation. No financial assurance is required for Notice-level activities such as exploration, and mining and mineral processing operations that disturb less than 5 acres.

Several states require or authorize agencies to require financial assurance for long-term protection of water quality. Some mines on federal lands have provided similar assurance, negotiated on a case-by-case basis with state and federal agencies, even where such requirements are not explicitly required in regulations.

The total amount of financial assurance necessary to guarantee that there will be financial resources to properly close the mine depends upon the current cost of reclamation activities to be completed. The cost for reclamation is calculated based on the basis of cost to bring a third party to complete the work with regulatory agency and/or consultant oversight and not on the operators cost to complete the work. Regulations also require periodic review and, if necessary, adjustment of reclamation costs and financial assurance instruments.

The regulations for each state and the federal land management agencies dictate what types of instruments are acceptable. These include: cash, letters of credit, corporate guarantees, and bonds. A corporate guarantee is a guarantee by the company that it will complete the reclamation work and is based on passing a "financial" test that determines a company's resource by looking at its assets and liabilities. There is no exchange of money with regulatory agencies. Less common are deeds of trust for real estate or liens on equipment. Publicly or privately held trust funds established by companies have been used to guarantee long-term water treatment and monitoring.

Financial assurance for reclamation is required until reclamation is complete and successful, i.e., physical stabilization and revegetation is successfully completed. Most state and federal agencies will consider partial release of bonds as portions of work are completed. For example, when regrading is completed the portion designated for that activity is released. Trust funds for long-term treatment or monitoring are not released but used to provide revenues for continuous funding.

Many mining companies have depended on insurance companies to provide financial assurance to various regulatory agencies that mined lands would be reclaimed and restored to the desired land use. The rapidly escalating costs of bonding requirements resulting from terrorist attacks, economic downturn and corporate bankruptcies have resulted in a climate where most insurers were unwilling or unable to issue reclamation bonds. Moreover, states and regulatory agencies have significantly increased the bonding requirements based on their experience where they had to step in to clean up mine sites and found that reclamation bonds were inadequate to address the contamination, such as at the Summitville mine in Colorado and the Zortman mine in Montana. For example, the State of Montana increased the bond for Asarco's Black Pine mine from $70,000 to $8,000,000, and the bond for Asarco's former Troy mine (now owned by Revett Silver Mining Company) is likely to increase from $10,500,000 to between $20,000,000 and $25,000,000. The state of New Mexico is attempting to increase the reclamation guarantee from $110,000,000 to $900,000,000 for three Phelps Dodge Company copper mines in that state. Alaska and Nevada have formed bond pools. However, such pools cannot provide "full-cost," financial guarantee for all operations they cover. Providing financial assurance on such a scale could adversely affect the viability of mining companies, as it makes raising capital for the industry even more difficult. However, these challenges and issues in mine reclamation, closure, and post-closure need to be considered at the very onset and mining and reclamation activities should be planned in a phased manner for economic viability and sustainability of operation.

The following section provides a case history of a well planned mine closure.

## 3.2.2   HOMESTAKE MINING COMPANY'S McLAUGHLIN GOLD MINE

The McLaughlin gold mine, now owned by Barrick Gold Corporation is located in the costal ranges of northern California, about 70 miles north of San Francisco. Barnes [2] describes the development of a reclamation plan for the McLaughlin mine and the selection of methods for reclamation of waste rock deposition areas, and Krauss [3] discusses mined land reclamation and contiguous lands management plan as an integral feature of environmental management at the McLaughlin mine.

The area was identified as an exploration target in 1978 and confirmed by drilling to be a commercial deposit in 1980. The ore body contained 20 million tons of gold ore averaging 0.152 troy ounces per ton, or approximately 3 million troy ounces of gold. The mine commenced production in March 1985. Mining was by open pit method and involved movement of about 50,000 tons per day of waste rock, low-grade and high-grade ores to provide a millfeed of 6,200 tons per day. The mine produced a total of 3.4 million ounces of gold. The waste rock deposition area was designed for 160 million tons covering 385 acres.

The mine is located on a ridge at an elevation of 2,000 feet above sea level, at the headwaters of two drainages. The climate is Mediterranean, with an average rainfall of 30 inches with precipitation occurring predominantly between October and April. Temperatures are in the 90°F range (peak exceeding 105°F) in summer. In winter daytime temperatures are in the 40°F with seasonal lows in the 20's. Snowfall occurs several times a year, with average precipitation less than 3 inches.

The project site is remote and sparsely populated, with the closest community 15 miles to the north. The economy and land use consists of seasonal tourism and agriculture, primarily pears, walnuts and grapes. The Napa Valley is located to the south of the project with the primary land use centered on the premium wine grape industry. The Sacramento Valley lies to the east of the project with the predominant land use being agriculture, mainly field crops and orchards. Vegetation at the site is a mixture of plant communities including Cismontane Introduced Grasslands, Blue Oak Woodland, Serpentine Chaparral and Northern Interior Cypress Forest. The latter two are associated with soils derived from the serpentinitic parent rock. Such soils have physical and chemical characteristics, such as calcium deficiency, excess magnesium, elevated levels of nickel and chromium, and clay consistency that results in soil instability and limits plant growth.

Reclamation is regulated by the California Surface Mining and Reclamation Act (Act) and is administered by the local county and city governments. The Act requires designation of a post-mining land use and preparation and approval of a mining and reclamation plan prior to commencement of mining. The reclamation plan must specify the final land form and topography; post-mining hydrology; top soil salvage, storage and replacement; seedbed preparation and fertilization; species to be planted and planting methodology; erosion and sedimentation control, and maintenance of revegetated areas. The Act also mandates that local governments require the determination of reclamation costs and the posting of financial assurances.

The McLaughlin mine reclamation plan specifies a unique post-mining land use, which provides for the conversion of the approximately 10,000-acre site to an environmental studies field research station when mining ceases. The mine site has been converted to the Donald and Sylvia McLaughlin Reserve (Reserve), a unit of the University of California Natural Reserve System. The field research station serves as the science support center for the surrounding 600,000-acre Blue Ridge/Berryessa Natural Area. It was felt that the mine's accumulated environmental and monitoring data would provide a valuable educational tool. The environmental baseline data and comprehensive monitoring program provide over twenty years of data, including aerial photography updated annually, ground and surface water quality data, aquatic ecology, wildlife surveys, and vegetation and sensitive plant surveys. In addition, physical facilities including laboratories, power, roads, water supply and sanitary facilities have been retained for Reserve use. The field station is being used by schools, colleges and universities. In excess of $2,000,000 of independent funding has been secured by Reserve researchers in each of the last three years.

The reclamation plan provides specific reclamation goals and the methods to be used to reach those goals. These goals are to:

- minimize erosion;
- stabilize disturbed areas with permanent diverse vegetative cover;
- maximize productive land use; and
- protect water quality.

The goals were met through preconstruction engineering and planning to facilitate implementation of the reclamation plan. Research and literature review identified reclamation methods best suited for the site and soils suitable for reclamation and stockpiling soils during construction. The plan provides for the reclamation of waste rock deposition areas concurrent with mining, rather than at the end of mine life. Annual monitoring data was used to evaluate and modify reclamation methods for future years.

Species selection for grass seed mixes were based on research conducted at the University of California at Davis, utilizing twenty years of pertinent revegetation experience at the Geothermal Steam Field Station electric generating facility and state highway revegetation work by the

California Department of Transportation, and research on grass species at the site, supported by the California Department of Mines and Geology. Native woody species indigenous to the site were evaluated for use in reclamation taking into consideration soil types, seed collection potential from surrounding native plants, direct seeding potential, nursery propagation potential and expected survival rates after transplanting into disturbed soils as tublings.

Topsoil was identified and salvaged during project construction and stockpiled at various locations for use in reclamation. The soils from the Davis Creek freshwater reservoir location were stockpiled and used for the reclamation of waste rock deposition areas. Additional topsoil was salvaged during open pit development for placement on final slopes and benches of waste rock deposition facilities during mining. Over one million cubic yards of soil were stripped from the tailings basin for use in closure of the tailings facility.

Topsoil is placed on the reshaped slope at an average depth of 1.4 ft and then trackwalked to provide a firm soil surface. Fertilizer additions were based on specifications for northern California soils of similar types. Hydrated lime was applied as a calcium supplement for serpentine soils by mixing lime with water in a hydroseeder and spraying the mixture on the areas to be reclaimed prior to application of seed and fertilizer.

Annual monitoring has demonstrated successional patterns in the reclaimed areas during the almost twenty years since initial seeding. In general, reclamation success has been excellent and all performance criteria have been met, with the exception of occasional rilling in limited areas. Woody species establishment has also met expectations. Approximately, 1,325 acres were reclaimed since the project commenced in 1983 and approximately 136 acres of waste rock deposition areas were reclaimed between 1985 and 1990.

Reclamation and waste disposal plan amendments approved in 1994 allow for the use of waste rock to backfill the mine pit. Approximately, two-thirds of the north pit and one-third of the south pit have been backfilled with waste rock. The remaining pit volume is projected to be filled with groundwater and surface water up to the 1,700–1,730 foot elevation. It will also serve as a containment and evaporation facility for waste rock leachate. Bedrock highwalls will remain while the remainder of the pit rim will be groomed and revegetated. The east and west rock deposition facilities and in-pit backfills have been capped and revegetated. Final slopes have been reduced to 2.5:1 or less and determined to be stable under projected seismic loading. Engineered drainage channels are armored where needed. Initial planting of woody species has been completed and will continue in the post-closure phase. Leachate collection and pumpback facilities will be maintained and operated by Homestake. Figures 3.1 and 3.2 show the progressive reclamation of the dumps in 1991 and 1998.

The mine area will remain fenced for public safety and upper benches revegetated to minimize visual impact. The Davis Creek fresh water reservoir remains near the maximum normal pool elevation and serves as an aquatic habitat, as well as a principal research facility for the Reserve. Homestake will continue to own and maintain the dam under the jurisdiction of the State Division of Dam Safety. The basic limnology and mercury transport studies undertaken at the reservoir over the past sixteen years have been widely reported in the scientific literature and provide the basis for much of the current work on the fate and management of mercury in California.

Process and mine area facilities have been dismantled and salvaged or scrapped. Their foundations have been broken and covered in place. These areas will be regraded for proper drainage, topsoil added and revegetated. Native trees and shrubs will be introduced to provide habitat and visual diversity. Fences, gate, access roads and utilities needed for Homestake's use, as well as the Reserve's use will be maintained. Administration and other ancillary buildings are now being used by the Reserve.

The tailings disposal facility will be maintained as a zero discharge facility and will be reclaimed as mixture of wetland and grassland habitats. Sediment ponds and other hydrological structures have been retained.

A financial guarantee for the reclamation of the site is provided by a Financial Assurance Agreement between Homestake and a Trust consisting of Lake, Napa and Yolo counties, the Bureau of Land Management (BLM) and the Regional Water Quality Board. The cost of site closure in each

Figure 3.1.   Progressive reclamation of the dump (1991).

Figure 3.2.   Progressive reclamation of the dump (1998).

of the years of operation has been calculated and the amount is payable to the Trust if the Company does not fulfill its reclamation commitment. The calculated amount is guaranteed by a renewable letter of credit payable to the Trust. Failure by the Company to renew the letter of credit prior to its expiry date causes the bank to pay the calculated amount for that year to the Trust. An agreement among the trustees provides for the expenditures of the paid funds. In addition, the agreement allows the Regional Water Quality Control Board to draw a portion of the guaranteed funds for pollution clean up, if the Company does not conduct the necessary cleanup. The assurance amount is reduced each year as successful reclamation is completed.

The majority of lands surrounding the McLaughlin mine are public lands jointly administered by the BLM and the California Department of Fish and Game (Fish and Game). A plan for the management of approximately 7,500 acres of Homestake owned non-mined lands, the Contiguous Lands Management Plan (CLM Plan) has been prepared, in consultation with Fish and Game and the California Water Resources Board. The CLM Plan has several integrated elements including a fire hazard reduction program, a sensitive plant restoration program, a pasture use program, a habitat restoration and improvement program and an education and research program. The CLM Plan provides for the involvement of the regional academic community.

Post-closure security will be maintained as during operations. Existing perimeter fences and signs will be maintained to prevent unauthorized public entry to the site. On-site Homestake and Reserve personnel are able to respond to incidents of trespass. On-site safety has been enhanced by the addition of locked gates across roads leading to areas of potential hazard and by placing berms at overlooks and along roads near high-walls and other exposures. Post-closure water quality will be monitored, as per Regional Water Quality Board requirements.

In summary, every mine leaves a legacy. In some cases the legacy is intended and in other cases a mine's legacy includes impacts and consequences that were not intended. Good reclamation and closure planning can make a difference in the kind of legacy left behind. Historically, the industry has been primarily concerned with a mine's economics, that is, did it turn a profit for the company and what benefits it provided the employees and the surrounding community while it operated. In the last decade, mining companies have also become increasingly concerned with the environmental and social legacies of their operations. The reclamation and closure of the McLaughlin Mine provides one example of a company's response to these post-closure environmental and social concerns in a developed country setting. Mining and mine closures in developing nations often present even greater opportunities to provide long term economic, environmental and particularly social benefits, while posing their own challenges in mineral discovery, mine development and eventual closure. Successful mine closures with positive legacies can provide important support for new mining ventures. In the future, environmental, social and economic considerations have to be carefully balanced for mining to be a sustainable enterprise.

REFERENCES

1. National Research Council, Hardrock Mining on Federal Lands, National Academy Press, Washington D.C., 1999, pp. 82–86; pp. 118–120; pp. 215–217.
2. Barnes, P.C., "Development and implementation of reclamation practices at Homestake's McLaughlin mine", Presented at the National Meeting of the American Society for Surface Mining and Reclamation, Durango, Colorado, May 14–17, 1991, p. 12.
3. Krauss, R.E., "Environmental Management at Homestake's McLaughlin Mine," Proceedings of the Gold '90 Symposium, Hausen, D.M., Halbe, D.N., Peterson, E.U., and Tafuri, W.J. (editors), Salt Lake City, Utah: Society of Mining, Metallurgy and Exploration Process Mineralogy X, pp. 509–519.

# 3.3  Economics of sustainable mining

Subhash Bhagwat
*Illinois Department of Natural Resources, State Geological Survey Division, Urbana/Champaign, IL, USA*

## 3.3.1  INTRODUCTION

"Sustainable development is the pattern of development that meets the needs of the present generation without compromising the ability of the future generations to meet their own needs"
(Cordes [1]).

The commonly used definition of sustainable development quoted above, when applied to mining, raises the question: What do we want to sustain when we mine a mineral, fuel or water resource? The answer cannot be the resource itself, given the finiteness of all physical resources. We cannot mine a mineral and sustain it at the same time. Thus, while we satisfy our needs for resources, we hope also to sustain the ability of future generations to meet their needs for the same resources. Furthermore, mining or extracting a resource affects the environment physically and esthetically and thereby also the ability of future generations to meet their needs for a healthy and pleasing environment. Mining affects the society and the environment locally when a specific deposit is targeted, but it also concerns the nation and the world when the entirety of reserves of particular minerals, fuels or water are taken into account. In the latter case, one exhausted deposit is replaced with another exhaustible deposit.

Two forms of sustainable development are defined in the literature: weak sustainability and strong sustainability, Tietenberg, 1996 [2]. The weak form of sustainability requires that we sustain the total of natural and man-made forms of capital. The strong form of sustainability sets an inherent limit on depletion of each form of capital, both natural and man-made. Substitution of man-made capital for natural capital is permitted in weak sustainability, whereas it is not permitted in the strong sustainability.

Applied to mining, the weak form of sustainability would permit near total depletion of a mineral resource if a man-made capital form is created using the mined mineral deposits such that future generations could continue to fulfill their needs. For example: Construction aggregate resources (sand, gravel and crushed stone) are used to build housing and infrastructure such as highways, schools, shopping malls, etc. The weak form of sustainability would consider the man-made capital such as the infrastructure as a substitute for the natural capital of the construction aggregate resources. In the strong form of sustainability, enough resources must be available to future generations to continue the maintenance and growth of infra-structure development at the rate deemed necessary by the future generations. The strong form of sustainability, thus, sets stricter limits on resource depletion than the weak form of sustainability. The extent to which materials substitution is possible, without negatively affecting the environmental, social and esthetic well being of future generations, would determine whether strong sustainability is possible in mining.

## 3.3.2  PATH TO SUSTAINABLE DEVELOPMENT

In some ways mining has, in fact, promoted the overall sustainability of natural capital of various kinds. For example, the use of firewood declined in countries that were able to mine coal. The discovery and exploitation of crude oil replaced coal to an extent, especially in the space heating. Increasing availability of natural gas has promoted its use in electricity generation in place of coal, thus promoting sustainability of coal resources. New technologies, such as fuel cells, could make increased use of natural gas to substitute for gasoline in automobiles and thereby promote sustainability of oil resources. On the other hand, each new fossil fuel created new needs and wants such as electricity and automobile that created depletion pressures on these resources and created hitherto unknown pressures on other natural capital forms such as the clean air and water. Each successive "fuel displacement" was possible because the new fuel was cheaper and/or more convenient to use than the old fuel. Convenience of use, in economic terms, represents a cost saving even if the price of the new fuel is higher than that which it replaces.

The comparison of actual or perceived costs of materials and their substitutes raises the question of how costs are defined and whether the consumer is required to pay the real cost of the material mined or extracted. Traditional accounting has been restricted to mining costs that are directly attributable to the mining operations, such as mine labor and overhead, equipment depreciation, maintenance and energy consumption. Historically, mined land was not reclaimed or restored. Post mining land reclamation according to pre-approved plans is now required. Therefore, the cost of reclamation became directly attributable to mining and mining cost accounting was modified accordingly.

Tietenberg, 1996 [2] proposed two principles to facilitate progress towards sustainable development: the Full Cost principle and the Carrying Capacity principle. Although originally proposed in the context of environmental resources, these principles are equally applicable to mining.

Full cost accounting helps in the realization of the monetary cost of the resource. For example, the laws requiring mine owners to restore mined land to its original contour and took mining a step closer to Full Cost accounting. Other steps toward full cost accounting include the costs of treatment of acid mine water, mitigating the effect of mining on groundwater levels and quality, etc. Including the costs of protecting the environment in the cost of production and delivery of mined resources provides an incentive to the consumers of the product to use it sparingly and/or to find a cheaper alternative. On the mining side, this consumer behavior provides the incentive to develop cost saving techniques and technologies as competition from substitutes increases. Together, the actions by the consumers and the producers contribute to the sustainability of the resource and to meeting the needs of future generations for the resource a little longer.

Developing countries frequently offer mining companies exemptions from the principle of full cost accounting in order to attract investments. Although such exemptions create more new jobs in the short run, they are associated with costs to be paid in the long run by the developing country and thus encourage unsustainable conditions with regard to the extracted minerals, as well as the over-dependence on revenues from the mineral resource. Special interest groups in developing countries also attempt to create legal frameworks that prevent full implementation of the Full Cost accounting principle. To the extent that such efforts fail, the cost of the materials mined is increased and producers as well as buyers of the mineral commodities seek lower cost opportunities in other countries. As mining moves from the more developed to the less developed countries, an increasing share of the real cost of the minerals is being borne by the less developed countries, thus increasing social inequality on a global scale.

The "Carrying Capacity" principle implies that there is a limit to nature's capacity to absorb the strain that pollution causes in natural habitats. When stretched beyond these limits, nature changes in "dramatic and unanticipated ways" making sustainability difficult or impossible. Mineral resources are a natural resource like clean air or clean water. The implication of carrying capacity to mining is that there is a natural limit to not only how much mineral is extracted over the lifetime of a mine, but also how much is extracted annually. "Bigger is better" does not always apply, even if it may appear to offer economies of scale in the short run. In the long term, across generations, the cost could be greater because future generations are forced to resort to far more expensive sources of minerals or their substitutes. The optimal rate of depletion of a mineral resource may not be the same when calculated by the conventional criteria of Net Present Value (NPV) maximization as by the sustainability criteria.

When time is not an important criterion, economic decisions are made on the basis of static efficiency, i.e., on the basis of maximization of the benefits to consumers as well as producers in the present time period. Benefits in a static case are maximized when marginal cost and marginal benefits are identical. The optimal quantity produced and sold corresponds to the point of intersection

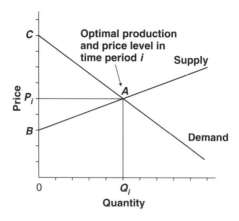

Figure 3.3.   Static efficiency of production in a given time period *i*.

of the demand and supply curves, as shown in Figure 3.3. In any given time period *i*, the optimal level of production $Q_i$ is obtained at the price level $P_i$. This level of production maximizes the consumer surplus $AP_iC$ as well as the producer surplus $AP_iB$ in the time period *i*. The total benefits from the decision to produce the quantity $Q_i$ are thus maximized at $ABC$.

However, the long term future of benefits and costs must also be considered when making business decisions. Financial investment decisions commonly are based on the analysis of the Present Value (PV) of future expected net profits (NPV). A discount rate "*r*" is used to determine the present value of future net benefits. This approach is called the dynamic efficiency. The problem is summarized and mathematically explained below:

$$\text{Maximize} \sum_{i=1}^{n} \frac{ABC}{(1+r)^n} \text{ , where } n = \text{number of time periods.}$$

*Mathematical analysis of dynamic efficiency*
A downward sloping demand schedule can be described mathematically as, $P_t = a - bq_t$, where $P$ = price and $q$ = quantity in year *t*; *a*, *b* = constants.
    The total benefits from the production of *q* are,

$$\text{Total benefits} = \int_0^t (a - bq)dq = aq_t - \frac{b}{2}q_t^2$$

If marginal cost of extraction is a constant *c*, the total cost of extraction is $= cq_t$.
    Let $\bar{Q}$ be the total available resource quantity and *n* be the number of years over which to extract it. The problem is to,

$$\underset{q}{Max} \sum_{i=1}^{n} \frac{aq_i - bq_i^2/2 - cq_i}{(1+r)^{i-1}} + \lambda\left[\bar{Q} - \sum_{i=1}^{n} q_i\right]$$

Subject to,

$$\frac{a - bq_i - c}{(1+r)^{i-1}} - \lambda = 0, \qquad i = 1,\dots, n.$$

and

$$\overline{Q} - \sum_{i-1}^{n} q_i = 0$$

The implication is that when the demand schedule is stable over time and the marginal cost of extraction is constant the present value of the marginal user cost increases at a rate $r$, the discount rate. This results in declining consumption over time, Tietenberg, 2000 [3].

The discount rate "$r$" depends on the preference for benefits in the current period over benefits in future periods. A greater discount rate strongly reduces the present value of future benefits. A smaller discount rate increases the present value of future benefits. Obviously, a zero discount rate places equal value on present and future benefits.

Sustainability considerations in mining require the choice of a production schedule and a discount rate that would satisfy the needs of current and future generations while maximizing the NPV. An important factor in choosing the discount rate is the perception about the public or private nature of the resource under consideration. At any given mineral deposit, the mineral is considered to be a private good. The owner of the deposit is free to exploit it in a manner that maximizes the NPV of profits. However, at the societal level, minerals, water and fossil fuels are also considered to be public goods. Thus a conflict of interests exists between the private and public perceptions in the choice of an appropriate discount rate. Whereas the private discount rate is determined by the opportunity costs of the best alternative investment, the public discount rate is subject to considerations of long term societal welfare. Public discount rates generally small, often no more than 1 or 2 percentage points above the estimated long term inflation rate. Some economists argue for a zero real discount rate for public goods.

How do we persuade the present generation to forego some consumption today so that future generations will be able to satisfy their need for the resource? Hartwick [4] proposed that "constant level of consumption could be maintained perpetually if all the scarcity rent were invested in capital." Such a policy would "be sufficient to assure that the value of the total capital stock would not decline". This proposition has come to be known as the Hartwick Rule. The Hartwick Rule thus represents the weak form of sustainability described earlier in this chapter. The capital stock created by such investment of the scarcity rent would substitute for the resource which future generations would have used to satisfy their needs.

### 3.3.2.1   *Measuring scarcity rent*

"Scarcity Rent" is the marginal opportunity cost imposed on future generations by extracting one more unit of the resource today. It can also be understood as the marginal cost imposed on future generations by hastening the exhaustion of the resource by one unit of the resource or by one unit of time. The cost thus imposed on future generations is equal to the benefit gained by the present generation. Figure 3.4 explains how "scarcity rent" can be conceptually measured.

As depicted in Figure 3.4, the marginal cost of an exhaustible resource increases with time until its exhaustion in the year "$T$", at which time a switch to a new source must be made. The new source generally is more expensive than the old one. Hence the instant increase in the marginal cost at "$T$". The cost curve continues to increase after "$T$". If we succeed in postponing the time of switching from "$T$" to "$T + 1$" we "earn" a rent equal to the area "$A$". "$A$" is the "marginal opportunity cost" and its present value is the "Scarcity Rent".

*Mathematical analysis of measurement of scarcity rents for exhaustible resources*
Let $C$ be the marginal extraction cost and $\theta$ be the scarcity rent. The efficient price of the resource, then, is: $P = C + \theta$.

Consider time to be divided into two periods, one before "$T$" and the other after "$T$". Further, let $C_1$ and $C_2$ be the marginal costs at the beginning of the two time periods, and let $g_1$ and $g_2$ be

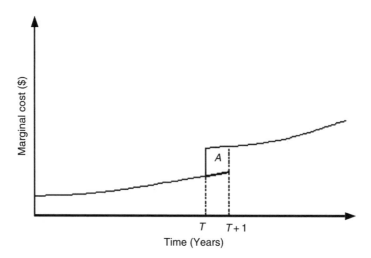

Figure 3.4.   Marginal cost of an exhaustible resource and the shift to a new resource.

the rates of cost growth in the two time periods. Assuming that costs grow exponentially, the marginal costs in the time periods before and after "*T*" are:

$$C_1 = K_1 e^{g_1 t} \text{ for } 0 < t < T \text{ and } C_2 = K_2 e^{g_2 t} \text{ for } t \geq T,$$

where $K_1$ and $K_2$ are constants (Fig. 3.4).

The total present value of the future marginal costs is determined by integration of the above cost functions over time, using a discount rate "*r*". If the current resource were unlimited, the integration of the cost function $C_1$ would suffice to obtain the present value of future marginal costs because there would be no need to switch to a new source at "*T*" and the cost curve $C_1$ would continue indefinitely. The present value of the future marginal costs would in this case be:

$$\int_t^\infty K_1 e^{g_1 t} \cdot e^{-r(q-t)} dq$$

where "*q*" corresponds to *T*.

When the current source is exhaustible and a switch to a new one occurs at "*T*", the total present value of future marginal costs is:

$$\int_t^T K_1 e^{g_1 t} \cdot e^{-r(q-t)} dq + \int_T^\infty K_2 e^{g_2 t} \cdot e^{-r(q-t)} dq$$

The present value of the additional (marginal) cost ($C_{PV}$) of resorting to a new source is the difference between the above two equations:

$$\int_t^T K_1 e^{g_1 t} \cdot e^{-r(q-t)} dq + \int_T^\infty K_2 e^{g_2 t} \cdot e^{-r(q-t)} dq - \int_t^\infty K_1 e^{g_1 t} \cdot e^{-r(q-t)} dq$$

which is the same as:

$$C_{PV} = \frac{K_2 e^{g_2 T - r(T-t)}}{g_2 - r} - \frac{K_1 e^{g_1 T - r(T-t)}}{g_1 - r}$$

The derivative of $C_{PV}$ with respect to time, $T$, measures the present value of savings in costs from postponing the switch to the higher cost source by one (1) time period, and thus represents the Scarcity Rent ($\theta$):

$$\theta = \frac{dC_{PV}}{dT} = K_2 e^{g_2 T - r(T-t)} - K_1 e^{g_1 T - r(T-t)}$$

Hartwick's Rule requires that the "Scarcity Rent A" should be invested in capital stock that would serve the same purpose for the future generations as the exhaustible resource being consumed by the current generation. It is assumed that such capital assets neither depreciate nor become obsolete. Neither assumption is strictly valid although they help simplify the understanding of Hartwick's Rule. A more realistic proposal is to strive to actually postpone the time to switch to a new source and thereby leave more of the current resource for the future generations. Postponement of consumption of anything is, however, counter-intuitive to human nature. Therefore, it is necessary to make a collective social decision to set a very low discount rate for exhaustible public goods resources. The correct price of an exhaustible resource, then, is equal to its marginal cost plus the scarcity rent, Moncur and Pollock [5].

Postponement of the exhaustion of a resource is facilitated when alternatives to the resource are currently available, for example natural gas or ethanol substituting motor gasoline or solar or wind energy supplying electricity. When postponement of consumption appears feasible, discount rates are lowered because leaving some of the resource for future generations appears to be economically acceptable. Alternatively, if the cost of current consumption of a resource rises, for example because environmental costs have to be internalized, incentives are created to find substitutes and/or postpone consumption, thereby, in effect, lowering the discount rate. This contributes to the sustainability of extraction of the resource. Obviously, a price is paid for sustainability. However, when the benefits that accrue to a society from resource conservation, improved or undegraded environmental quality, and improved health of the population are taken into account, the greater price may be acceptable. Such an accounting system does not currently exist but would be useful in promoting sustainability considerations.

### 3.3.2.2   *Triple bottom line (TBL)*

The "triple bottom line" (TBL) concept in sustainability involves the addition of the environmental and social values to the economic value to form the decision-making "bottom line". The TBL concept requires companies to minimize any harm resulting from the economic activity and to increase environmental and social values or, at the least to not diminish them. This involves costs as well as benefits, called "externalities". Negative externalities are costs not currently borne by the consumer of the product. Positive externalities are benefits enjoyed but not paid for by the consumer. Some examples of mining externalities are:

*Negative externalities:*
- Disturbed landscapes due to surface mining
- Vibrations from explosions serving mining operations
- Increased truck traffic in and around communities
- Effect of dust or other forms of pollution on the health of humans, animals and plants
- Possible land subsidence from underground coal mining
- Possible effect on levels of ground water
- Possible costs of future import dependence due to domestic resource exhaustion
- Decrease in property values as a result of mining.

*Positive externalities:*
- Employment created in mining and in auxiliary/service industries
- Tax revenues from mining and employment

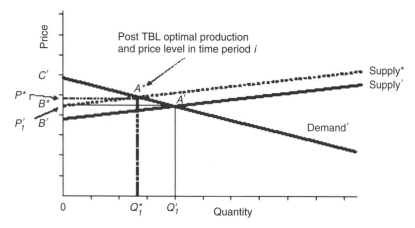

Figure 3.5.   Effect of TBL consideration on consumer/producer surplus.

- Increase in property values as a result of land reclamation
- Economic and political benefits of reduced imports from foreign sources.

The above listed negative and positive externalities are reflected in the price of the mineral only to the extent to which they are mandated and, therefore, included in the cost accounting. It is easy to visualize that the internalization of negative externalities shifts the supply schedule upward and the internalization of positive externalities shifts it downward (Fig. 3.5). The "net" effect on the supply schedule resulting from the consideration of the externalitics i.e., from the application of the TBL concept, is presumed to be a net upward shift of the supply schedule resulting in a reduction in resource consumption from $Q'_i$ to $Q^*_i$ as its price increases from $P'_i$ to $P^*_i$. Unless the demand curve also shifts upward, the consumer and producer surplus denoted by the area $A'B'C'$ is reduced to $A^*B^*C'$.

When more than one resource deposit is available for extraction, preference is given to the one with the lowest marginal cost to the user. The exhaustion of this deposit leads to its replacement with the next more expensive deposit. The replacement occurs before the physical exhaustion of the first deposit if its marginal user cost exceeds that of the replacing source. In the case of mining, the cost increase related to the location of the replacing source is incorporated in the decision. The replacing source may be another non-renewable deposit or a renewable source. Preference is given to the source with the lower cost when both are available. Developments in the technology of exploration, mining and mineral processing require a continued reassessment of costs because technology can lead to cost reduction and make a new source more attractive. Societal decisions can also create new market conditions. For example, community development projects may result in rezoning land use from mining to urban development and remove otherwise economically viable mineral deposits from mining consideration due to increased property value or outright prohibition of mining. Even when private property rights are well defined, changes in the profitability of land use for mining vs. its use for non-mining purposes can result in less than optimal use of the mineral deposit and impose a cost on the society in terms of current prices as well as future sustainability.

### 3.3.3   DISCUSSION

Deciding how much of the resource should be extracted in the current time period and how much should be saved for the future requires the efficient inter-temporal allocation of the resource. The allocation between current and future time periods is said to be efficient when any change in the allocation leads to a reduction in the consumer/producer surplus (*Pareto Optimality*). However, efficiency in inter-temporal allocation alone would not solve the sustainability problem because the environmental and social considerations are not amenable to strictly economic analysis. The optimal inter-temporal allocation is rather an ethical issue. Moving toward sustainability requires ethical and moral foundations in addition to the willingness to internalize the economic externalities. The key to its realization is "restraint" – restraint in how much we consume, how much economic growth we need and want, and restraint in how much damage can be inflicted upon our environment, Cordes [1]; Tietenberg, 2000 [3].

REFERENCES

1. Cordes, John A., In: "Sustainable Development and the Future of Mineral Investment". James M. Otto and John A. Cordes, Editors, United Nations Environment Program (UNEP), May 2000.
2. Tietenberg, Thomas, H., Managing the Transition to Sustainable Development: The Role for Economic Incentives, in Pricing the Planet – Economic Analysis for Sustainable Development, Peter H. May and Ronaldo Seroa da Motta, Editors, Columbia University Press, New York, 1996.
3. Tietenberg, Tom, Environmental and Natural Resource Economics, Fifth Edition, 2000.
4. Hartwick, John, M., Intergenerational Equity and the Investing of Rents from Exhaustible Resources, The American Economic Review, Volume 67, Issue 5 (December, 1977).
5. Moncur, J.E.T. and Pollock R.I., 1988, Scarcity rents for water – A valuation and pricing model; Land Economics, Volume 64, No. 1.

# 3.4   Managing environmental impacts

Raj Rajaram[1], Subijoy Dutta[2] and Krishna Parameswaran[3]
[1]*Complete Environmental Solutions, Oak Brook, IL, USA*
[2]*S&M Engineering Services, Crofton, MD, USA*
[3]*ASARCO LLC, Phoenix, Arizona, USA*

## 3.4.1   INTRODUCTION

In many circumstances, an environmental impact assessment is submitted and reviewed by government authorities before a mining permit can be issued. The permit specifies the discharge limits for waters from the mine, air quality emissions, allowable surface subsidence (for underground mining), and requirements for managing groundwater and waste generated from the mining operation. A proposed underground mine in Wisconsin submitted their environmental impact assessment report to the Department of Natural Resources in 1995, and had not received approval of the report as of November 2002. During this seven-year period, the Department requested additional information from the company and the company had to make several revisions to the environmental management plans to satisfy the Department. Once the permit is received after the arduous review process, the government monitors the operation to ensure that the company is following all the approved plans.

The following environmental impacts should be managed during and after the mining operation to obtain the mining permit:

- blasting impacts
- dust control in the mine and air quality in the surrounding area
- surface water management
- groundwater management
- surface subsidence control
- wetland impacts
- biological diversity impacts
- compatible land use (post-mining)
- surface stability (erosion and tectonic).

Land use is discussed in Section 3.5 of this chapter and blasting impacts, dust control, surface water and groundwater management, surface subsidence control are discussed in Chapter 4 in the context of using sustainable mining systems and technologies to manage these impacts. The following section focuses on managing biological diversity impacts.

## 3.4.2   BIOLOGICAL DIVERSITY

This section is based largely on the discussion of biodiversity conservation and minerals development discussed by Warhurst and Franklin (OUP [1]).

Biological diversity is defined as the variability among living organisms from all sources including terrestrial, marine and other aquatic ecosystems. It includes the diversity within species, between species and associated ecosystems. In the context of sustainable mining practices, biological diversity can be conserved if biological resources are used in a manner and at a rate that does not lead to long-term decline of biological diversity. This will maintain the biological resources to meet the needs and aspirations of presentand future generations. The keys to sustainable development and conserving biologicaldiversity are:

• Minimizing impacts
• Mitigating impacts
• Maintaining a diverse and healthy economy.

Minerals are concentrated in economically viable ore deposits in relatively limited areas around the globe. Known ores constitute less than 0.01% of the metal content of the upper 1 km of the continental crust. In most ore deposits many elements, besides the ones of economic interest are enriched, potentially contributing to revenue producing by-products or potential contaminants in downstream processes. Thus, mines can only be located in relatively few places where economically viable mineral deposits are formed (NAP [2]). Products of mining are not only essential for the economy but are essential in our daily lives as well. The National Mining Association (NMA) estimates that an average American uses 3.8 short tons of coal each year and 47,000 pounds (23.5 short tons) of newly mined minerals each year. These minerals include metals such as aluminum, steel, copper, lead, zinc and platinum used in cars and at least 29 minerals used to make the computer screen, chips, circuitry and battery for a home computer (NMA [3]).

The impact of minerals development on the long-term health of ecosystems and biodiversity has long been a public concern. The 1972 Club of Rome report, *The Limits to Growth*, predicted the imminent depletion of fossil fuels and metals. The discovery of new oil, coal and mineral reserves, improved recycling of metals and technological advances have alleviated the fear of running out of nonrenewable resources. The debate has since moved towards controlling depletion and degradation of renewable resources such as water, air, land and biodiversity. There is a growing emphasis on "sustainable development," aimed at striking a balance between economic development, social activities and environmental quality. The approach is to minimize impacts on ecosystems and biodiversity as a result of minerals development.

Although mineral extraction itself may occur on a fairly small land area, associated infrastructure development and pollution from extraction activities have the potential to impact the health of ecosystems and their ability to provide goods and services necessary for human and environmental well being. These ecosystem goods and services include: air and water purification, conservation of soils and control of the hydrologic cycle. The extent of these impacts is influenced by management strategies or technologies employed by the operator in minerals development, as well as the local geology and geography of the ecosystem itself. The direct effects from minerals development arise from use of the land resulting in removal of biota, soils and minerals. The disruption, removal, erosion or contamination of the soil environment can impact soil microbes and fungi that have associations with forest plant and trees, which in turn, could affect their productivity. This could result in loss of agricultural productivity and impact the success of mine site reclamation. While reclamation can go a long way towards replacing soil, flora and fauna, it may take many years or decades for the ecosystem to reestablish associations amongst plants, trees and microorganisms. This fact has to be taken into account in developing mine reclamation plans. Mining and mineral processing and associated waste management activities have the potential to adversely impact ecosystems through acid rock drainage, and the possible release of contaminants such as cyanide and heavy metals.

Many of the potentially adverse effects of mineral development on biodiversity can be minimized or avoided by proper planning, design and innovation. The ecosystem itself has some capability to withstand the effects of emissions and effluents. Excess concentrations of emissions and effluents can cause more significant effects and need to be controlled. To ensure that management strategies for conserving biological diversity are implemented effectively, an understanding of the physical, social, economic and cultural context in which they are implemented is essential. Biodiversity management requires a multi-disciplinary, intersectoral approach. The National Environmental Policy Act (NEPA) in the United States requires that federal permitting agencies study and document potentially significant effects of a project on wildlife. In addition, the Endangered Species Act (ESA) requires that before a federal agency grants a permit or approves a project, it must determine if the action may affect listed or proposed threatened and/or endangered species and their critical habitat. If a federal action involves a "major construction activity," NEPA and ESA require that a biological assessment be prepared to determine if the action causes jeopardy to listed species or adversely affects critical habitat. The scope of this assessment is quite broad requiring identification of species and critical habitat discussion of proposed action and available alternatives and mitigation measures.

Discussion in Chapter 8 on the sustainable practices employed during exploration on the Camp Caiman gold project, a prospect explored by ASARCO LLC (Asarco), notes that the exploration permits for that project were adjacent to the Kaw Biotope zone and roughly coincident with the Kaw Wetland System, a Ramsar site, i.e., designated for inclusion in the Ramsar List of Wetlands of International Importance. These wetlands are identified for specific protective measures in the Convention on Wetlands held in Ramsar, Iran in 1971. Asarco withdrew an exploration permit based on advice from the Environmental Advisory Committee (Committee) that the lands could be within the Biotope zone. Also on the Committee's recommendation, two separate baseline studies were conducted, one for the natural environment and another for the social environment. As part of the baseline review a citizen's group from a town of 500 people was invited to meet with the Committee. The group understood the potential benefits of the project and the need to protect the rainforest but expressed concern that social issues may be overlooked (OUP [4]).

At the other end of the spectrum, passive biological treatment systems have been applied to the treatment of effluents from active metal mines and drainages from inactive metal mines. This innovative technology was used at Asarco's former West Fork Unit, an underground lead-zinc mine in Missouri, U.S.A. (now owned by the Doe Run Company) (Gusek et al. [5]) and at Asarco's inactive mine sites at the Mike Horse mine in Montana, Gem mine in Idaho and Buchans mine in Newfoundland, Canada. In fact such technologies have been employed to treat acid mine drainage from coal mines since the mid-1980s. Wetlands have long been recognized as a natural means of improving water quality. Contaminant reductions are achieved through the precipitation of metal hydroxides, sulfates and carbonates. Passive treatment systems can be designed to mimic the cleansing action of wetlands and engineered to optimize the biochemical processes that occur in a natural wetland system. The principal advantages over a conventional wastewater treatment system is that large volumes of sludge are not produced as the metals are precipitated as oxides, sulfides and carbonates in the treatment system substrate and reagent use is considerably reduced. There are also examples where mining has resulted in the creation of habitats. The tailings pond at the Sweetwater lead mine in Missouri, formerly owned by Asarco (now owned by the Doe Run Company, is a widely appreciated bass fishery). Mine adit discharges in the Silverton mining district in Colorado have resulted in creation of a trout and beaver habitat.

In their discussion of innovation for remediation and reuse in mining and downstream operations, Warhurst and Franklin (OUP [6]) note Asarco's use of hazardous wastewater treatment sludge produced at the California Gulch, Colorado, site at its lead smelter in East Helena as a lime flux substitute. Prior to that, the wastewater treatment plant sludge was sent to the National Zinc smelter in Bartlesville, Oklahoma for the recovery of zinc. Not only was reuse of the material achieved but also ecosystem impacts were prevented by avoiding land disposal.

REFERENCES

1. Oxford University Press (OUP), Warhurst, A. and Franklin, K., "Biodiversity Conservation, Minerals Extraction and Development Towards a Realistic Partnership," Footprints in the Jungle, Editors Ian A. Bowles and Glenn T. Prickett, Oxford University Press, 2001, pp. 183–203.
2. National Academy Press (NAP), Hardrock Mining on Federal Lands, National Research Council, Washington D.C., 1999, p. 23.
3. National Mining Association (NMA), 2002, Facts about Coal and Minerals, pp. 2–56.
4. Oxford University Press (OUP), Graybeal, F.T., 2001, Evolution of Environmental Practice During Exploration at the Camp Caiman Gold Project in French Guiana, In: Bowles, I.A., and Prickett, G.T., ed.,

Footprints in the Jungle. Natural Resource Industries, Infrastructure, and Biodiversity Conversation: New York City, Oxford University Press, and pp. 227–229.

5. Gusek, J.J., Wildeman, T.R., Miller, A. and Fricke, J., "The challenges of designing, permitting and building a 1,200 gpm passive bioreactor for metal mine drainage," West Fork, Missouri, Proceedings of 15th Annual Meeting of American Society of Surface Mining and Reclamation, 1998, pp. 203–212.

6. Oxford University Press (OUP), Warhurst, A. and Franklin, K., "Biodiversity Conservation, Minerals Extraction and Development Towards a Realistic Partnership," Footprints in the Jungle, Editors Ian A. Bowles and Glenn T. Prickett, Oxford University Press, 2001, p. 191.

# 3.5   Land use: the geomorphic and land use impacts of mining

Ian Douglas and Nigel Lawson
*School of Geography, University of Manchester, Manchester, UK*

## 3.5.1   INTRODUCTION

Since people began to remove stones to create tools, then metals and other materials to build shelters, villages, towns and eventually modern cities, the deliberate manipulation and alteration of the earth's surface has increased exponentially. Mining and quarrying activities have, in many places, gradually become the principal agent of geomorphic change, often, save for areas of great tectonic activity, glaciation or high mountain ranges, far outstripping the natural forces of erosion. In the earliest scientific attempt to demonstrate the magnitude of changes to the earth's surface by human intervention, George Perkins Marsh set out:

> "... to indicate the character and, approximately, the extent of the changes produced by human action in the physical condition of the globe we inhabit; to point out the dangers of imprudence and the necessity of caution in all operations which, on a large scale, interfere with the spontaneous arrangements of the organic or the inorganic world; to suggest the possibility and the importance of the restoration of disturbed harmonies..."
>
> (Marsh [1]).

A second milestone in evaluating people's impact on the physical landscape, R.L. Sherlock's *Man as a Geological Agent* (1922) emphasized the contrast between natural and human denudation and provides many illustrations of the quantities of material involved in mining. Whilst general in character, it concludes "man is many times more powerful, as an agent of denudation, than all the atmospheric denuding forces combined" (Sherlock [2]). More recently, Roger le B Hooke estimated that anthropogenic material movements are more than double the approximately 24 Mt sediment load delivered to the world's oceans and interior basins by rivers annually and that humans are arguably the most important geomorphic agent currently shaping the earth. He also points out that whereas, in most instances, it is possible to identify a line such as the coast across which material is moved by natural methods, human movement of material is more random and has a greater visual impact on the landscape (Hooke [3]).

Now concern over making land use and urbanisation more sustainable has led to much greater attention than before to the flows of materials (Fischer-Kowalski [4]), the ecological footprints of cities (Rees [5]) and the ecological "rucksack" of mining (Bringenzu and Schütz [6]). Already analyses of materials fluxes have been produced for China, the USA, Japan, Germany, the Netherlands, Austria and Italy (Chen and Qiao [7], De Marco et al. [8], Mathews et al. [9]). Girardet has estimated the ecological footprint of Greater London as 125 times the area it occupies (Sustainable London Trust [10]). Comparative data on the ecological footprints of many cities are now available e.g., Macau (Lei and Wang [11]), Barcelona (Prat and Noguer [12]), and Vancouver (Rees [13]).

## 3.5.2   METHODOLOGY FOR ASSESSING TOTAL MATERIAL MOVEMENTS THROUGH MINING AND QUARRYING

The International Union of Geological Sciences (IGS) and the Scientific Committee on Problems of the Environment (SCOPE) have brought together earth scientists concerned about the human dimensions of geological processes in a project entitled "Earth Surface Processes, Material Use and Urban Development" (ESPROMUD) to evaluate how the growth of cities and extractive industries has affected the rates and nature of geomorphic change (Cendrero and Douglas [14], Douglas and Lawson [15]). A key step in the programme has been to establish the materials flows due to mining by accounting for the annual masses of rock and earth surface materials extracted, including overburden and mineral processing residues. Another major step is to assess the flows of materials in the urban construction, life-support, maintenance and waste disposal processes.

National mining and quarrying statistics record the net amounts of material produced for each commodity extracted, the run of mine production. However, these figures do not account for all the overburden removed before extraction can start and the waste created during mineral processing. Together these latter quantities are the hidden flows associated with mining and quarrying (Douglas and Lawson [16]). Whilst it can be argued that during open pit mining, especially coal mining, much of the overburden and locally stored waste is eventually replaced in the hole from which it came, even that local temporary shift of materials changes the nature of the ground surface. In addition, there is always a risk that some of the stockpiled overburden will be eroded or lost to the people-modified, natural drainage system and that the final land restoration will result in a somewhat different landscape from that which existed before mining. The quantities of material shifted by mining and quarrying will vary with the mineral being extracted, the technology used to extract it, the geological situation of the deposit, the age and life of the mine, and the precise management policies of the mine operator. Nevertheless, these hidden flows of overburden removal and production waste from mining and extractive industries affect the sustainable land use and ecosystem recovery. Estimation of their magnitude is therefore important. Many researchers, such as Sherlock [2], Hooke [3], Warhurst [17], Von Weizächer et al. [18] and Adriaanse et al. [19], have tried to gain an idea of these "hidden" flows using estimated multipliers of the mineral production to obtain a figure for the total amount of materials shifted. In the ESPROMUD project, case-study examples of mine overburden quantities cited in the literature were examined to obtain the multipliers for all materials mined and quarried globally. These multipliers for global production figures were established for all 49 base extractive minerals in the United Nations Industrial Commodity Statistics (Douglas and Lawson [16]). The resulting approximations were then extensively revised through consultation with a wide range of experts involved with the mineral industries (Douglas and Lawson [20]). However, the application of these global multiplier values to national statistics may be less safe, particularly if a single mine, or group of mines, in an unusual geologic formation is responsible for virtually all the production of a particular mineral in that country.

### 3.5.2.1 *Global material movements*

20,933 Mt (Million tons) of run of mine production came from the world's mines and quarries in 2000. This was 47.7% more than in 1975. However, the gross total mass of materials, including waste and overburden, moved in the extraction process in 2000 was over three times greater, 62,265 Mt, which is 48.2% more than 25 years earlier (Table 3.1).

96.2% of the gross material movement (95.7% net run of mine) in 2000 was confined to building stones, aggregates and just 7 minerals. The inclusion of waste and overburden clearly demonstrates the relative impact on geomorphological change made by the production of gold, copper, and coal (Table 3.2).

### 3.5.2.2 *Material movements during informal and unrecorded mining*

National mineral production statistics vary in accuracy, and there is probably a certain amount of unrecorded, and often illegal, removal of gravel and rock in every country. Some countries have a large amount of small-scale artisanal mining, some of which may be unregulated and unrecorded. Gold Fields Mineral Services of London's estimate of 175 t of gold being produced in 1975 by informal miners world-wide may be lower than the actual total (Veiga [24]). Three million carats of diamonds may be mined informally each year in Africa (Holloway [25]). The low recovery rates obtained by informal artisanal gold and gem miners working sites with low yields mean that these operators probably shift more than twice the mass of material per net tonne of mineral extracted than do large mining companies. Jéjé [26] suggests a ratio as high as 3.7 M to 1 for artisanal gold mining in the Amazon, as compared with a ratio of about 1 M to 1 for large mechanised gold

Table 3.1.    Global production of land won minerals, including waste and overburden.

| Year | Production net weight, 1,000 tons | % increase since 1975 | Materials moved gross weight, 1,000 tons | % increase since 1975 |
|---|---|---|---|---|
| 2000 | 20,933,070 | 47.7 | 62,265,038 | 48.2 |
| 1998 | 20,610,531 | 45.4 | 60,015,452 | 42.9 |
| 1995 | 19,735,291 | 39.3 | 57,548,678 | 36.8 |
| 1988 | 18,607,294 | 31.1 | 56,864,321 | 35.4 |
| 1985 | 16,942,603 | 19.5 | 52,028,252 | 23.9 |
| 1980 | 16,196,957 | 11.4 | 47,407,590 | 12.8 |
| 1975 | 14,172,463 | | 42,002,725 | |

Sources: United Nations, 1975–1998 [21]; United States Geological Survey, 2002 [22]; International Energy Annual, 2002 [23]. (For methodology see Douglas and Lawson [16]).

Table 3.2.    Global production in 2000 of the principal land won minerals responsible for total materials moved, including multipliers for waste and overburden.

| Commodity 2000 | Net weight, 1,000 tons | % World | Multiplier | Gross weight, 1,000 tons | % World |
|---|---|---|---|---|---|
| Coal, Hard | 4,151,598 | 19.83 | 4.87 | 20,218,282 | 32.47 |
| Coal, Brown + lignite | 915,079 | 4.37 | 9.9 | 9,059,282 | 14.55 |
| Petroleum, crude | 33,56,968 | 16.03 | 1.016 | 3,410,679 | 5.48 |
| Iron ores – Fe content | 621,803 | 2.97 | 5.2 | 32,333,765 | 5.19 |
| Copper ores – Cu content | 13,200 | 0.06 | 450 | 5,940,000 | 9.54 |
| Nickcl orcs – Ni content | 1,250 | 0.01 | 560 | 700,000 | 1.12 |
| Gold ores – Au content | 3 | 0.00 | 950,000 | 2,422,500 | 3.89 |
| Aggregates and Building Stones | 10,976,488 | 52.43 | 1.36 | 14,928,023 | 23.97 |

Sources: United Nations, 1975–1998 [21]; United States Geological Survey, 2002 [22]; International Energy Annual, 2002 [23]. (For methodology see Douglas and Lawson [16]).

mining. These operations alone could easily result in the movement of over 400 Mt of material per annum, equivalent to nearly 1% of the global mass moved during mineral extraction.

In many parts of the developing world, building materials are worked as small enterprises. People seek the clay, sand, gravel and stone that they need in the closest convenient place and it is inconceivable that these activities are accurately recorded. In Vietnam, for example, at headlands along the coast, individual entrepreneurs quarry granite, while around towns and cities, small brick clay pits leave a series of derelict hollows and degraded soils. In inland China, villagers operate small crushers making aggregates from rock quarried from tiny rock outcrops near their fields. Many hundreds of thousands of tonnes of material are worked in this way and in clay pits (Edmonds [27]) and inefficient unauthorised mines (Qu and Li [28]). Almost uniquely, many brick clay pits around Chinese villages, especially near Beijing, are restored to agricultural production some 2 m below the original ground level. Elsewhere, however, such pits are usually left until filled with urban waste and reclaimed for urban development. Road construction in remote areas, such as logging roads in Borneo, uses up to 370 t of gravel per kilometre (km), and may involve the unrecorded quarrying of over 4,000 t of rock per year in each major Borneo logging concession (assuming some 10 km of road construction or repair each year). The global amount of material moved by these informal, and largely unrecorded, mining activities is difficult to assess. Such mining probably involves 250 t per person per year (Nötstaller [29]). In India alone about 200,000 informal miners (Chakravorty [30]) shift about 50 Mt of material each year, about 2.5%

of all mining related materials displacement in India (Lawson and Douglas [31]). In all, these activities probably add no more than 1 to 5% to the global mass of materials moved during the extraction of minerals (Douglas and Lawson [32]). However, they result in substantial amounts of earth movement and land degradation. Furthermore, much of the land affected by these informal mining activities is situated in ecologically sensitive areas or in areas where land for agriculture is scarce.

### 3.5.3   MULTIPLE URBAN DEMAND FOR LAND

In terms of sustainable development, the paradox is that urban growth usually creates a double demand for land: land on which to build houses and land from which to extract the materials needed for house building, infrastructure construction and waste disposal. This double demand for land may be divided into the direct land demand and the hidden land use of urban construction and the disposal of urban waste. This hidden land use is analogous to the hidden flows created by over-burden removal and waste discussed above.

Societies vary in their ability to restore degraded mining land. Restoration may be expensive in terms of the local economy, especially where mining has been the only or the major source of paid employment. In a few cases, such as the upper Thames gavels workings near Swindon, England (Fig. 3.6), mineral extraction may be designed as part of a long-term resource and land use management strategy. In most cases, however, plans for restoration were not made at the outset and the development of new uses for the land, or ways of preventing further land degradation may not be undertaken until several years, or decades, after mining has ceased. The examples below set out some of the problems that are encountered.

### 3.5.4   IMPACTS ON LAND USE AND ON GEOMORPHOLOGY

#### 3.5.4.1   *General*

Mining and quarrying are temporary activities on any given site. When minerals or earth surface materials are worked out, the extractive activities are resulted elsewhere leaving behind disturbed areas that may, or may not, have had some degree of reclamation or restoration work completed. Few data are available for the derelict land created by mining, although its restoration ecology is relatively well understood (Hall [33], Holl et al. [34]). Britain's long industrial history has created a legacy of mineral workings that locally is so significant in the land cover pattern that considerable data and reclamation histories are available.

Some 60,000 ha of active mineral workings or former workings not yet reclaimed, and including land used for associated buildings and machinery, occupy 0.46% of all land in England (Department of the Environment [35]). Most of the areas mined or quarried in past centuries but now used as grazing land, whose landforms still reflect the effects of past activity are excluded from this total. The legacies of these extractive activities are vividly visible in the landscape, especially from the air, with the gravel pits along the flood plains of rivers in the south-east and Midlands of England (Figs 3.6 and 3.7) and the large limestone quarries of the Peak District (Fig. 3.8) standing out clearly. The mountains of the Lake District are still scarred by the slate and copper mining activities of previous centuries. The former coalfields retain the traces of spoil heaps, despite the restoration of areas like the Central Forest Park in Stoke on Trent (Kivell [36]).

Even though there are still so many legacies of past activity, the amount of restored, remodelled land is high. Between 1982 and 1994 the average rate of land restoration was 3,310 ha per annum. In England as a whole, the restoration of former mineral workings sites between 1974 and 1994 affected no less than 0.52% of the total land area of the country (Department of the Environment [35]). While statistics of derelict land are available for England, that the total area once worked for

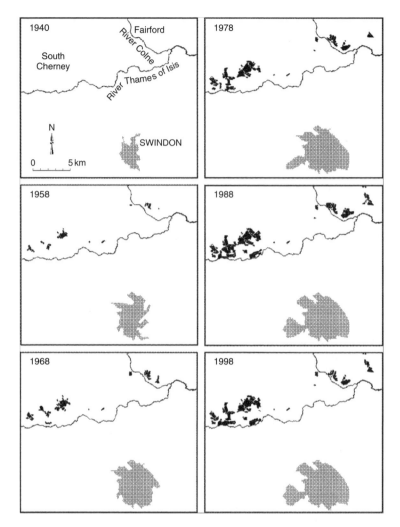

Figure 3.6.   Land transformation by gravel extraction in the Colne Valley and urban expansion of Swindon, UK 1940–1998.

minerals and then given over to some other use cannot be calculated. Building stone quarrying, aggregate extraction and brick clay removal have left their mark on the periphery of most large cities, as changes in the Colne Valley west of London and the Trent valley south of Nottingham (Figs 3.6 and 3.7) show. Conversion of these old mineral workings to new uses, be they water parks, nature reserves, or restoration after landfilling involves further investment.

Not all restoration is returns land to productive uses. Poor quality restoration may inhibit plant regeneration and slow soil improvement. For example, during opencast (strip) coal mining the rock overlying the coal (overburden) is fractured and broken before being removed. Hence, problems such as compaction and acidification mean that the agricultural productivity of restored sites is often low (Walsh et al. [37]). Many of British ancient upland mining sites are still unstable, releasing contaminated sediments to nearby streams (Lewin [38], Macklin et al. [39], Hudson-Edwards et al. [40]). Forestry is one of the more cost-effective soft end-uses for derelict land schemes, with many new woodlands being established on former mineral extraction sites that were subsequently

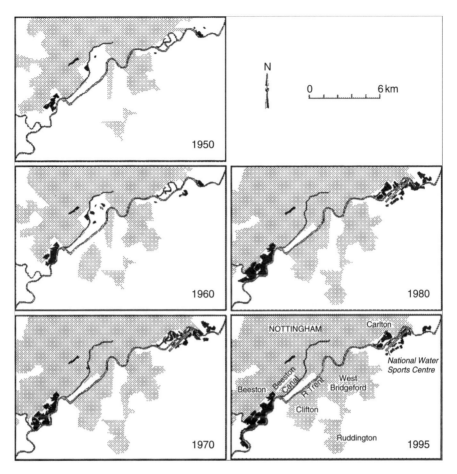

Figure 3.7.   Land transformation by gravel extraction in the Trent valley and urban expansion of Nottingham, UK 1950–1995.

reused for landfill, although there is potential for much further forestry development (Perry and Handley [41]).

### 3.5.4.2   *Case studies of coal, limestone, oil shale and tin mining*

#### 3.5.4.2.1   *Coal mining in South Yorkshire (UK)*

Delays in land remediation always create unnecessary problems. Detailed site infrastructure plans may be lost over time and empirical knowledge of pollution "hot-spots" may no longer readily available. New concrete platforms laid over old and polluted demolition material can become contaminated from below. The indiscriminate spreading of the "clean" construction and demolition waste as screed means that separation of contaminated and clean materials becomes impossible. As a result, less uncontaminated construction and demolition waste is available for recycling and more material has to be disposed to landfill. The former Grimethorpe colliery site in the Dearne Valley (South Yorkshire) illustrates the consequences of having no decommissioning programme to carry out remediation work as soon as mining ceases. Coal mining in the area began in the thirteenth century. By the end of the eighteenth century, a well-established coal industry existed. In 1894, the 140 ha Grimethorpe and Ferry Moor colliery site opened. Eventually it became the largest coal

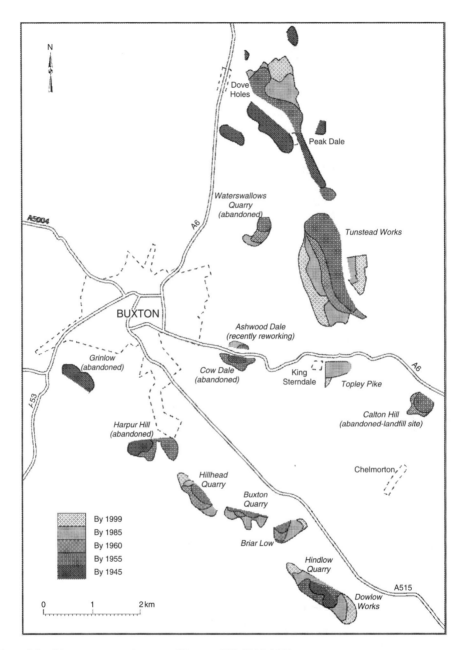

Figure 3.8.    Limestone extraction around Buxton, UK, 1945–1999.

producing and processing unit in the UK with a coking plant, chemical works, a power station and associated lagoons and spoil heaps occupying 480 ha (Fig. 3.9). After several changes to the complex, the plants were all closed by 1993. However, various political and economic factors delayed land remediation work until 2000.

The abandoned buildings and infrastructure were demolished, the site covered with new concrete bases and buildings, and the remaining debris left on site as fill. Despite this demolition,

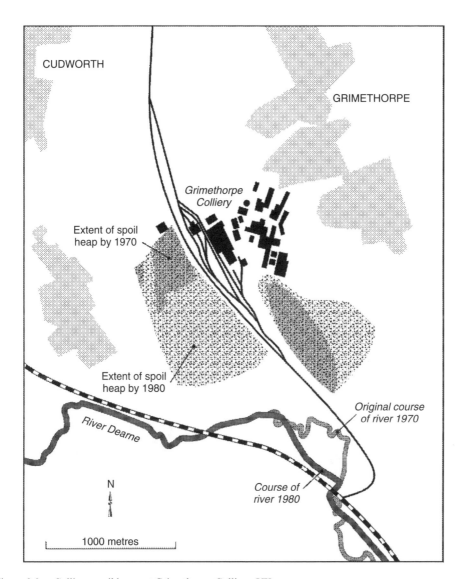

CUDWORTH

GRIMETHORPE

*Grimethorpe Colliery*

Extent of spoil
heap by 1970

Extent of spoil
heap by 1980

*River Dearne*

*Original course
of river 1970*

N

*Course of
river 1980*

1000 metres

Figure 3.9.   Colliery spoil heaps at Grimethorpe Colliery, UK.

by 1994 all the foundations and below-ground structures, including under-ground tanks, remained in situ.

Events at Grimethorpe were part of only 15 of the 24 sites covering some 11,000 ha inherited by Yorkshire Forward, the regional land improvement agency, which had seen any regeneration commence by 2000 (yorkshire-on.net [42]). These delays in commencing land remediation work mean that the former coalfields remain marginalised places (Bennett et al. [43]). Grimethorpe is now suffering from severe socio-economic deprivation with out-migration of population and increases in local unemployment rates totally unrelated to national trends (Table 3.3).

*3.5.4.2.2   Coal mining in Poland*
Poland is the 5th largest producer of coal in the world, extracting 110 Mt coal and 62.5 Mt lignite in 2000. The impact is particularly recognisable in the Upper Silesia Coal Basin where coal has

Table 3.3.   Population change and unemployment in Grimethorpe and the United Kingdom, 1981–1999.

| Year | Population Grimethorpe | Unemployment (%) Grimethorpe | Unemployment (%) United Kingdom |
|------|------------------------|------------------------------|----------------------------------|
| 1981 | 5,456 | 11.7 | 8.1 |
| 1991 | 4,978 | 25.1 | 8.0 |
| 1999 | 4,420 | 33.0 | 3.9 |

Source: Crompton, 2000 [44].

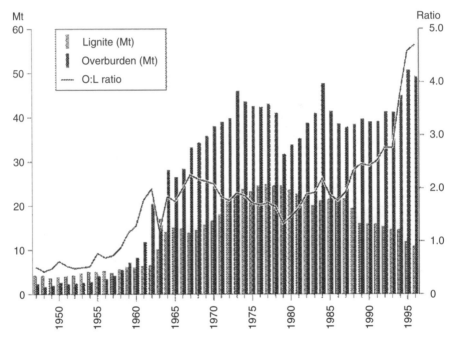

Figure 3.10.   Ratio of overburden to lignite in Turów Mine, Poland, 1947–1995.

been extracted for over 160 years, affecting over 1,500 km$^2$ and radically changing the landscape. 1999 mine waste production was over 44 Mt which, when added to the 6,750 million m$^3$ waste and overburden accumulated in dumps, despoil an area of 210 km$^2$. In addition, some 860 million litres of water are pumped out of deep mines each year and in places the land has subsided by as much as 30 metres (Rybicka and Rybicki [45]). Lignite mining in particular is likely to severely impact on land use in Poland for the foreseeable future. Production is expected to rise to over 70 Mt per annum between 2010 and 2017 before falling back to 67 Mt by 2021 and to 48 Mt by 2040. Conversely, in many Polish mines, the ratio of overburden removal has increased steadily from 0.50 in 1947, to 2.0 in 1980 and to as much as 4.70 in 1995 as technical advances in extraction equipment render the exploitation of increasingly deep deposits economical (Fig. 3.10) (Rybicka and Rybicki [45]).

### 3.5.4.2.3   *Coal mining in China*
The problem of land degraded by mining is particularly severe in a country such as China undergoing rapid urbanisation and development and where the mountainous terrain and eroded soils create a shortage of agricultural land. Edmonds [27] estimated that around 1990, China had some

2 million hectares of land already eroded or exploited by mining that could be reclaimed. China, with the world's largest coal reserves, produced 1,380 Mt in 2000 to fuel the rapid expansion of the nation's economy. There are about 500,000 ha of derelict land from past coal workings in China, with 0.2–0.3 ha being damaged for each 10,000 ton of coal mined (Bian [46]). Thus every year, coal mining degrades 34,500 ha more land. Underground coal mining, accounting for 96% of Chinese coal production, causing 1–3 ha of land subsidence per million tonnes produced, has already led to subsidence over approximately 80,000 ha (Chen et al. [47]). In Tongshan County in the north of Jiangsu Province, the Xuzhou coal mines have caused the loss of approximately 10,000 ha of arable land, or 6.7% of the total cultivated land area in the county. This has restricted land ownership to only 0.033 ha of land per head for about 210,000 people, and has left a further 100,000 people with no land at all (Chen et al. [47]).

### 3.5.4.2.4  *Limestone: geomorphic change caused by limestone around Buxton, Derbshire Peak District (UK)*

Limestone is the most important non-fuel mineral extracted in Britain transforming key parts of the upland landscape (Fig. 3.8). Gunn and Gagen [48] calculate the natural rate of erosion, or rock removal, by solutional processes in the Peak District as approximately $70\,m^3\,km^{-2}\,y^{-1}$, equivalent to an annual loss of some 83,500 tonnes of limestone per year. The amount of limestone removed annually by quarrying rose from 2 million tonnes per year in the period 1900 to 1920, to 5 Mt by 1949 to a peak of 22.5 Mt in 1973, since when it has fallen back to a more or less steady out put of about $18\,Mt\,y^{-1}$. Extrapolating from these figures, the tonnage of limestone removed by human activity in just 100 years equals the amount lost by natural solution processes over the whole 10,000 years of the Holocene (Gunn and Gagen [48]). These quarries adjoin the Peak District National Park and are scenically important. One indicator of scenic value is the expenditure of people travelling to and visiting areas of outstanding natural beauty which can indirectly contribute to the cost of landscape restoration. These scenic values are difficult to quantify, although many attempts have been made to do so either in terms of calculating tourist expenditures or developing means of quantifying scenic attributes, such as the uniqueness value (Leopold [49]) or LNC (landscape, natural and cultural-historical) values (Lenders et al. [50]). In the case of these Peak District limestone quarries, this restoration function has involved trying to simulate natural scree slopes on the faces of abandoned quarries as well as in worked out areas of active quarries (Department of the Environment Transport and the Regions [51]).

### 3.5.4.2.5  *Oil shale/sand mining: an example of major land degradation*

While oil shales and sands could solve many future liquid fuel problems, their exploitation involves much land disturbance. The petroleum potential of oil sand deposits in Alberta, Canada exceeds that of the proven oil reserves of Saudi Arabia. The oil sand mining, extraction, utilities and upgrading operation at Syncrude's Mildred Lake facility currently covers an area of 40,000 ha. Syncrude hold leases to exploit up to 68,600 ha, an area larger than 25% of the world's countries and containing an estimated 1.7 trillion barrels of bitumen, of which 300 billion are recoverable by current mining methods (Syncrude [52]). Opencast mining of oil sand and shale deposits up to 50 meters below the surface is commercially viable. This necessitates, on average, the additional material movement of 22 times the net quantity of oil produced (Douglas and Lawson [16]) and consumes 1.09 million kJ energy per barrel of crude oil produced (Syncrude [53]).

Estonia, the major producer of oil shale in Europe, has been mining oil shale since 1916. The mined-out area already covers $330–335\,km^2$ ($220\,km^2$ underground workings and $110\,km^2$ open pit mining) and the area being damaged is increasing by $5–7\,km^2$ per annum (Toomik and Liblik [54]). Underground workings in the Ordovician oil shales are so damp that $10–20\,m^3$ water have to be pumped out per ton of oil shale produced (Ministry of the Environment of Estonia [55]). Subsidence and changes to surface and underground water regimes have damaged some $100\,km^2$ of arable land in the deep mined areas. The consequent soil compaction and water logging make reclamation for agriculture an expensive alternative to reforestation (Toomik and Liblik [54]). Whilst over 80% of the $110\,km^2$ of land disturbed by open pit mining has been reclaimed, the difficulty in removing, handling and replacing overburden horizons during the mining cycle has resulted in general remediation to sparse woodland, with only a very small amount returned to agricultural use (Levene [56]). Furthermore, 70 years of thermal processing of oil shale has resulted in a chain of "mountains" of solid waste up to 100 meters high producing about 26 km$^2$ of new landforms in north-east Estonia. Approximately 1.3 Mt of coke-ash and other wastes contain pollutants such as sulphides and phenols are added to these dumps annually (Toomik and Liblik [54]). Political pressure to reduce dependency on oil imported from the Middle East will clearly drive North American exploitation of oil sands, regardless of the impact on future land use. Planned investments between 2000 and 2007 in additional oil shale extraction facilities in Alberta, Canada, exceed $6 billion (Syncrude [57]). However, environmental and financial considerations have led Estonian production to decrease from a high of 30 Mt p.a. in the early 1980's to 11.7 Mt in 2000.

### 3.5.4.2.6   *Tin: restoration of humid tropical tin mine land*

The restoration of derelict industrial and mining land in humid tropical countries poses particular problems due to heavy rainfall and rapid breakdown of chemical compounds. With the rapid industrialisation of many countries, these problems become even more important in the future. The nutrient poor sand banks and clay-rich slime ponds of former tin mining land in Indonesia and Malaysia provide a good example of these problems. Generally encouragement of plant growth on these areas has been difficult. 150 years of alluvial tin mining left large areas of disturbed ground worked by hydraulic pumps and sand dredges of low fertility and little agricultural potential. Investigations of tin mine tailings around Kuala Lumpur, Malaysia in 1967–1969 (Palaniappan [58]) showed that unstable sand banks were only colonized by pioneer species on flat sand plains. Such areas did not progress towards a forest succession. On the other hand, areas of former slime ponds had a succession from pioneer species, to *Phragmites* spp. which initiated a shrub stage that later evolved towards a forest with species normally found in lowland secondary forests, such as *Bridella, Macaranga* and *Ficus.* The shrub species normally develop after three years of pioneer plants and the forest types develop about 12 years later. Thus although the slime habitats are unique and unusual, they end up supporting a vegetation that replicates species found in the lowland secondary forests. Some land may thus become usable through natural regeneration, while other land requires costly interventions to be restored to productive use.

Few studies have examined how a minimal restoration effort may still achieve positive ecological results (Passell [59]). Several tin mine tailings restoration efforts were made in Indonesia and Malaysia in the twentieth century, but few went beyond minimal land regrading and the planting of species such as *Acacia mangium*. Analysing former tin mine sites on the 11,340 km$^2$ Indonesian island of Bangka, Passell [59] found significant increases after 3 years of restoration in bird species richness and diversity. Even rudimentary restoration produces positive increases in bird species numbers and diversity. Low cost, basic restoration methods are helpful and a great improvement over neglect of the abandoned mine sites.

### 3.5.4.3   *Chemical time bombs*

Many substances left by mining take time to be released to the environment. Sometimes the releases are catastrophic as with the collapse of tailings mounds at Ok Tedi (Fookes and Dale [60], Hearn [61]) or the release of slurry from the Aznalcollar mine into the Cota Donana National Park (Pain et al. [62], Meharg et al. [63]). In the end accumulations can become major hazards for future land use and prolonged soil contamination. These chemical time bombs (Ter Meulen et al. [64]) continue to build up, often as a result of uncontrolled waste disposal from informal mining (Lacerda and Salomons [65], Salomons and Lacerda [66]).

### 3.5.4.4   *Impacts specific to artisanal and informal mining*

The geomorphic and land use impacts of small-scale artisinal and informal mining activities are, relative to net mineral production, invariably far greater than those created by large scale high technology mining. Artisanal gold mining in Ghana provides a good example of the ways current small-scale mining activities affect the local environment and eventually human health. The majority of small-scale gold mining activities extract alluvial deposits of gold. Their processes of extraction are outdated and harmful to the surrounding environment. The major environmental impact results in the diversion of rivers. After the mining is completed, the rivers are not realigned along their original courses, which in turn results in the pollution of waters and destruction of surrounding flora and fauna. The mining activities also degrade the surrounding land by increasing air pollution, contaminating surface and ground water and increasing soil erosion and nutrient leaching. The pollution is, in the most extreme cases, leading to desertification and permanently changing land use from agriculture to waste rendering it useless to traditional inhabitants when the mining operations are completed. In the short run the inhabitants of the region are suffering from sickness and disease related to contaminated drinking water supplies such as dysentery and malaria.

Brick-clay pits are a common feature around the fringes of many towns in developing countries. For example in the Central Highlands of Vietnam near Dakto, clay pits cover an area as large as the town itself. Worked-out pits either remain as patches of bare earth which become severely eroded or become partially filled with water. None of the land appears to be restored or returned to agriculture. They are also subject to erosion and add to the accelerated transfer of materials to rivers. The same has happened with brick pits around Mar del Plata in Argentina. These hollows now form a blight on urban planning, restricting the growth of the town (del Rio [67]). In India, many commentators have complained that small scale brick clay workings around towns have taken up much valuable farming land which more sustainable planning policies would have retained for agriculture. In China, the ubiquitous expansion of rural brick factories, a third of which dry their bricks in the sun, have played a considerable role in the reduction of cultivated land (Larivière and Sigwalt [68]).

Township coal pits employing some 2 million people still produce about 40% of China's coal output (Pui-Kwan Tse [69]). However the recovery ratio of coal from these local pits is less than a third of that from state owned mines and these local pits have a large environmental impact as their wastes are disposed of haphazardly (Qu and Li [28]). In the gorges of the Jialing River at Chonquing, for example, adits run straight into the hillside and waste is allowed to slide directly into the river.

Small mining frequently leads to a single large deposit being owned and worked by many different operators, each wanting to maximize their own profits at the expense of rational exploitation of the whole ore body. Rapid exploitation of the highest grade ore often occurs, with the dumping of waste material on adjacent reserves, so making the remainder of the ore-body submarginal as far as small-scale mining is concerned (Berger [70]).

### 3.5.4.5 *Hydrogeological impacts*

Mining alters the natural hydrology over large areas. Pumping whilst mining is taking place alters water tables and creates depressions. Voids created by past mining activities, both at and below ground level become repositories of water. Sub-surface mineral extraction has considerable geomorphological impact, especially where it induces subsidence and surface landform change. In England alone, 660,626 ha of underground mineral workings from past and present mining were recorded in 1994 (Department of the Environment [33]). Abandoned flooded voids in deep mine can, with interconnected voids and major pit lakes, cover areas in excess of 2,000 km$^2$ in some of the larger European coal fields and be up to 1,500 m deep in places (Younger [71]).

Since the earliest days of mining, waste rock and processing debris have been left behind or been backfilled into mine voids. Acidic water in flooded abandoned coal mines can eat in to adjacent lime and calcitic sandstone thus creating sinkholes and subsidence. Many alkaline water discharges from old metal workings are still sufficiently rich in iron and or zinc to be highly contaminated. Contaminated surface water leads to soil degradation and increased rates of erosion. Whilst the worst pollution tends to emanate from the most recently flooded voids, mines abandoned thousands of years ago in Spain, Greece and other European countries still release highly polluted waters (Younger [71]). Large tracts of land remain degraded.

Table 3.4.   Summary of impacts of mining on land use.

| Causation | Impact |
|---|---|
| Working pit specific | • Subsidence by groundwater extraction<br>• Waste and overburden production<br>• Soil contamination by water pollution<br>• Land degradation by air pollution<br>• Land degradation by tailing dams<br>• Slumping due to groundwater depletion<br>• Depletion of arable land due to higher land values for mineral extraction<br>• Land degradation by draining of large mines also drying local springs and streams |
| Abandoned pit specific | • Subsidence due to voids<br>• Geomorphic change (pits and depressions)<br>• Loss of scenic value through mining and quarrying in areas of natural beauty<br>• Slope instability by spoil disposal<br>• Water filled gravel pits along flood plains<br>• Land degradation by waste and overburden dumps<br>• Loss of arable land by soil contamination from pollutants leaching from mine waste dumps<br>• Loss of arable land by compaction of replaced overburden<br>• Loss of arable land by increased erosion due to acidic water from abandoned deep mines degrading soil<br>• Cross contamination of wastes decreases resource substitution by recycling<br>• Danger to humans and animals |
| Chemical time bomb specific | • Loss of land use due to long term release of contaminants<br>• Loss of fertility in arable land due to rainfall induced release of chemical compounds<br>• Contaminated alkaline water from ancient abandoned metal workings degrades arable land<br>• Land degradation by acidic water affecting lime and calcitic sandstone from long abandoned coal workings |
| Artisanal and informal mining specific | • Water and land contamination due to uncontrolled use of harmful chemicals such as mercury<br>• Land use change by uncontrolled alterations to natural drainage systems<br>• Loss of arable land due to increased overburden removal and waste production during informal operations in areas of deprivation<br>• Loss of arable land due to irrational and haphazard waste and overburden disposal |

## 3.5.5   DISCUSSION: MITIGATING THE GEOMORPHIC IMPACTS OF MINING

Mining and quarrying for building materials and fossil fuels creates more material flows and changes in landform than the combined total of all other mineral extraction and processing (Table 3.4). However the industries to which these materials are supplied have many opportunities to cut down waste through traditional and innovative ways of recycling, and thereby to cut down the geomorphic impacts of mining.

### 3.5.5.1   *Recycling construction and demolition waste*

Greater use of construction and demolition (C&D) waste for aggregates in concrete manufacture and the re-use of components of buildings, such as reclaimed bricks or timber would have the double benefit of reducing the amount of land need for both landfill and new aggregate extraction.

In 1999, England and Wales produced some 72.5 Mt of C&D waste, including excavated soils, but excluding both road planings and other materials re-used without processing on the site where they first arose (Environment Agency [72]). 27.0 Mt $y^{-1}$ of this C&D waste were deposited in landfill. 9.5 Mt $y^{-1}$ of this material were employed in engineering works on site (haul and access roads, construction of cells, cover, etc.). 20.3 Mt $y^{-1}$ were exempt from licensed disposal and were used in unprocessed form or coarsely crushed for use on demolition/construction sites and for sale/disposal off site for land modelling during the construction of projects such as golf courses and equestrian centres. $25.1 \, \text{Mt} \, \text{y}^{-1}$ were recycled by screening and/or crushing (Environment Agency [72]). However, only $5 \, \text{Mt} \, \text{y}^{-1}$ of this recycled material is either crushed to produce a graded product or directly recovered (Department of the Environment [73]). The per capita consumption of cement in the England and Wales is around $250 \, \text{kg} \, \text{y}^{-1}$, with some 4 t of aggregate being used per capita per year. Modest increases in the recycling of C&D waste could reduce the demand for newly mined and quarried materials significantly. Overall, large quantities of C&D waste are being used for low grade recycling activities, as a result of the cost of disposal to land-fill; transport costs; the commercial need to re-develop sites as quickly as possible; or the low value of recycled material.

Where C&D waste that might be contaminated cannot be recycled and reused, it gets incorporated in contaminated land and thus adds to the problems of contaminated land remediation. The use of possibly contaminated C&D waste as an alternative to newly quarried aggregates depends on the development of appropriate risk assessment methodologies for materials that might be recycled (Lawson et al. [74]). Whilst major structures can remain both viable and desirable for many centuries, the majority of the urban infrastructure (domestic housing, factories, retail developments, offices, transport infrastructure, etc.) is generally renewed in less than 100 years. Much of this material could be reused and thus replace the extraction of aggregates as long as it is located in the right place, and is available at the right time, to meet market demands. Purposely established "borrow mounds" to stock re-useable C&D waste surplus to immediate requirements, at strategic locations on ring roads or major transport routes around urban areas, could become valuable sources of raw materials for future urban development.

### 3.5.5.2  *Construction materials from industrial wastes*

#### 3.5.5.2.1  *Colliery spoil and coal ash*

Coal mining waste is one of the most abundant and widely available by-products potentially suited for reuse as substitute to fresh aggregates. Coarse discard, largely sand and gravel sized pieces of stone and shale, forms the largest fraction of coal mining spoil. The fine discard made up of the clay, silt and fine sand sized pieces of stone and shale filtered from coal washing. Both burnt and unburnt colliery spoil are suitable materials for engineering fill in construction, but are unsuitable for use in concrete, unless processed to produce a synthetic aggregate (Aggregates Advisory Service [75]). In Japan, a dry flue gas desulphurizer using absorbents made from coal ash has been built at Tomato-atsuma Thermal Power Station No. 1. These absorbents remove more than 90% of the power station's $SO_2$ emissions. Now the company is working to put the coal ash used in this desulphurizing process to further use, for example, as a refrigerator deodorising agent (Hepco.co [76]).

#### 3.5.5.2.2  *Oil shale waste*

This material has similar chemical and physical properties to burnt colliery spoil and can be used as a substitute for primary aggregates. It has been widely and successfully used as bulk fill and as selected granular fill (Department of the Environment [77]).

#### 3.5.5.2.3  *Pulverised fuel ash*

Pulverised Fuel Ash (pfa) is ash derived from the combustion of hydrocarbon based materials and comprises mineral residues. The main components of pfa are oxides of silicon, aluminum and iron. As it is similar in colour to Portland Cement with pozzolanic and self-hardening properties, pfa can be used as structural fill, in block and lightweight aggregate manufacture, as a cement replacement and as an additive in concrete and in brick manufacture. In some parts of Europe, and in particular in France, all pfa is utilised. In the UK however, where coal-fired power stations generate around $7 \, \text{Mt}$ annually, only $3.5 \, \text{Mt}$ per annum are used, resulting in the present stockpile of some $250 \, \text{Mt}$ of potentially reusable pfa (Aggregates Advisory Service [75]).

### 3.5.5.2.4   *China clay waste*
Each ton of kaolin produced leaves some nine tons of waste in the form of unaltered overburden rocks, sand and mica. The United Kingdom production of 26 Mt kaolin per annum is confined to south-east of Cornwall. While locally, within Cornwall, china clay waste can meet 30–40% of the total aggregate requirements, some 450 Mt remains stockpiled (Aggregates Advisory Service [75]).

### 3.5.5.2.5   *Slate waste*
Slate has been worked in Britain for many centuries and production still generates some 7 Mt of waste annually. Slate waste is, in effect, a crushed rock and is therefore potentially suitable for all applications where crushed rock is specified. However, the use of slate waste as a secondary mineral is constrained by the cost of transporting it to the point of use. Thus the total consumption is insignificant in relation to the volume of waste being generated. It is estimated that there are over 500 Mt of slate waste on the 2,100 ha of derelict and current slate workings (Department of the Environment [78]).

### 3.5.5.2.6   *Glass*
Recent investigations have suggested that fragmented glass might be used as an aggregate substitute or additive to asphalt for road construction (Aggregates Information Service [79]).

### 3.5.5.2.7   *Paper sludge*
Experiments at the University of Salford have demonstrated that high quality plaster-board substitutes may be made from the sludge produced in paper manufacture, again offering the potential to reduce a major disposal problem (Webster and McNicholas [80]).

### 3.5.5.2.8   *Tyres*
Malaysia has for long used rubber in road surfacing materials, providing an outlet for shredded tyres. The springy surfaces of modern children's playgrounds are also made from used vehicle tyres. Tyres are also used as an additive in the manufacture of concrete (Ratnasamy and Radin [81]).

### 3.5.5.2.9   *Foundry sand*
To produce moulds in to which metal is cast, the UK foundry industry uses, and disposes of, over one million tonnes of silica sand per annum. Recycled foundry sand can be used for the manufacture of concrete blocks as a substitute for silica sand, in the production of asphalt and in roofing felt manufacture (Aggregates Advisory Service [75]).

### 3.5.5.2.10   *Incinerator bottom ash*
The ash from municipal solid waste incineration is another substitute for primary aggregates, having been used successfully in several European countries for over twenty years as a substitute for primary aggregates in embankment fill; road-base material; asphalt and concrete building blocks. In Britain, fear of dioxin emissions is still restricting the incineration of municipal solid waste and worries about cross-contamination limit opportunities to recycle incinerator bottom ash. Conversely, in the Netherlands, with practically no land suitable for waste disposal to landfill, all non-recyclable municipal solid waste is incinerated and almost 100% of the incinerator bottom ash produced is reused. Working together through an industry association, municipal solid waste companies have encouraged the introduction of regulatory standards governing the useful application of incinerator bottom ash. The Dutch government authorities have endorsed these regulatory measures (Aggregates Advisory Service [75]).

### 3.5.5.2.11   *Dredged silts*
Another possible source of building materials for the UK arises from the 40 Mt of material which is dredged annually from estuaries and ports and disposed in the sea. Most of the material is not contaminated and can be regarded as a resource (Burt and Cruickshank [82]). (Most of this

dredged material is silt sized). Although sand and gravel are needed for aggregates, silt could be used for making clay bricks and tiles and also in concrete tile manufacture.

Many estuaries become contaminated and relatively basic treatment of contaminated dredged material may make its reuse uneconomic. As transport costs are high, clean dredged material is likely to be an economically viable substitute for fresh aggregates only close to dredging sites and at some distance from other aggregate sources. Disposal of contaminated dredged material is costly and thus finding alternative uses for it is becoming more attractive. At present there are no legally binding standards affecting the use of dredged materials, but the local waste regulation authorities would have to approve its reuse. While re-use of dredged material would reduce the

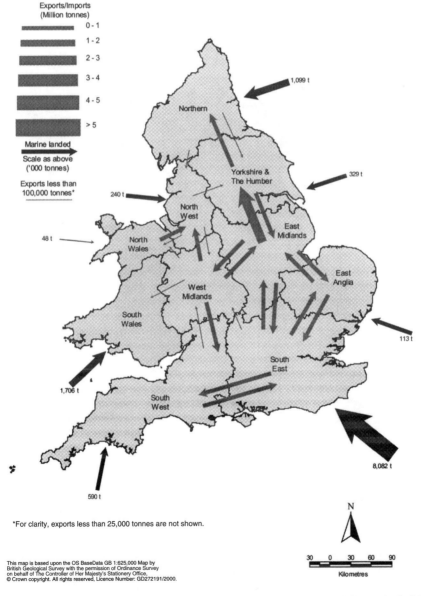

Figure 3.11.   Sand and gravel interregional flows in England and Wales, 1997 (British Geological Survey, 2000 [83]).

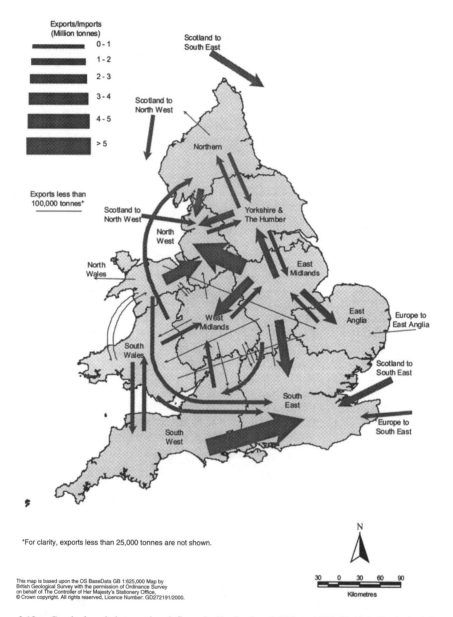

Figure 3.12.    Crushed rock interregional flows in England and Wales, 1997 (British Geological Survey, 2000 [83]).

amount of land required for both dumping spoil and extracting new aggregates, the potential for recycling is restricted by economics and acceptability (Burt and Cruickshank [82]).

The above examples show the great potential for use of recycled materials in place of freshly dug aggregates. However, their adoption depends on their geographic location and the economics of production and distribution. In England and Wales, the annual consumption of over 200 Mt of primary sand and gravel and crushed rock requires a wide range of inter-regional transfers of material (Figs 3.11 and 3.12). Use of environmental economics to analyse materials flows might demonstrate how modest charges for the disposal of used materials and taxation on the extraction of raw materials would make recycling becomes cost effective. The building industry's overwhelming concern with cost concerns and engineering and technical acceptability drives the potential for the use of secondary materials in construction. Rational policies to reduce raw materials extraction would encourage this.

### 3.5.6 CONCLUSIONS

In global terms, each ton of minerals produced disturbs 3 tons of earth (Table 3.1) and in Britain the extraction of minerals removes nearly 400 Mt of material each year, even assuming the total replacement of all hidden flows from the opencast extraction of coal. This represents approximately 8 times the $50 \, \text{Mt} \, \text{y}^{-1}$ accelerated sediment transferred by British rivers (Douglas and Lawson [84]). The significance of this 8 fold difference between erosion by earth surface processes and the deliberate shifting of material by direct human action lies in the recognition that the overall rate of change of the land surface of Britain is now more rapid than at any time since the last glaciation (i.e., in the last 12,000 years).

Just as the water industry now accepts that it has to provide water both for human and industrial consumption and to support aquatic ecosystems along rivers, so planners and regulators have to manage mineral extraction licences as far more than just a commodity to be bought, sold or leased. Mining and quarrying alters land. It removes land from alternative use, sometimes irreparably for long periods of time. Around rapidly expanding cities, aggregate and brick clay extraction takes land out of food or timber production and forces greater energy consumption to bring production from farms and forests further away. The changes to rivers caused by gravel extraction can have severe impacts on industrial and transportation infrastructure further downstream, often requiring costly, and materials hungry, remedial river training works such as flood prevention embankments and re-aligned channels. The stress on land availability caused by ever increasing extraction of primary minerals to meet the world wide drive to greater urbanisation and material consumption will only be reduced by greater efforts to restrict material use. Prospects for managing minerals extraction more carefully, for reducing the overburden mass and the off-site impacts are good. The technologies exist. Wealthy communities can afford to adopt them. Less affluent communities may substitute labour for energy costs and restore old mineral workings by sheer hard work, as with the re-establishment of cereal crops on the floors of old brick-pits in rural China. Careful design to reduce material requirements, increased efforts to promote alternative and renewable forms of energy to replace fossil fuels, and acceptance of mineral waste and other secondary products as an alternative resource can mitigate the land use and geomorphological impacts of mining.

### REFERENCES

1. Marsh, G.P., *Man and Nature or Physical Geography as Modified by Human Action.* Charles Scribner, New York, 1864.
2. Sherlock, R.L., *Man as a Geological Agent.* Witherby, London, 1922.
3. Hooke Le B.R., On the efficacy of humans as geomorphic agents. *GSA Today 4 (9)* (1994), pp. 223–225.
4. Fischer-Kowalski, M., Society's metabolism – origins and development of the material flow paradigm. In: S. Bringezu (ed.) *Regional and national material Flow accounting: from paradigm to practice of sustainability (Wuppertal Special 4).* Wuppertal Institute for Climate, Environment and Energy, Wuppertal, Germany, 1997, 16–23.
5. Rees, W.E., Pressing global limits: trade as the appropriation of carrying capacity. *University of British Columbia Centre for Human Settlements Policy Working Paper: Issues and Planning Responses* (1994), pp. 4–24.
6. Bringezu, S. and Schütz, S., Analyse des stoffverbrauchs der deutschen wirtschaft: status quo, trends und mögliche prioritäten für maßnahmen zur erhöhung der ressourcenproduktivität'. In: J. Köhn (ed.): *Neue ansätze in der weltökonomie.* Metropolis-Verlag, Marburg, 1996.
7. Chen, X. and Qiao, L., *Material flow analysis of Chinese economic-environmental system.* Paper presented at the SCOPE Open Executive Committee Meeting, Beijing, April 2000.

8. De Marco, O.L. and Pizzoli Mazzacane, E., Material flow analysis of the Italian economy: preliminary results. In: R. Kleijn (ed.): *ConAccount Workshop: Ecologizing Societal Metabolism: Designing Scenarios for Sustainable Materials Management, (CML Report 148)*. Centre of Environmental Science, Leiden University, Leiden, Netherlands, 1999, pp. 31–37.

9. Mathews, E., Amann, C., Bringezu, S., Fischer-Kowalski, M., Hüttler, W., Kleijn, R., Moriguchi, Y., Ottke, K., Rodenburg, E., Rogich, D., Schandl, H., Schütz, H., van der Voet, E. and Weisz, H., *The Weight of Nations. Material Outflows from Industrial Economies*. World Resources Institute, Washington, D.C., 2000.

10. Sustainable London Trust, *Creating a sustainable London*. The Trust, London, 1996.

11. Lei, K. and Wang, Z., The analysis about ecological footprints of Macao in 2001. In: *Ecoscape Eco-industry Eco-culture: Proceedings of the Fifth International Ecocity Conference*, 2002, pp. 70–72.

12. Prat, I. and Noguer, A., Application of the ecological footprint to Barcelona: summary of calculations and thoughts on the results. In: *La Petjada Ecologica deBarcelona*. Commissio de Medi Ambient IServeis Urbans, Ajuntament de Barcelona (Environment and Urban Services Committee of Barcelona City Council), Barcelona, 1998.

13. Rees, W.E., Ecological footprints and appropriated carrying capacity: what urban economics leaves out. *Environment and Urbanization,* 4 (2) (1992), pp. 121–130.

14. Cendrero, A. and Douglas, I., Earth surface processes, materials use and urban development; project aims and methodological approach. *Geological Society of America Abstracts with Programs*, 28 (7) (1996), p. A-79.

15. Douglas, I. and Lawson, N., An earth science approach to material flows generated by urbanisation and mining. In: S. Bringezu (ed.): *Regional and national material Flow accounting: from paradigm to practice of sustainability (Wuppertal Special 4)*. Wuppertal Institute for Climate, Environment and Energy. Wuppertal, Germany, 1997a, pp. 108–118.

16. Douglas, I. and Lawson, N., 1997b An earth science approach to assessing the disturbance of the earth's surface by mining. *Mining and Environmental Research Network Research Bulletin* 11/12 (1997b), pp. 37–43.

17. Warhurst A., *Environmental degradation from mining and mineral processing in developing countries: corporate responses and national policies*. OECD, Paris, 1994.

18. Von Weizächer, E., Lovins, A.B. and Lovins, H.L., *Factor four. Doubling wealth-halving resource use*. Earthscan, London, 1997.

19. Adriaanse, A., Bringezu, S., Hammond, A., Moriguchi, Y., Rodenburg, E and Schütz, H., *Resource flows: the material basis of industrial economies*. World Resources Institute, Washington DC, 1997.

20. Douglas, I. and Lawson, N., Problems associated with establishing reliable estimates of materials flows linked to extractive industries. In: R. Kleijn, S. Bringezu, M. Fischer-Kowalski, and V. Palme, (eds): *ConAccount workshop Ecologizing Societal Metabolism: Designing Scenarios for Sustainable Materials Management. November 21st. 1998, Amsterdam, The Netherlands. CML report 148 Section Substances & Products*. Centre of Environmental Science (CML). Leiden, The Netherlands, 1998, pp. 127–134.

21. United Nations, *Industrial Commodity Statistics Yearbook, 1975–1998. Production and Consumption Statistics*. United Nations, New York, 1975–1998.

22. United States Geological Survey, http://minerals.usgs.gov/minerals/pubs/myb.html, 2002.

23. International Energy Annual, http://www.eia.doe.gov/emeu/iea/contents.html, 2002.

24. Veiga, M.M., Personal communication, 6/7/1998.

25. Hollaway, J., Small-scale mining: how to combine development with low environmental impact. *UNEP Industry and Environment* 20 (4) (1997), pp. 44–48.

26. Jéjé, L.K., Personal Communication, 1998.

27. Edmonds, R.L., *Patterns of China's lost harmony. A survey of the country's environmental degradation and protection*. Routledge, London, 1994.

28. Qu, G. and Li, J., *Population and the environment in China*. Lynne Rienner, Boulder, Colorado, 1994.

29. Nötstaller, R., Personal communication, 13/7/1998.

30. Chakravorty, S.L., Economic dimensions of small-scale mining in India. In: A.K. Ghosi (ed.): *Small-scale mining: a global overview*. Mohan Bimlani, New Delhi, 1991, pp. 287–298.

31. Lawson, N. and Douglas, I., Urban metabolism, materials flows and sustainable development: a geoenvironmental perspective. In: Proceedings of the 5th International Symposium on Geoenvironmental Technology and Global Sustainable Development, August 9–13, 1998. CEEST – University of Massachusetts, Lowell, 1998, pp. 3–12.

32. Douglas, I. and Lawson, N., The contribution of small-scale and informal mining to disturbance of the earth's surface by mineral extraction. *Mining and Energy Research Network Research Bulletin* No. 15, 1999/2000, (2000), pp. 153–161.

33. Hall, I.G., The Ecology of Disused Pit Heaps in England. *Journal of Ecology*, 45 (1957), pp. 689–720.

34. Holl, K.D., Zipper, C.E. and Burger, J.A., Recovery of native plant communities after mining. *Virginia Cooperative Extension Publication* 460 (2001) p. 140.
35. Department of the Environment, *Survey of land for mineral workings in England 1994, Volume 1.* HMSO, London, 1996.
36. Kivell, P. T., Dereliction and environmental regeneration. In: A.D.M. Phillips (ed.): *The Potteries: continuity and change in a Staffordshire conurbation.* Alan Sutton, Shrewsbury, 1993.
37. Walsh, F., Lee, N. and Wood, C., *The environmental assessment of opencast coal mines.* EIA Occasional Paper 28, University of Manchester, Manchester, 1991.
38. Lewin, J., Contemporary erosion and sedimentation. In: J. Lewin (ed.): *British rivers.* Allen and Unwin, London, 1981, pp. 35–58.
39. Macklin, M.G., Hudson-Edwards, K.A. and Dawson, E.J., The significance of pollution from historic metal mining in the Pennine ore fields on river sediment contaminant fluxes to the North Sea. *Science of the Total Environment.* 194–195 (1997), pp. 391–397.
40. Hudson-Edwards, K., Macklin M. and Taylor. M., Historic metal mining impact to Tees river sediment." *Science of the Total Environment.* 194–195 (1997), pp. 437–445.
41. Perry, D. and Handley, J., The potential for woodland on urban and industrial wasteland in England and Wales – with a review of the dynamics of urban and industrial wasteland. Forestry Commission Technical Paper 29. Forestry Commission, Edinburgh, 2000.
42. yorkshire-on.net, www.yorkshireon.net/matters/progress/action%5Fplan/actionplan6.htm, 2000.
43. Bennett K., Benyon, H. and Hudson, R., *Coalfields regeneration. Dealing with the consequences of industrial decline.* The Policy Press, Bristol, 2000.
44. Crompton, M., Senior Planning Assistant Development Programme Area, Barnsley Metropolitan Borough Council: Personal communication, 2000.
45. Rybicka, E.H. and Rybicki, S., *Environmental impacts of coal mining in Poland.* Paper presented to the Workshop on Mine and Quarry Waste – the Burden from the Past, Lake Orta, Italy, May 27–28, 2002.
46. Bian, Z., Department of Mining Engineering, China University of Mining Technology, Jiangsu Province: Personal communications, 4/6/1998 and 11/8/1998.
47. Chen, H., Zheng, C. and Zhu, Y., Phosphorous: a limiting factor for restoration of soil fertility in a newly reclaimed coal mined site in Xuzhou, China. *Land Degradation and Development* 9 (1998), pp. 115–121.
48. Gunn, J. and Gagen, P., Limestone quarrying as an agency of landform change. In: D. Gilleson, and D. Ingle Smith (eds): *Resource Management in Limestone Landscapes: international perspectives. Special Publication No. 2.* Department of Geography and Oceanography, University College, Australian Defence Forces Academy, Canberra, 1989, pp. 173–181.
49. Leopold, L.B., Landscape aesthetics. In: G. Bell and J. Tyrwhitt (eds): *Human identity in the urban environment, edited by* Penguin Books, Harmondsworth, 1972.
50. Lenders, H.J.R., Huibregts, M.A.J., Aart, B.G.W. and Van Turnhout, C.A.M., Assessing the degree of preservation of landscape, natural and cultural-historical values in river dike reinforcement planning in the Netherlands. *Regulated Rivers: Research and Management.* 15 (1999), pp. 325–337.
51. Department of the Environment, Transport and the Regions: *The reclamation of limestone quarries using landform replication.* Department of the Environment, Transport and the Regions, London, 1997.
52. Syncrude: *Syncrude facts 1997.* Syncrude, Alberta, 1997a.
53. Syncrude: *1997 Environmental facts.* Syncrude, Alberta, 1997b.
54. Toomik, A. and Liblik, V., Oil shale mining and processing impact on landscapes in north-east Estonia. *Landscape and Urban Planning* 41 (1998), pp. 285–292.
55. Ministry of the Environment of Estonia: *Estonian mineral Resources.* Environmental Information Centre, Tallinn, 1994.
56. Levene, R.M., The mineral industry of Estonia. In: *US Geological Survey – Minerals Information.* http://minerals.usgs.gov/minerals/pubs/myb.html, 1997.
57. Syncrude, http://www.syncrude.ca/our_growth/growth.html, 2002.
58. Palaniappan, V.M., Ecology of Tin Tailings Areas: Plant Communities and Their Succession. *Journal of Applied Ecology,* 11 (1974), pp. 133–150.
59. Passell, H.D., Recovery of Bird Species in Minimally Restored Indonesian Tin Strip Mines. *Restoration Ecology,* 8 (2000), pp. 112–118.
60. Fookes, P.G. and Dale, S.G., Comparison of interpretations of a major landslide at an earthfill dam site in Papua New Guinea. *Quarterly Journal of Engineering Geology,* 25 (1992), pp. 313–330.
61. Hearn, G.J. (1995). Landslide and erosion hazard mapping at Ok Tedi copper mine, Papua New Guinea. *Quarterly Journal of Engineering Geology,* 28, 47–60.

62. Pain, D.J., Sanchez, A. and Meharg, A.A., The Doñana Ecological Disaster: contamination of World Heritage estuarine marsh ecosystem with acidified pyrite mine waste. *Science of the Total Environment* 222 (1998), pp. 45–54.
63. Meharg, A.A., Osborn, D., Pain, D.J., Sanchez, A. and Naveso, M., A challenge to ecotoxicology and environmental chemistry: The Doñana mining catastrophe. Letter to the Editor: *Environmental Toxicology and Chemistry,* 18 (1999), pp. 811–812.
64. Ter Meulen, G.R.B., Stigliani, W.M., Salomons, W., Bridges, E.M. and Imeson, A.C., Chemical Time Bombs. In: *Proceedings of the European state-of-the-art conference on delayed effects of chemicals in soils and sediments,* Veldhoven, the Netherlands, 1993.
65. Lacerda, L.D. and Salomons, W., *Mercury in the Amazon: a chemical time bomb?* Chemical Time Bombs Project, Foundation for Eco-development, Haren, 1992.
66. Salomons, W. and Lacerda, L.D., *Mercury from Gold and Silver Mining: A Chemical Time Bomb.* Springer, Heidelberg, 1998.
67. del Rio, L., *Environmental consequences of 120 years of urban development in Mar del Plata, Argentina.* Paper presented at the Latin American ESPROMUD Project Workshop, Bogata, Colombia, 29 November 1996.
68. Larivière, J.-P. and Sigwalt, P., *La Chine.* Masson, Paris, 1991.
69. Pui-Kwan T., The mineral industry of China. In: United States Department of the Interior Bureau of Mines *Minerals yearbook international review, Volume III.* US Government Printing Office, Washington DC, 1996, pp. 191–208.
70. Berger, A.R., The importance of small-scale mining: a general review. In: J.M. Neilson (ed.): *The Geoscience in international development AGID Report No. 8 Strategies for small-scale mining and mineral industries.* Association of Geoscientists for International Development, Bangkok, 1982, pp. 1–12.
71. Younger, P.L., *Don't forget the voids: impacts of emissions from abandoned mines in Europe.* Paper presented to the Workshop on Mine and Quarry Waste – the Burden from the Past, Lake Orta, Italy, May 27–28, 2002.
72. Environment Agency, *Construction and Demolition Waste Survey. R&D Technical Summary* PS368. Environment Agency, Swindon, 2001.
73. Department of the Environment: *Managing demolition and construction wastes. Report on the study on the recycling of demolition and construction wastes in the UK, for the Department of the Environment by Howard Humphries and Partners.* HMSO, London, 1994.
74. Lawson, N., Douglas, I., Garvin, S., McGrath, C., Manning, D. and Vetterlein, J., Recycling construction and demolition wastes – the case for Britain. *Environmental Management and Health.* 12 (2) (2001), pp. 146–157.
75. Aggregates Advisory Service: http:www.planning.detr.gov.uk/aas/index.htm, 2002.
76. Hepco.co.: http://www.hepco.co.jp/english/profile/topics/topics13.html, 2000.
77. Department of the Environment, *Use of waste and recycled materials as aggregates: standards and specifications. A report prepared by the Building Research Establishment (BRE) for the Department of the Environment.* HMSO, London, 1995a.
78. Department of the Environment, *Slate waste tips and workings in Britain.* HMSO, London, 1995b.
79. Aggregates Information Service *Digest 72 Glasphalt – the use of recycled glass in Asphalt by RMC Aggregates (UK) Ltd.:* http:www.viridis.co.uk/ais, 2002.
80. Webster, J. and McNicholas, P., From negative cost materials to high strength Boards. *ICE Research Focus,* 38 (1999), p. 3.
81. Ratnasamy, M. and Radin Umar, R.S., Modified rubberized stone mastic asphalt mix for Malaysian roads. *Road Engineering Association of Asia and Australasia Journal,* 6 (1997), pp. 67–78.
82. Burt, T.N. and Cruickshank, I.C., *Uses of dredged and other materials in construction. HR Wallingford Report SR555.* H.R Wallingford, Wallingford, 1999.
83. British Geological Survey: *Collation of the results of the 1997 aggregate mineral survey.* British Geological Survey, Keyworth, Nottingham, 2000.
84. Douglas, I. and Lawson, N., The human dimensions of geomorphological work in Britain. *Journal of Industrial Ecology* 4 (2), 2001, pp. 9–33.

# 3.6   Energy use management

Krishna Parameswaran
*ASARCO LLC, Phoenix, Arizona, USA*


## 3.6.1   INTRODUCTION

Mining natural resources consumes large amounts of energy in the mining and processing tasks. In addition, large mining projects are in remote areas where access to an electricity grid may not be available. Diesel generators or hydroelectric power plants are used at mining sites to produce the energy required to run equipment, and provide lighting. The impact of inefficient energy use on the environment and the profitability of the mine have to be carefully considered. Large excavators and hauling equipment use a lot of diesel fuel, and emit diesel emissions. In underground mines, the impact of diesel particulates is regulated by the mine safety and health agencies, and mining companies are finding ways to comply with the regulations.

Energy audits conducted can result in reduced energy use and increased profitability. The audit should consider the following key areas of a mining operation:

- large excavating and hauling equipment
- ore crushing and grinding
- mineral processing equipment
- pumping systems, both for open pit and underground operations
- ventilation system for underground mines
- lighting and heating, ventilation, and air conditioning of buildings
- material conveying systems.


## 3.6.2   ENERGY USE MANAGEMENT

The mining industry consumes about 1 to 2 quadrillion BTUs annually, or about 3.2% of the energy consumed by all industries in the United States of America. Energy costs are an important constituent of mining costs, estimated to be about 5% of the value of all commodities produced (DOE, 1999 [1]).

In primary nonferrous metals production such as refined copper, approximately two-thirds of energy used to produce refined metal is consumed in extraction (mining) and beneficiation (mineral processing) to produce a concentrate that can be smelted and refined. An important energy consumer is the energy consumed in comminution or size reduction especially grinding. A National Materials Board Study (NMB [2]) in the year 1981 estimated that U.S. mining industry consumes about 29 billion kWh of electrical energy (or approximately 99 trillion BTUs) for size reduction. It is highly likely that this figure has escalated by 20% in the last 22 years. These operations have low efficiency in that only 1–2% of the input energy is utilized. Most is expended as losses, particularly as heat and metal wear. Energy used could be reduced through improved blasting techniques, more efficient electrical and pumping equipment, and reduced wear and tear on equipment. Finally, implementation of new modeling and simulation technologies should result in proper design of size reduction equipment.

Over the years, modest improvements have been made in energy efficiency of size reduction equipment. The main trend has been the increased use of Semi-Autogenous Grinding (SAG) mills. These mills have replaced secondary and tertiary crushing and rod mills. SAG mills utilize the impact provided by falling ore to replace part of the grinding media. Other developments include improved computer simulation of SAG mills that has led to better design of shell (Sherman and

Rajamani, [3]) and pulp lifters in SAG mills. Other areas of research being pursued include mine to mill concept, where improvements in drilling techniques and blasting design have lead to smaller size inputs to crushers. Shifting part of the size reduction to blasting itself improves energy efficiency in crushing and grinding, as well as reduced maintenance and wear. An added benefit of more efficient blasting and drilling is reduced dust emissions (DOE, 2000 [4]). Advances in computer technology allow for increased data logging capabilities and advances in data mining have resulted in considerably improved analytical capabilities, especially in the correlation between mine site variables and concentrator production.

In-situ mining or solution mining offers the potential to extract metals from deep seated deposits because the energy consumed in mining and beneficiation is saved, as well as due to the more effective use of the intrinsic heats of reaction. However, this technique can be utilized where the mineralized rock is fractured to allow contact with the extracting solution. Proper collection and containment of the metal-bearing pregnant solution is critical to success and reduction of environmental impacts.

In smelting operations, energy management issues relate to the cost and availability of fossil fuels. In copper smelting, for example, smelting has moved towards autogenous flash smelting (Outokumpu, INCO) and bath smelting processes (Isasmelt, Ausmelt, Mitsubishi, Noranda and El Teniente) in which the sulfur in concentrates is oxidized with enriched air or pure oxygen to replace some or all of the fossil fuel usage. In some of the developing world, older reveberatory furnaces burning fossil fuels are still being used. In these cases, improvements in energy efficiency can be obtained through oxygen enrichment and waste heat utilization.

For copper, one important development over the past several decades is the increased adoption of leaching and solvent extraction-electrowinning (SX-EW) technology. The share of total copper output in the USA accounted for by SX-EW technology has increased from 6% in 1985 (Tilton and Landsberg [5]) to approximately 50% in 2002. Worldwide SX-EW represents roughly 20% of total copper output.

The SX-EW process involves leaching of existing mine dumps and prepared ore heaps with dilute sulfuric acid solution. In the solvent extraction step, the pregnant leach solution is mixed with an organic solvent (extractant), which selectively removes the copper. The copper loaded solution is mixed with an aqueous acid solution, which strips out the copper. This resulting electrolyte is relatively concentrated and relatively pure and in the electrowinning step is converted to cathode copper.

Non-sulfide copper minerals (carbonates, oxides, hydroxy silicates and sulfates) are the easiest to leach. Sulfide minerals (chalcopyrite, bornite, chalcocite and covellite) require an oxidant be added to the acid leach solution. The process takes longer with sulfide minerals and is less complete. Bioleaching shortens the leaching time for sulfide ores. SX-EW technology has evolved from extraction of copper from waste rock dumps in the 1970's and early 1980's to ore mined specifically for leaching. Efforts continue to extend the range of technology to processing sulfide minerals, as an alternative to traditional pyrometallurgical technology. Phelps Dodge's Morenci mine converted from traditional mining and milling of sulfide concentrates to solely leaching and SX-EW in 2002, and produces about 375,000 metric tons of copper. Also, at its Baghdad mine, Phelps Dodge is operating a 16,000 metric tons per year demonstration plant for pressure leaching of copper concentrates, followed by SX-EW to produce electrowon cathodes (Marsden et al. [6]).

The leaching of existing mine dumps does not require additional mining. Heap leaching, on the other hand, requires mining and some crushing of ore. Thus, the SX-EW process has the potential to reduce the cost and energy used in mining. The energy use in milling, smelting and refining is replaced by the energy used in leaching and SX-EW operations. Also, in the SX-EW process, the time copper remains in process is shortened. One of the disadvantages of SX-EW technology is that by-product precious metals, if present, are difficult to recover. With more plant operating experience from Morenci, better energy consumption and cost comparison with conventional technology can be made.

REFERENCES

1. Department of Energy, February 1999, Mining Industry Roadmap for Crosscutting Technologies, Mining Industry of the Future, p. 3.
2. National Materials Advisory Board, 1982, Comminution and Energy Consumption, Report no. NMAB-364, National Academy of Sciences, Washington, DC USA, pp. 16–24.
3. Mark Sherman and Raj Rajamani, 1999, The Effect of Lifter Design on the Alumbrera SAG mill performance, Canadian Mineral Processor's Conference Proceedings, Ottawa, Canada, pp. 255–265.
4. Department of Energy, September 2000, Mineral Processing Technology Roadmap, Mining Industry of the Future, pp. 8–9.
5. John E. Tilton and Hans H. Landsberg, 1999, Innovation, Productivity Growth and Survival of the U.S. Copper Industry, Productivity in the Natural Resource Industries Improvement through Innovation, Ed. R. David Simpson, Resources for the Future, Washington DC, pp. 109–139.
6. J.O. Marsden, Robert E. Brewer and Nick Hazen, 2003, Copper Concentrates Leaching Developments by Phelps Dodge Corporation, *Hydrometallurgy 2003* – Fifth International Conference in Honor of Professor Ian Ritchie Volume 2: Electrometallurgy and Environmental Hydrometallurgy, Ed. C.A. Young, A.M. Alfaniazi, C.G. Anderson, D.B. Dresinger, B. Harris and A. James, TMS (The Minerals, Metals & Materials Society), pp. 1429–1446.

# CHAPTER 4

## Sustainable mining systems and technologies

Raj Rajaram
*Complete Environmental Solutions, Oak Brook, IL, USA*

### INTRODUCTION

The World Summit on Sustainable Development (WSSD [1]) was held in Johannesburg from August 26 to September 4, 2002. This summit reviewed the progress made by governments and other groups, including the private sector, in the implementation of the Agenda 21, the declaration that emanated from the United Nations Conference on Environment and Development (UNCED [2]), the so called "Earth Summit", that was held in Rio de Janiero in 1992. Prior to the WSSD, the International Council on Mining and Metals (ICMM [3]) prepared a Working Paper that provides a brief overview of the progress over the past decade of the mining and metals industries' contribution to sustainable development. The ICMM working paper was discussed at the Global Mining Initiative (GMI) in Toronto in May 2002.

The GMI is an industry initiative involving about 28 mining and mineral companies, sponsorship by governments in Canada, Australia and United Kingdom, labor unions, UN agencies, the World Bank, and some academic and civil society organizations. The Mining, Minerals and Sustainable Development (MMSD [4]) project is the major component promoted by the GMI. The GMI is assisted by the London-based International Institute for Environment and Development (IIED) in implementing the MMSD project.

The GMI framework recognizes that:

- decisive and principled leadership is required by the mining companies
- accountability, transparency, and credible reporting is essential
- mining companies must do business in a manner that merits the trust and respect of key constituencies, including the communities in which they operate
- constructive and value-adding engagement among constituencies at the local, national, and global levels is essential
- mining companies must move beyond a regulatory-compliance-based mindset to effectively manage the complex trade-offs of economic, environmental, and social issues
- industry requires additional capacity (trained personnel) to be effective in advancing sustainable development
- the roles and responsibilities of the diverse parties comprising governments, civil society, and business are different and must be respected
- small-scale mining and orphan site legacy issues are important and complex; governments and international agencies should assume the lead role in addressing them (MMSD, 2002).

The growing public interest in a clean environment and the new governmental regulations are forcing companies to give more attention to environmental performance. The companies are also realizing that improving environmental performance is also resulting in energy savings, waste reduction, and process efficiency improvements. Thus, implementing environmental improvements is leading to improved profitability (UNEP, DTIE, 1997 [5]). The environmental management aspects of the sustainable mining framework set forth above, and emerging technologies available to implement sustainable mining are discussed in this chapter.

REFERENCES

1. United Nations, 2002, World Summit on Sustainable Development (WSSD), Johannesburg, South Africa, September.
2. United Nations Conference on Environment and Development (UNCED), 1992 Rio de Janiero, Brazil, September.
3. International Council on Mining and Metals (ICMM), 2002, Global Mining Initiative, Toronto, Canada, May.
4. International Institute of Environment and Development, 2002, Mining, Minerals and Sustainable Development Report (MMSD), London, England.
5. United Nations Environment Programme, Division of Technology, Industry and Environment (DTIE), 1997, Mining and Sustainability, New York, NY, USA.

# 4.1 Cleaner production

Jim Altham
*Curtin University, Western Australia, Australia*

Turlough Guerin
*Shell Distribution Oceania, Rosehill, NSW, Australia*

## 4.1.1 INTRODUCTION

An important means for making sustainable development operational is through the use of cleaner production approaches. Cleaner production is defined by the United Nations Environment Program as the "continuous application of an integrated preventive environmental strategy to increase eco-efficiency and reduce risks to humans and the environment." Cleaner production is about making more efficient use of materials and energy in business:

- for ongoing environmental improvement
- for minimizing waste generation and air emissions, energy and materials use and toxics use
- for maximizing ecological and economic benefit
- can be applied to processes, products and services.

Cleaner production is a preventive strategy and to improve its success, should be linked to the core activities of the business. It includes introduction of revised processes, management and housekeeping practices from the beginning to the end of the business process, including redesign of products, with the emphasis on reducing waste and pollution at source. Cleaner production is about working efficiently and efficiency is the cornerstone of business success (Van Berkel [1]).

Cleaner production is an environmental improvement strategy, which leads to specific solutions applicable to any given business. The options found generally fit under one or more of five preventive practices i.e. product modification, input substitution, technology modification, good housekeeping, and onsite recycling.

Thirteen case studies from the Australian mining and minerals industries are presented to showcase cleaner production. Each of these demonstrates the commitment of the individual organizations to improving their triple bottom line. These case studies were selected from a much larger number of case studies, many of which are available on the Internet, and identify proven or best practices environmental management in the minerals industry across Australia. These case studies cover a cross section of operations. The case studies cover the following aspects:

- air emission
- dust management
- energy and materials efficiency
- integrated sustainable development
- process wastewater management
- waste minimization
- water efficiency.

## 4.1.2 AIR EMISSIONS REDUCTION

### 4.1.2.1 *Comalco's Bell Bay aluminum smelter – dry scrubbing technology*

Comalco, a subsidiary of Rio Tinto, is a major bauxite mining and aluminum smelting company with operations in Australia and New Zealand. With a power supply agreement in place to extend the life of the Bell Bay operation to 2014, the investment of $AUD44 million was made to provide the world's best practice fume scrubbing technology.

Table 4.1. Case studies of sustainable development and cleaner production in the Australian Minerals Industry.

| Case study | Minerals sector | Company & location | Description of case study | Reference |
|---|---|---|---|---|
| Air emission reduction | Aluminum | Comalco Aluminum Limited – Bell Bay Smelter, Tasmania | Introduction of dry-scrubbing technology to initiate $AUD11 million in savings. Contributed significantly to the local community in Tasmania, Australia. | Center of Excellence in CP [2] |
| Dust management | Lime | Blue Circle Southern Cement – Marulan, NSW | Addition of minor changes to plant has enabled the removal of fine dusts from the process and the collection of the limestone dust for sale as a by-product. | Aus. Center for CP [3] |
| Energy and materials efficiency | Industrial minerals | Tiwest joint venture – Pigment plant, Kwinana, Western Australia | A new process to recover synthetic rutile uses waste hydrochloric acid from a neighboring company to produce ammonium chloride for use in pigment production at another one of the first (i.e. rutile plant) company's operations. | Center of Excellence in CP [4] |
| | Industrial minerals | Iluka Resources Limited Synthetic Rutile Plant | The company investigated alternatives to waste gas streams to wet scrubbing, adopting waste heat power and an electrostatic precipitator. This technology required an additional $AUD11 million to install. Modifications generate 6.5 MW of energy returning 16% on capital. | Center for Excellence in CP [5] |
| | Coal | BHP Coal – Illawarra, NSW | Capturing coal seam methane and piping it to surface where it generates 94 MW of energy through electricity generation. | Aus. Center for CP [6] |
| | Energy | Mount Isa Mines – Queensland | A program of innovations has enabled the company to open a new mine and add new electricity-using activities while cutting total annual electricity use, carbon dioxide emissions and delaying the demand for a new power station. | EM Ind. Assn. of Aus. [7] |
| Integrated sustainable development | Aluminum | Alcoa World Alumina – Various locations in Western Australia | Alcoa World Alumina has implemented a wide range of cleaner production initiatives at its Western Australian bauxite mines and alumina refineries. For example, its dust control measures have been saving the company approximately $AUD500,000 a year. | Center of Excellence in CP [8] |

| Category | Commodity | Operation | Description | Reference |
| --- | --- | --- | --- | --- |
| | Gold | Joint venture between Delta Gold and Placer Dome – Granny Smith, Laverton, Western Australia | This mining venture has developed and introduced a unique blend of sustainability practices, taking a holistic approach to mining activities including creation of opportunities for remote indigenous communities and reducing waste to landfill. | Miles [9] |
| | Diamonds | Argyle Diamond Mine (Rio Tinto) – Kununurra, Western Australia | This mine was threatened with closure in 2001. Instead, it is now a thriving as a result of efforts to create a new future implementing a range of sustainable development measures. | Stanton-Hicks [10] |
| | Gold | Newmont – Golden Grove Operations, Western Australia | Cleaner production initiatives have reduced pollution and waste, improved energy efficiency and reduced greenhouse gas emissions. Improved hydrocarbon management, rehabilitation and solid waste management have halved the volume of waste requiring landfill disposal. | Tyler [11] |
| Process wastewater | Steel | OneSteel Whyalla Steelworks – South Australia | The company introduced artificial reed beds for treatment of its industrial waste water to reduce water consumption and increase quality of discharged water. | Bus. Council of Aus. [12] |
| Waste minimization | Aluminum | Alcoa Portland Aluminum – Portland, Victoria | The site has achieved a significant reduction in the amount of waste going to landfill. Although driven primarily by business needs, waste reduction has been achieved by evaluating processes, gaining the commitment of its workforce and combining these efforts with cleaner production and waste minimization concepts. | EM Ind. Assn. of Aus. [13] |
| Water efficiency | Copper-Uranium | WMC Olympic Dam Mine – Olympic Dam, South Australia | Production processes have been modified so that less water is used in flotation and separation of the minerals from the ore including use of high density thickeners to reduce water passing to the tailings system; recycling the acidic liquids from mine tailings that historically had been evaporated, and using highly saline water which seeps from the mine for drilling and dust control. | EM Ind. Assn. of Aus. [14] |

4.1.2.1.1   *Pre-existing and new process*
The process of smelting aluminum is a continuous operation. Fumes produced during smelting need to be "scrubbed" in order to remove fluoride. Prior to the commissioning of the dry scrubbing technology, the smelter relied on wet scrubbing for the treatment of potroom emissions, whereby alkaline liquor was brought into contact with hot gases and the fluoride chemically removed. The wet scrubbing process had two stages. The first stage included a cyclone for the removal of coarse particles. These particles were recycled back into the process by blending with the primary alumina. Removal of particles was not very efficient. The fume passed through the cyclones and then passed through the second stage consisting of a series of water sprays. Water with alkaline chemicals added contacts the fume and absorbs the hydrogen fluoride. The hydrogen fluoride laden water was then piped into a treatment plant, where fluoride materials were removed by precipitating cryolite using expensive chemicals. The residual water was later discharged into the Tamar River. The precipitated cryolite was dried to a solid using a rotary kiln that burnt fuel oil before being recycled back into the smelting process. Cryolite is used as an electrolyte in the chemical bath of the reduction cell or pot. With wet scrubbing, water consumption for the site was 90 ML per month. Up to 60 hours per month of downtime was required for maintaining each of the six scrubbing systems. It is a complex process requiring large amounts of chemicals to neutralize the hydrogen fluoride.

4.1.2.2   *Dry scrubbing*

4.1.2.2.1   *Description of the technology*
Dry scrubbing is the most technologically advanced fume scrubbing system available for the aluminum industry. It is a simple process, whereby hydrogen fluoride is captured in a gaseous form to be combined with alumina in a reactor. The alumina absorbs the fluoride and returns fluoride rich alumina to a silo ready for recycling into the smelting process. In the early 1990's, existing dry scrubbing technology required the same alumina to be recycled many times in order to remove sufficient quantities of the hydrogen fluoride. This resulted in higher operating costs and scaling, where alumina with small amounts of hydrogen fluoride and water stick to steel surfaces, causing blockages and flow problems. The existing technology was complex. Getting good contact between the hydrogen fluoride and the alumina, moving the fume from the potlines to the scrubber and transporting the alumina around the site, proved challenging.

In 1995 Comalco finished the first installation of its own dry scrubbing technology at its New Zealand aluminum smelter. The Comalco owned Research Center in Melbourne developed this technology. The pilot plant and the first full-scale commercial installation at the New Zealand smelter had proven excellent contact between the alumina and the hydrogen fluoride and the alumina only needed to pass through the system once, (as compared to several passes for alternative methods). Dry scrubbing was installed at the Bell Bay smelter in 1997. In 1999, Comalco implemented plans to construct two more industrial fume scrubbers in its drive to continually improve environmental performance. The two projects involving green carbon fume scrubbing and carbon baking furnace fume scrubbing are estimated at $AUD18–$AUD20 million. With the installation of dry scrubbing, Comalco has achieved world's best practice for potroom fume scrubbing technology and reduced fluoride emission by 33%.

4.1.2.2.2   *Advantages of the process*
The $AUD44 million dry scrubbing project has not only delivered a significantly improved environmental performance, it is an inherently cleaner production process and the specific benefits are described in the following paragraphs. The installation of dry scrubbing has resulted in significant savings in operating costs:

• $AUD5 million from reduced chemical usage
• $AUD4.5 million from recycling fluoride rich alumina, reducing aluminum fluoride costs
• $AUD0.5 million from reduced maintenance
• $AUD1 million in miscellaneous savings as a result of dry scrubbing including $AUD250,000 savings in water consumption.

Total annual operating and maintenance cost savings are estimated at $AUD11 million, with an investment payback of 4 years. Environmental benefits have been:

- 95% reduction in potroom ducted fluoride emissions
- 70% overall reduction in the operation's fluoride emissions
- 70% reduction in the operation's water consumption
- emissions significantly better than fluoride regulation requirements
- negligible particulate emissions
- reduced discharge of water into the Tamar River
- cleaner working environment
- substantial reduction in chemical usage
- overall improved process efficiencies from simple handling procedures
- increased potroom fume removal rates and significantly reduced fugitive emissions
- improved occupational health and safety (OH&S) performance.

The benefits to the Tasmanian community have included:

- emissions of fluoride per ton reduced by more than 95%
- capital expenditure for the project saw $AUD18 million invested directly in Tasmania of which $AUD4 million was spent in the Tamar Valley alone
- aesthetic improvements as there are no more plumes from multiple stacks
- improved business viability through efficiency gains and lower cost performance
- significant increase in recycling of materials with the operation having more than halved the use of aluminum fluoride
- eliminated the use of chemicals required for wet scrubbing
- peak construction force of 200 during project commissioning.

All of the improvements to environmental performance and the savings in operation and maintenance costs ensure the continued and viable operation of the smelter into the future. This means continued employment and the resulting economic benefits to Tasmania.

### 4.1.3   DUST MANAGEMENT

#### 4.1.3.1   *Blue Circle Southern Cement*

The company supplies limestone to cement plants at Berrima and Maldon as well as the BHP steelworks at Port Kembla, all of which are located in NSW, Australia. The mine has a total production of three Mt per year. The mine has been in existence since 1929. This dust either settles on the plant floor and equipment, which requires regular cleaning, or is discharged to the atmosphere. The airborne dust poses a health hazard and maintenance concerns.

#### 4.1.3.1.1   *Pre-existing process*

Limestone rock is drilled, blasted, loaded and hauled from the open-cut mine to a processing plant where it is crushed and then dispatched in rail-trucks. The crushing plant consists of primary, secondary and tertiary crushers. The limestone is transported between crushing stations and to dispatch silos by conveyors. As a result of the crushing and conveying processes a significant amount of dust is generated. The primary and secondary crushers are connected to baghouses to minimize direct dust emissions to the atmosphere. Both baghouses have collection hoppers which hold up to 6 tons of fine limestone dust. The collected dust is discharged periodically back onto the conveyor system via an automatically controlled rotary valve and chute. However, as the collected dust is very fine, a large portion of it becomes airborne. Additional dust is generated from this fine material at conveyor transfer points or at conveyor return rollers, where material that has become lodged in small indentations in the conveyor surface is shaken off.

#### 4.1.3.1.2   *Cleaner production program*
The cleaner production initiative implemented at Blue Circle was relatively simple. A bypass chute and second rotary valve were fitted to the collection hoppers on the primary and secondary crusher baghouses. The bypass chute was then connected to one ton capacity bulker bags. This enabled the collected fines to be discharged directly to a contained system and has eliminated the substantial generation of dust associated with the former practice. In addition the collected material is now sold as a lime fertilizer.

#### 4.1.3.1.3   *Advantages of the process*
The three main benefits from this cleaner production initiative are:

• a reduction in airborne dust particles
• a consequent reduction in plant cleaning requirements
• a saleable product.

Sale of the dust as lime fertilizer generates income of $AUD25,000 per year (data for 1997). The costs of the modifications were only $AUD3,500, giving a payback period of approximately two months. The drivers for the process change came from the staff on the plant floor who wanted to reduce the hours involved in plant and equipment cleaning.

A further upgrade involved the enhancement of a limestone treatment process. This involves the conversion of limestone (calcium carbonate) into quicklime (calcium oxide) by heating the crushed limestone in a kiln. Prior to the installation of the pre-heater, the limestone feed entered the kiln at ambient temperature. The pre-heater, which is fueled by waste gases from the kiln, raises the temperature of the limestone to 800°C prior to entering the kiln. The benefits of the change in the processing at the operation are that:

• the limestone requires less time in the kiln
• the oxidation reaction occurs more fully resulting in superior product quality
• the fuel efficiency of the kiln is improved by 25%
• the capacity of the kiln is increased by 40% and hence production capacity is also increased.

An initial obstacle encountered was the occasional difficulty experienced by customers in unloading the bulker bags due to moisture retention which caused the limestone dust to set hard. This problem was overcome through bulk transport of the dust in specially designed trucks.

### 4.1.4   ENERGY AND MATERIAL EFFICIENCY

#### 4.1.4.1   *Tiwest's Kwinana pigment plant*

Tiwest Joint Venture (Tiwest) is an equal joint venture between Ticor Resources Pty Ltd. and KMCC Western Australia Pty Ltd. The operations include: a titanium mine and wet processing plant at Cooljarloo producing titanium minerals concentrate; road transport to a dry separation plant at Chandala 60 km north of Perth, which produces ilmenite, rutile, zircon and leucoxene; a synthetic rutile plant at Chandala to upgrade ilmenite to synthetic rutile; a pigment plant at Kwinana, south of Perth, converting synthetic rutile to titanium dioxide pigment; warehouses at Henderson, south of Perth, for storing the pigment; and exporting and other facilities at the Kwinana port. The principal end-product, titanium dioxide pigment, has broad application in, for example, paint, plastics, paper, ink and pharmaceutical products.

##### 4.1.4.1.1   *The pre-existing and new process*
The process is based on chloride technology adopted by many companies during the 1980s and 1990s to improve pigment quality and reduce effluent rates. Using feedstock from the Chandala complex, the operation reacts synthetic rutile with petroleum coke and chlorine in fluidized bed reactors or chlorinators. The reaction process produces titanium tetrachloride which is purified by

condensation and fractional distillation. The remaining gases are systematically treated by scrubbing and incineration. Liquid effluent goes to the wastewater treatment plant and treated effluent goes to ponds where further settlement takes place. The treated water from the ponds is discharged to the sea under strict environmental controls.

The next stage of the process uses a special process for reacting titanium tetrachloride with superheated oxygen and support fuel to produce base titanium dioxide pigment. The base pigment goes through a finishing process which involves milling, classification, surface treatment, filtering, drying, micronizing and bagging. Various grades of pigment are produced to meet market requirements. Residue from the operation is separated and returned to Cooljarloo where it is encased in specially constructed clay lined pits and used as part of the mine rehabilitation program.

#### 4.1.4.1.2 *Cleaner production program*
Various initiatives have been implemented under the broad headings of energy, materials and water efficiency. These are discussed in the following subsections.

##### 4.1.4.1.2.1 Energy efficiency
A major initiative has been installing a cogeneration plant, commissioned in 1998 and owned by Western Power. A gas turbine generates electricity and the exhaust gases which would have otherwise been vented into the atmosphere, are used to generate superheated steam for the micronizer, the last part of the production process. The plant generates all of Tiwest's power requirements plus surplus electricity for the South West Western Australia interconnecting grid. It also reduces steam demand from the package boilers and reduces greenhouse gas emissions. Some of the other energy efficiency initiatives implemented include:

* implementing a procedure to prevent excessive temperature of superheated steam which is wasted by subsequent cooling of the excess heat
* tuning to prevent excessive burning of natural gas on the waste gas incinerators when on minimum fire
* tunnel driers tuned to prevent unnecessary over-heating of pigment
* commissioning of a second waste gas incinerator, producing steam and thereby reducing demand from the central plant boilers
* replacing the effluent pond transfer pump with an overflow weir
* a variable speed drive on the new lime pump and fan, to reduce flow when not required
* non-return valves on all sump pumps to eliminate back flow and reduce the frequency of operation
* reduced air puffer frequency on non-critical bag filters in the pigment finishing section
* improvements in process monitoring and control, reducing the requirement to re-process off-specification material
* initiatives to reduce petroleum coke consumed by the chlorination reaction including tightening feed control and improving reaction efficiency
* initiatives to reduce consumption of LPG as support fuel for the oxidation reactors including improving LPG flow metering, monitoring oxidation reaction stability and improving combustion efficiency in the oxidizers
* improving furnace energy efficiency
* reducing oxidizer heat losses.

##### 4.1.4.1.2.2 Materials efficiency (Production of hydrochloric acid)
Dilute hydrochloric acid (HCl), generated from scrubbing the gas stream from chlorination, was previously neutralized in the waste treatment plant. Two initiatives were realized during 1997 to recover the HCl, first as acid recovery for sale and secondly for use as ammonium chloride at the Chandala operation. For this purpose, a second scrubber was installed to produce HCl at a higher concentration that enabled reuse as a low quality acid. In this conversion process to ammonium chloride, waste HCl is transferred to neighboring Coogee Chemicals which converts it to ammonium chloride and tankers it to Chandala for use in the production of synthetic rutile. The cost of the ammonium chloride to Tiwest is significantly cheaper than that previously imported.

4.1.4.1.2.3   Rutile recovery plant
The aim of this initiative has been to recover synthetic rutile from process effluent. After chlorination, all metallic chlorides go to a sump and to the wastewater treatment plant. Overflow from the fluidized bed reactor is rich in unreacted synthetic rutile and petroleum coke but this went with the metal chlorides to effluent treatment. A new plant was installed and commissioned in 2000 to recover synthetic rutile from the effluent using hydrocyclones. The titanium-rich fraction is filtered on a belt filter, washed, dried in a fluidized bed drier and returned to the chlorinator with the normal input material. The titanium-poor fraction continues to the wastewater treatment plant.

4.1.4.1.2.4   Use of supplementary fuel
In 2001, a switch was made to using an alternative fuel that produces less water to react with the chlorine and form HCl. This has improved overall chlorine efficiency significantly.

4.1.4.1.2.5   Water efficiency
In 1995, Tiwest conducted an in-house water audit, which identified opportunities for reducing water consumption and identified areas for savings and reuse. Successful projects have included the commissioning of counter-current washing in pigment filtration, the reuse of micronizer condensate in pigment filtration, and the installation of a recovery tank for water reuse. The use of groundwater and reprocessed water are currently being pursued as part of the Kwinana Waste Water Recycling Plant. Specific water consumption has been cut by 50% since plant start-up.

4.1.4.1.2.6   Production of hydrochloric acid
Various benefits have been achieved including:

* savings in the costs of ammonium chloride to Chandala
* reduced quantity of lime for neutralization
* reduced waste for disposal.

In this example of "industrial waste exchange," there is a mutual benefit for Tiwest and Coogee Chemicals. Tiwest benefits by having a local means for "recycling" much of its waste HCl to Coogee, and Coogee Chemicals benefits from being able to produce ammonium chloride cheaply and having an ongoing local supplier (for HCl) and customer (for ammonium chloride).

4.1.4.1.2.7   Rutile recovery plant
The rutile recovery plant is designed to recover up to 21,000 tons per year of unreacted synthetic rutile and coke (based on 180,000 tons per year) which equates to a net savings of about $AUD31,000 per day on an investment of $AUD6 million. At current pigment production rates, the recovery potential is about 53%.

4.1.4.2   *Iluka Resources Limited – waste heat recovery for power generation*

Iluka is an international mining and mineral processing company. The company has operations in Australia, the US and Indonesia and employs over 2,500 people. The company has been mining and processing mineral sands in Western Australia for over 40 years. A major project to more than double the capacity of the company's synthetic rutile plant in North Capel, Western Australia, was commenced in 1995. A significant component of this project was to be the handling and treatment of hot waste gas from the plant, which ultimately resulted in the construction of a waste heat recovery plant. The project was commissioned in 1997.

4.1.4.2.1   *The pre-existing and new process*
Synthetic rutile is produced by removing iron from ilmenite in order to increase the titanium content. Iluka uses the Becher process which involves feeding the ilmenite ore into a rotary kiln to reduce the iron oxides to metallic iron. The iron is precipitated as hydrated iron oxide, and along with other impurities, is removed from the synthetic rutile. The reductant used in the process

is coal, which also acts as a fuel for the kiln. The process results in a hot, dirty waste gas stream (primarily $CO_2$), which needs to be treated before it can be released to the atmosphere.

### 4.1.4.2.2  *Cleaner production program*

The traditional pollution control method of dealing with a waste gas stream, which is high in both temperature and particulates, would be to install a wet scrubbing system. While such a system would cool the gas and remove the particulates there are a number of environmental and economic impacts associated with it. These include:

- high water consumption with water converted to steam and released to the atmosphere resulting in loss of heat energy and water
- generation of high particulate content, acidic liquid waste which requires removal of solids and addition of lime for pH neutralization
- high energy consumption resulting in consumption of non-renewable fossil fuels and generation of air emissions, including greenhouse gases
- high maintenance and operating costs as a result of pumping of water and liquid waste and disposal of waste solids from neutralization plant.

Iluka investigated alternatives to a wet scrubbing system in order to determine whether there was a more effective way of dealing with the waste gas stream. The company decided to adopt a major technology modification in the form of a waste heat power generation facility and an electrostatic precipitator for the removal of particulates. This waste heat power generation facility is recovering the heat energy in the waste gas to produce electricity, which is used on site, reducing the amount of power purchased.

The company completed the design, construction and commissioning of a super-heater, boiler and economizer capable of producing about 30 tons/hr of steam. This super-heated steam drives a fully condensing steam turbine capable of producing about 6.5 MW of electricity (after allowing for power to run plant auxiliaries). As this was the first time a plant such as this had been installed on a synthetic rutile plant in Australia, a number of challenges were encountered. These were addressed through the innovation, originality and dedication of the team in solving the corrosion problems resulting from dirty waste gas, the input heat variations, and maximizing electricity production.

### 4.1.4.2.3  *Advantages of the process*

The total cost of the waste heat recovery plant was just over $AUD20 million and the expected rate of return on the investment was 16%. This compared favorably with the traditional wet scrubbing system that was expected to cost around $AUD9 million but had no financial return on investment. The plant now generates up to 6.5 MW of electricity, with an average of 5.5 MW. Of this 4 MW is used in the new synthetic rutile plant, 0.7 MW is used to run the waste heat recovery plant auxiliaries and any excess is used in other parts of the North Capel operations. By avoiding the need to purchase electricity and taking into account operational savings, the company has reduced operational costs by over $AUD1.5 million per year. With a payback time of 8 years and an expected operating lifetime of over 25 years, savings will continue to accumulate well after the plant has been paid back.

The main barrier to the implementation of this project was the perceived risk of adopting a new technology, outside of the company's core business. Most businesses would agree that it is far easier to adopt a traditional, well established process. However, this means that the benefits of newer technologies are frequently overlooked. Iluka had to look closely at what financial and environmental benefits the project offered, and at how they could minimize their exposure to risk.

### 4.1.4.3  *BHP – coal seam methane recovery*

BHP Coal Illawarra operates four underground coal mines in and around the Illawarra region of NSW, which is situated 75 km south of Sydney, NSW. Three of these mines, the Appin, Tower and

West Cliff mines, produce approximately 3.5 Mt of coal per year. The coal is primarily used for domestic steel making, although some coking and energy coal is also exported.

#### 4.1.4.3.1   *Pre-existing process*

Gaseous methane is contained within subterranean coal seams and is a potential explosion hazard. Furthermore, methane is also a greenhouse gas with high global warming potential and was recognized as a wasted resource.

#### 4.1.4.3.2   *Cleaner production program*

In 1995, BHP, in conjunction with Energy Developments Limited and Lend Lease Infrastructure, developed a power generation plant that uses waste methane to generate up to 94 MW of electricity. This is sufficient to energize 60,000 homes. Supply of the fuel for electricity generation is achieved by capturing methane from within and below the coal seam (approximately 250 million cubic meters per year). It is piped to the generation plants on the surface where it is distributed to a series of modular gas engines that drive electrical generators. Natural gas supplied by pipeline is used as supplementary fuel in the event of a shortfall in methane supply from the mines. The capture and utilization of methane from coal seams provides a major environmental benefit through reduced release of greenhouse gases in the form of methane emissions. This was a significant consideration in determining the benefit of the project.

BHP pays a fee to Energy Developments Limited to operate the generation plant, however, the energy that is generated is sold by BHP to the electricity grid. Some of the gas collection costs incurred by BHP, which must be met to allow mining to continue, are recovered in this way. Methane drainage of the mines is required to allow mining to continue safely. Utilization of the methane provides an important energy resource while reducing greenhouse gas emissions by approximately 50%. This represents a reduction in greenhouse gas output of the equivalent of approximately 3 Mt of carbon dioxide per year. In addition to providing an independent source of electricity for the community and the mines, the utilization of this otherwise wasted resource reduces the amount of coal consumed by the state's power stations.

Difficulty in estimating future electricity prices due to deregulation of the power industry was a key consideration in determining the economic viability of the project. The power generation plant also emits oxides of nitrogen as a by-product of the combustion process and these can contribute to the formation of photochemical smog. A management plan for reducing nitrogen oxide emissions when ambient ozone levels approach target levels is now in place.

#### 4.1.4.4   *Mount Isa Mines (MIM) (now X strata) – energy efficiency*

Local discoveries of many new ore-bodies and expansion of current operations meant the inevitable increase in power demand for MIM operations in the Mount Isa Region of Queensland, Australia. It also meant that MIM had either to invest in new generating capacity or utilize existing generating plant more effectively. More effective use meant that MIM was required to undertake rigorous energy management within its own operations.

#### 4.1.4.4.1   *Cleaner production initiative*

Major reductions in energy consumption, peak demands and greenhouse emissions were achieved by implementing initiatives in a number of areas:

1. A 1,000 kW impulse turbine and generating set was installed 1,000 m underground. Chilled water at 1°C is discharged at around 100 L/s down a vertical pipe from the surface to underground. Prior to the installation of the set, the water gained around 2.5°C between the surface inlet and the underground outlet, resulting in a temperature rise of 3.5°C. The installation of the set not only generated emission free electrical energy but recooled the water by 2°C down to 1.5°C, reducing the chilled water requirement by 11%. Running the set during times of peak demand also lower required generating capacity.

2. Many of the mine crib rooms were subsequently fitted with dedicated refrigerated air conditioners, reducing pumping costs and the need for chilled water.
3. The twelve 1–2 MW axial ventilation fans on the surface are mounted over vertical shafts, which are typically 1,000 m deep. Operation of the fans is to either extract or supply air to the underground workings. The pitches of the fan blades are automatically changed at regular intervals during the day by a process controller installed on the surface. Fan blades are driven at minimum pitch during times such as change of shifts, when ventilation of the whole mine is not required.
4. Dispersed throughout the mine are approximately 1,000 smaller ventilation fans, each fan having an average connected load of 11 kW. These fans increase general air movement underground and direct ventilation to priority areas. Many of the fans have been fitted with ripple frequency controllers.

Two ripple frequency transmitters inject electrical reticulation control pulses at 750 Hz. Initially these transmitters were only used to control domestic hot water heaters in the city and surrounding properties, which used off peak electricity supplied at a lower tariff. The transmitters are now used to control both the hot water heaters and fans underground. Fans are turned off at the end of each shift.

The hoisting control systems of the R62 (lead mine) and U62 (copper mine) ore skips operated independently. Coincidently, during their hoisting cycles, full copper and full lead skips would be accelerated from rest at the same time. The mass of each skip and contained ore is 40 tons, the acceleration time for both is about 20 seconds. This meant that there were random occurrences of high current, short duration demands on the generators. To allow for these occasions, the maximum sustainable load of the mine's power station has been set at 5 MW below the station's maximum steady state generating capacity. The two winders are now controlled by an interconnected Programmable Logic Controllers (PLCs) such that only one skip can be accelerated from rest at a time.

It was realized that greater energy efficiency and reduced emissions would result if operators were made more accountable for energy use. In early 1997, a company wide personal computer based energy and emission management system (PC: EMS) was instituted. The PC: EMS was designed to allow easy data input together with meaningful displays. Plant operators are the key to PC: EMS operation. All plant operators were supplied with a PC. Each PC was connected to an area network which covered a surface area of about 28 km by 2 km and it also reaches deep underground. Plant operators have to enter daily operational forecasts at half hour increments via their terminals. In the central database, the forecast demand and energy requirements are calculated by multiplying the proportion of plant estimated to be operating by the full load rating (MW) of the plant. Each plant operator, therefore, is able at any time to see the energy and environmental costs of running their plant, and any other plant in dollars, energy consumed in MWh, peak demand in MW, and tons of carbon dioxide emitted to generate the necessary power. At present the target is for forecasts to be within a band between 110% and 90% of actual. Because local operators may not be immediately aware of influences outside their control, their forecasts are subject to adjustment by the PC: EMS Administrator. These adjusted or final forecasts are then issued to the power station. Final forecasts are usually within 105% and 95% of the actual and are typically 104% of the actual or measured value.

Energy management has resulted in reduced energy consumption and the postponement of further capital outlay for generating plant. As a direct result, carbon dioxide emissions have been substantially reduced. The company was been able to open the deepest mine in Australia, with all the additional power requirements that it entailed, and still reduce total energy use. Carbon dioxide emissions related to metal produced have fallen by 11% since 1995/96.

As a result of the initiatives, including cost savings (Table 4.2), a number of conclusions can be made:

- *Accountable Operations*: Increased efficiency cannot be attained without operators appreciating the costs of their actions.
- *Reduced Energy Consumption*: Operators must own reductions.

Table 4.2.   Summary of cost benefits, MIM operations.

| Initiative | Investment ($AUD) | Savings ($AUD pa) | Deferred expenditure ($m) | Payback period[2] |
|---|---|---|---|---|
| Turbine/Generating set | 1,000,000 | 450,000 | 2.5 | <1 year |
| Air conditioning | 500,000 | 2,000,000 | | Weeks[2] |
| Mine processor | 500,000 | 400,000 | 10.0 | <1 year |
| Underground | 20,000 | 600,000 | | |
| Ventilation | | 500,000 | | 7 weeks |
| "Ripple control" | | | | |
| Hoisting | 20,000 | | 13.0 | – |
| PC: EMS[1] | 300,000 | | | – |

[1] PC: EMS reduced energy costs by 5% in its initial year of operation (1997/1998).
[2] Estimates only.

- *Improved Operating Practices*: Usually result not only in lower operating costs but also improved safety awareness.
- *Lower Maximum Demand*: Better utilization of generators ending in lower capacity charges.
- *Lower Emissions*: Energy management ensures a company does not exceed limits and allows involvement in future emission reduction trading.
- *Long Term Survival*: Any industrial concern lacking energy management will have higher production costs than its more efficient competitors, lower public opinion, and suffer penalties for exceeding greenhouse emission limits.
- *Enhanced Energy Efficiency*: Energy management leads to a profitable well run organization.

### 4.1.4.4.2   *Cleaner production drivers*
MIM was subjected to increased market competition with lower prices from the sale of metals. In this context, a reduction in operating costs was essential. MIM was also looking to meet growing environmental concerns internationally by reducing the carbon dioxide emissions from its coal and oil-fired generators and postpone or remove the need for additional energy generation capacity.

### 4.1.5   INTEGRATED SUSTAINABLE DEVELOPMENT

#### 4.1.5.1   *Alcoa World Alumina*

Alcoa World Alumina Australia is a trading name of the unlisted public company, Alcoa of Australia Limited. The company owns and operates alumina refineries at Kwinana, Pinjarra and Wagerup with a combined capacity of 7.3 Mt a year, equivalent to 15% of world demand. Alumina is exported world-wide from shipping terminals at Kwinana and Bunbury. Alcoa employs approximately 3,700 people in Western Australia. Apart from refining, the company operates bauxite mines at Huntly and Willowdale in the Darling Range south of Perth, which supply the three refineries.

Environmental management is a high priority for Alcoa in all aspects of its operations. In 1990 the company's work on mine rehabilitation earned it a listing on the United Nations Environment Program's "Roll of Honor" for environmental achievement – the only mining company in the world to have been so recognized. Sustainable development is the basis for Alcoa's 2020 Global Environmental Strategy. At Alcoa, cleaner production is a tool used within the Alcoa Business System (ABS) to assist the company to move towards sustainable aluminum production. Business units have adopted challenging waste reduction, cleaner production and cost improvement targets as well as strategies aimed at increasing community and social contribution. The ABS was initiated in 1996 and is being implemented throughout all 350 Alcoa locations in 37 countries.

4.1.5.1.1   *Processes and activities*
Access to bauxite is the keystone of Alcoa's activities in Australia. Darling Range bauxite is a low-grade resource to which value is added through refining and smelting. Bauxite is defined as any ore in the lease which has a content of more than 27.5% aluminum oxide. Typically, it takes seven tons of Western Australian bauxite to yield one ton of aluminum.

4.1.5.1.2   *Mining*
At current rates of alumina production, 26 Mt of ore are mined each year from the Huntly and Willowdale mines. Darling Range bauxite is the lowest grade ore mined on a commercial scale anywhere in the world, and requires substantial investment in a fleet of loaders, excavators, trucks and crushing and conveying equipment. At each new mining area, about half a meter of topsoil and overburden is removed and conserved for later reclamation, and the top one to two meters of cemented caprock bauxite is drilled and blasted so that it can be extracted along with the more friable bauxite below. Alcoa has developed a sophisticated computer-based blast acoustic model to ensure that blasting noise is kept below acceptable levels. Four bulldozers successfully ripped caprock at pits near neighbors where blasting would otherwise have been unacceptable. Once the ore has been broken, it is loaded onto haul trucks by excavators or front-end loaders and transported to primary crushers at the mines. Ore mined at Huntly is transported by conveyor to supply the Pinjarra refinery and the Kwinana railhead stockpile. From the stockpile, bauxite is railed to the Kwinana Refinery. Bauxite from the Willowdale mine is conveyed to the Wagerup refinery.

4.1.5.1.3   *Refining*
Alcoa operates a three-refinery system in Western Australia between the capital city, Perth, and the port of Bunbury 200 km to the south. The bauxite is fed via conveyors to refineries where it is subjected to the Bayer process to produce alumina. Aluminum must first be refined from bauxite in its oxide form. The Bayer refining process used by alumina refineries worldwide, involves (1) digestion, (2) clarification, (3) precipitation and (4) calcination. Alcoa's Kwinana refinery, which began operating in 1963, has a current rated capacity of 1.9 Mt a year. The Pinjarra refinery is one of the world's biggest with a capacity of 3.2 Mt, and Wagerup has a capacity of 2.2 Mt.

4.1.5.1.4   *Cleaner production program*
The key elements of the cleaner production program were as follows:

• Eco-efficient design including environment and safety risk assessment and minimization, reduction or substitution of process inputs, elimination of waste, and pollution prevention.
• Integration of environmental considerations into all business decision-making processes.
• Standardized environmental management systems (to ISO 14001) emphasizing continuous improvement.

Detailed examples of these initiatives are included in the following sections.

4.1.5.1.5   *Improved vessel descaling practices*
In the past explosives, jackhammers and high-pressure water have been used to remove scale from inside process tanks. These techniques are expensive, affect equipment availability, and impose serious injury and health risks. They produce large volumes of waste product with associated soil and groundwater contamination and waste disposal risks. Caustic and acid wash methods have now superseded these older techniques.

4.1.5.1.6   *Improved heat exchanger maintenance*
Heat exchangers in the process become contaminated and require regular maintenance. In the past, prior to de-scaling, the contents were emptied to the concrete floor resulting in temporary loss of process solutions (75 kL each time), degradation of the concrete, safety issues and risks of soil and

groundwater contamination. Now, cool water is used to push the process solution through the heat exchangers prior to drain down. This saves money and avoids the previous problems.

### 4.1.5.1.7   *Improved bauxite quality*
The contamination of the refinery liquor stream by organic compounds is a major constraint on production. These organic compounds originate with the bauxite. During mining, the topsoil and overburden is removed using heavy equipment. Previously, pockets of overburden were mined along with the bauxite. Research indicated that this was a significant source of organic compounds. Mining practices were modified to carefully remove all overburden, thereby reducing the input of these organic compounds into the refining process and subsequent reduction of waste.

### 4.1.5.1.8   *Reduction in fine alumina waste*
Fine alumina is produced by uncontrolled precipitation and product breakdown during materials handling. It is captured in electrostatic precipitators. In the past, product quality specifications resulted in the recycling of over 200,000 tons per year of fine alumina in three refineries. More than half of this was lost to residue. Technology and product modifications led to waste reduction refinements in precipitation control and calciner design, as well as reuse of some of the superfines as liquor burning feed medium. These modifications have significantly reduced the waste. After discussions with customers, some relaxation in the product specifications further reduced the superfines recycle.

### 4.1.5.1.9   *Oxalate management*
Sodium oxalate, an organic impurity, is removed from the liquor stream as a crystalline material. In the past, it was either treated with lime and disposed of to residue areas, or incinerated in a rotary kiln. While recovering some of the sodium value with the oxalate, these processes impose a range of health, safety and environmental risks. The sodium oxalate has proved to be a useful reagent for vanadium processing and it is being transported from Kwinana and Wagerup to Windamurra. Oxalate kilns at Kwinana and Wagerup have now been shut down resulting in cost avoidance, energy savings and reduced emissions.

### 4.1.5.1.10   *Benefits*
- vessel descaling
  - reduced capital costs of $AUD0.8 m, and benefits of greater than $AUD5 million per year
  - reduced safety and environment risks
  - major reductions in scale waste disposal
- improved heat exchanger maintenance
  - reduced costs of product loss
  - reduced concrete maintenance
  - reduced safety hazards
  - significant reduction in the high floor maintenance costs
- improved bauxite quality
  - improved production and reduced organic removal, treatment and disposal
  - reduction in fine alumina waste
  - net savings from waste reduction are estimated at $AUD14 million per year
- oxalate management
  - cost avoidance
  - energy saving
  - reduced emissions from shut down of oxalate kilns.

### 4.1.5.1.11   *Cleaner production drivers and barriers*
The principle incentives for these initiatives were reduced costs. Barriers to the project were primarily that it was "too hard" and that "everything had already been tried". In both cases, the impetus

by the team, and backing by management, led to a successful conclusion, which has stimulated ongoing drivers for further improvements.

### 4.1.5.2 *Granny Smith Gold mine*

The Granny Smith Gold mine is a joint venture between Delta Gold and Placer Dome. It is located approximately 25 km south-southwest of the township of Laverton in the north-eastern goldfields region of Western Australia. This mining venture has developed and introduced a unique blend of sustainability practices. The gold mining industry was particularly weak with respect to the employment of Aboriginal people. Through considering the well being of local communities, in addition to the accepted needs for proper environmental and economic stewardship, an isolated operation such as Granny Smith has made significant gains towards sustainability in a relatively short period of time.

Efforts to facilitate an increase in local employment opportunities for the indigenous people, in both the town itself and on the mine site, have become a significant gauge of social progress. Cultural initiatives that seek to encourage and support opportunities for local artists to display and sell their work have also become a normal part of the mine's development strategy. The evolving process of consultation and implementation continues, with the result that the mines sustainability strategy becomes increasingly multidimensional. The environmental, socio-cultural, and economic dimensions of this program at Granny Smith is discussed below.

#### 4.1.5.2.1 *Environmental*

The current development of a revegetation program incorporates the tried and tested seed, save and sow method, which is used extensively on mining sites. The revegetation program involves planning and design for both operations and closure at the outset. As progressive decommissioning of sites occurs over the life of the operation, revegetation follows in phases. The revegetation strategy includes final forming of disturbed land, planting schemes for tailings areas and general rehabilitation of the Granny Smith location as extraction operations shift over to the Wallaby site. An emphasis has been placed on ensuring that a wide range of local species are seeded onto rehabilitation areas. The original plant species at the dig site are de-seeded for propagation and eventual reseeding. Finally, when mining operations are completed, the pits can be backfilled or graded, depending on cost.

Granny Smith initiated the "Ruggies" recycling program in 1997 to recycle waste materials previously disposed of into landfills. Several mines have since joined the Ruggies program and thousands of tons of waste have been recycled. The program has also succeeded in cleaning up mine sites. Steel from mill balls, copper from cables and aluminum from drink cans are just some of the waste items that are now being recycled. Trucks that once traveled to the mine sites with full loads and returned back to Perth empty are now taking saleable cargoes back with them. The money collected from the Ruggies program benefits Western Australia's only children's hospital and other charities in Western Australia. All the people working in the Ruggies Recycling initiative are doing so on a voluntary basis, reflecting the community spirit of the program.

#### 4.1.5.2.2 *Social and cultural*

The Laverton Leonora Cross Cultural Association (LLCCA) is working in conjunction with other key supporters to achieve its commitment to improve employment, retention and training for Aboriginal people and to assist in any community initiatives that seek to address the socio-economic disadvantages suffered by Aboriginal people. A central theme is maintaining the flexibility necessary to accommodate cultural differences and initiating mentoring courses. Selected employees with superior people skills are trained in communication and mentoring to help guide new trainees. Since 1997, the LLCCA has placed approximately 60 people per year, both Aboriginal and non-Aboriginal, into meaningful full-time employment. Pre-employment training has been provided for many others, as well as assistance with resume writing and preparation.

### 4.1.5.2.3   *Economic*

Economic sustainability of a mining community is a difficult issue. Ultimately, no matter how many years hence, mines do close and this leaves communities that haven't found other means of generating income with few options. Presently employment opportunities and income filter through to the community. Efforts to increase the penetration of this capital into the local economy are helpful to both short-term economic and longer term education and other social prospects. However, innovation of other means of harnessing previously undeveloped local potential, peculiar to an area, is essential to providing a truly sustainable vision for an area. With this in mind, the potential for olive farming, tourism, and crafts sales are being investigated as part of the overall project to provide real diversification of the local economy. Another project is the current experimentation with various aquaculture techniques. It is envisioned that experiments currently underway will see discarded open pits utilized for either commercial or recreational aquaculture activities. Species including trout, silver perch, black bream, barramundi, yabbies and marron are currently being studied in order to determine their suitability to local climatic factors. All of these projects are in early stages of development, however, they are reflective of the larger sustainability effort being undertaken by Granny Smith.

### 4.1.5.2.4   *Argyle Diamond Mine*

The Argyle Diamond Mine case study is a good example of sustainable development in mining, setting a standard for how the synergistic achievement of economic, social and environmental goals can show us the power of this new approach.

The Argyle Diamond Mine is located in the remote East Kimberley region of Western Australia, upstream from Lake Argyle. Diamonds are usually found in kimberlite "pipes" that have extruded and cooled rapidly during a volcanic event. These "pipes" can move around during geological time, and when exposed to the surface can have their diamonds washed off into creek beds.

The mine employs approximately 750 people, of whom three quarters work at the mine site. In 2001, 88 Mt of rock yielding 10 Mt of ore were excavated using massive earthmoving equipment. The ore was then transported to the primary crusher to begin the refinement process, while the waste rock was hauled to the waste dumps. Within the processing plant the ore is crushed, scrubbed and screened before gravity separation. The final step is a unique Australian innovation that separates the diamonds from the ore by X-ray fluorescence, which picks out the diamonds from their light flashes and fires an arrow of air knocking the diamond into the clear and collecting it.

There are approximately 150 people employed in the Perth office, some of whom sort and polish the most valuable diamonds. The remaining diamonds are shipped to Antwerp for sale, 90% of which are then sent to India for cutting and polishing. Argyle helped to develop this Indian industry, which currently employs some 750,000 people. From here the diamonds are sold around the world to various markets. Some of these markets, such as the market for champagne and cognac diamonds, were created by Argyle through marketing displays of attractively set diamonds of different colors, created by world-class master jewelers. The result is that over 90% of Argyle diamonds are now sold as various grades of gemstones. The local Kimberley component of this is projected to increase as local employment, education and contracting schemes contribute more to growth over time. So, with all this development history and potential, why was the mine going to close down? The diamond pipe was becoming harder and harder to access, meaning that the mining operation was becoming less and less viable. In the late 1990's management began to prepare the mine for closure in 2001.

### 4.1.5.2.5   *The process*

In 1998, a new management team was given an open brief to see if a future could be created for the diamond mine. A process was begun entitled "Creating a Future." The starting premise shifted at this point from an acceptance that economics determined the mine would close in 2001, to asking what it would take to prolong the mine's life beyond that date. Rapid action was needed to ensure that the window of opportunity was not lost. Once the technical feasibility of future mining was proven through to 2007, a positive attitude within the new management's ranks was required to guarantee that project delivery dates were met. Mobilizing the workforce to implement this vision achieved a streamlined and successful transition process.

Another innovation emerged as a spin-off benefit once the future of the project was ensured. A huge amount of earth around the ore was cut back to create the necessary access for the new phase of mining. Previous to the extension of the life of the mine, the focus had been on final rehabilitation plans of the mine site. It had been determined that the waste dumps would need shoring up at an estimated cost of approximately $AUD50 million. With the new cutback creating additional waste rock it could be sorted and used to achieve this end at no real cost greater than ordinary operations. This kind of synergy is a hallmark of the new thinking that ensured the continuation of mining at Argyle. However, it also required substantial restructuring across the entire mining operation as the next phase of mining was going to produce less diamonds from the same amount of rock.

To make its future operations viable, Argyle needed to find productivity improvements and efficiencies across the business. Among these was the installation of new rolls-crushers, which enabled greater profitability through capturing a much higher percentage of Argyle's smaller diamonds through more precise separation of materials. Many of these had previously slipped through with the tailings.

An enhanced water management regime has also created the potential for greater efficiencies, as well as increased reuse in the future. Greenhouse-friendly hydropower from Lake Argyle currently provides 98% of Argyle's power, with diesel generators providing back-up power during peak periods of use in Kununurra and other regional applications. Some efficiency has been gained in diesel usage from earthmoving equipment, through the introduction of newer models, however, this remains a major environmental cost for the mine.

The ecological and economic efficiency gains were matched with human efficiency gains in the workforce. Improved understanding of the concept of sustainability, continuation of the previous management's improvement of work practices, and a different management emphasis, have together combined to produce significantly more cost effective diamond extraction. Thus, plans have changed to enable diamonds to be mined up to 2007, while creating the time needed to assess the added possibility of extending the life of the mine further through underground operations. A rigorous underground feasibility study is also underway. An underground operation has the potential to add another decade or more to the mine's life. All of these managerial and efficiency achievements are significant advancements in sustainability, but they are only part of the set of innovations that followed the decision to "create a future."

### 4.1.5.2.6   *Waste recycling innovations*
Programs are in place to manage, use and recycle waste resulting from the mining operation, as well as the mining village. Waste oil from the lubrication of machinery and vehicles is recycled. The market for the oil is found in nearby communities and as far away as Darwin, allowing the operation of the plant, after initial costs are factored out, to continue at no cost.

Waste from the Argyle settlement is separated in color-coded bins as a normal part of life in the isolated mining camp and back loaded to Perth at minimal cost. A culture of greater care among employees has begun to take hold at the same time that ecological gains are made through recycling and greater transport efficiency. Further studies into the complete waste profile of the mine and settlement at Argyle are being conducted including composting, worm farms, and using organic compost to rehabilitate tailing dams.

### 4.1.5.2.7   *Rehabilitation innovations*
Perhaps the most globally innovative approach that has been adopted in the mining operations is the rehabilitation of land that has been stripped for alluvial mining. This process began in 1988. The goal is to rehabilitate the area so that there is a landscape with all the biodiversity of the micro-ecosystems of the area re-established. Over 100 hectares a year is rehabilitated. The process involves:

- immediate soil replacement leading to self seeding
- contouring and deep ripping to ensure water retention on the areas

- gathering seed stocks through local harvesting
- developing nursery stocks of important species
- planting those areas where these particular species are low
- monitoring to ensure the areas are restored.

A breakthrough in this aspect of the mine's operation came when it was discovered that many of the important and rare species in the rehabilitation sites were "bush tucker" species that the local Aboriginal community were keen to see included. Thus, they were invited to play a role in seed collection to help in the development of nurseries and to ensure that a full representation of these important species was grown throughout the sites. There are more than 50 plants of ethno-botanical significance in the area and these have been documented with the assistance of Aboriginal traditional owners, so that the value of the plants for nutrition, medicine and other household applications can be passed on to the community. The result of this approach has been an important development of the relationship with the local Gija people.

### 4.1.5.2.8   *Greater Aboriginal involvement*

The level of trust within local communities surrounding the mine has grown cautiously, through Argyle's involvement in many community development projects across the Kimberley. This is a direct result of a deliberate policy that attempts to improve the proportion of Aboriginal people employed at the mine. With 10% indigenous representation at present, Argyle is aiming to increase this to reflect Kimberley demographics at 15%, as part of an overarching local employment scheme that has a goal of 30% local employment by 2005. This change has not come easily. It meant that the management team had to become aware of many subtle changes in the culture of how people were hired in the company. The approach of using written resumes and direct interviews is not used, as it was evident that this drove away many good applicants in the past. The process now involves a recruiting workshop, where people are brought in for up to a week and slowly are able to adjust to living in the community. They are given tasks by an Aboriginal instructor including training in truck driving at a mini-mine site. This process gives managers and superintendents the opportunity to observe potential employees involved in a range of team and individual activities, and to assess their skills in this way. These same people have become the "champions" of Aboriginal employment, working to mentor and support new Aboriginal employees in the workplace. Training is also undergone continuously once people are employed. The culture gap is now being bridged rather than ignored and/or accommodated, resulting in greater trust and retention rates.

### 4.1.5.2.9   *Health and safety innovations*

The development of a safe and healthy work environment is an essential part of a sustainable operation as it is basic for retaining the workforce. Argyle has instituted a range of programs including:

- Fit for work (a set of rules that includes monitoring for substance abuse and alcohol as well as healthy lifestyle commitments)
- Healthy Lifestyle Program (promoting healthy eating habits and alternative fitness programs)
- Employee Assistance Program (psychological and emotional wellbeing services, monitoring and promotion)
- Plant safety (detailed courses and training for management in recognizing patterns of unsafe work practices)
- Emergency response (four fully trained teams are at Argyle, who also assist with local emergencies outside the mine site)
- Health information (e.g. on skin care for cancer prevention).

The results for the mining operation have been dramatic reductions of total injuries, severity of injuries, and time lost to injury since 1998. Employee retention rates remain very high relative to the industry average.

#### 4.1.5.2.10 *Integration of common goals between mine departments*

Health and safety, environment, economic efficiency and community relations become increasingly sustainable when integrated. This is one of the key process elements discovered by Argyle in the process of "creating a future." This means that there is a structure to ensure that these elements are all part of the one operation. Each is considered with respect to the others before any steps are taken, to maximize the benefits and avoid any potential problems. It also means that the process of decision-making must be able to recognize innovation at all levels of involvement in the company.

#### 4.1.5.2.11 *Conclusion*

Argyle now reports their achievements under the banner of sustainability. The mine has reaped multiple benefits from integrating environmental and social benefits into their economic performance. However, it has not been an easy process. They needed to be innovative and forward-thinking in their ideas, their technology, and their management systems.

These are the long-term benefits from these innovations. Sustainability improvements will outlast the life of the mine and will be used by Argyle people in other mining operations. Meanwhile, the community and the company gain from the investment in better practices in real economic, social and environmental terms. The company gains a reputation as a true asset to any community in the future, a longer economically viable lifetime, as well as a strategy that has its own economic value. The community gains the economic support to continue to develop into the future by diversifying their activities along different lines with company assistance, while minimizing environmental and social impacts.

### 4.1.5.3 *Golden Grove*

#### 4.1.5.3.1 *Cleaner production program*

Newmont Australia is committed to cleaner production and over the past two years has convened a number of group eco-efficiency workshops to facilitate eco-efficiency improvements at all operational mine sites. Corporate eco-efficiency targets have been set and site general managers are accountable for ensuring these targets are met. The corporate commitment of Newmont Australia to eco-efficiency and sustainability is supported at the site level.

#### 4.1.5.3.2 *Containment of spilt product*

The loss of product and revenue from spillage and emissions has historically been high at Golden Grove. Prior to the construction of a storage shed for concentrates in 1998 for example an estimated 1,000 tons of concentrates with a value exceeding $AUD1 million were lost each year as wind and airborne emissions. This gave a financial pay back period for the $AUD1.5 million shed of 18 months. More importantly, it greatly reduced the contamination of surrounding bush-land and soils with metal sulfides. This contamination had caused significant vegetation die back and the acidified soils would not support plant growth. The environmental and financial benefits of the storage shed helped to justify further expenditure on spillage containment including a $AUD1 million upgrade to spillage containment structures in the process plant completed in 2001. It also enabled operational staff to spend more time on operating the plant efficiently without the distraction of constant spillage clean up.

#### 4.1.5.3.3 *Underground hydrocarbon management*

Golden Grove discharges over 2 billion liters of water to a salt lake each year under rigorous license conditions, including a requirement that oil and grease contamination be less than 10 ppm. The maintenance of clean discharge water is, therefore, fundamental to the company's license to operate. Mine discharge water is also used as process water in the plant and oil contamination will dramatically inhibit mineral flotation. Persistent oil contamination will result in millions of dollars worth of potentially recoverable product being lost to the tailings dam. The main source of underground oil contamination is hydraulic hose failure. Hydraulic hose failures on loaders are common and can result in the immediate loss of 500 L of oil. In most cases, oil will spill directly into water being pumped out of the mine. When monitoring of oil and grease in discharge water commenced in 2000, contamination levels up to 30 ppm were common. This contamination is thought to be at least partially responsible for the intermittent poor performance of the plant at that time. An improvement program for underground hydrocarbon management commenced in 2000. Poor practices such as depositing oily wash pad sediment in drives receiving groundwater inflow ceased as did the practice of pumping oily wash-pad water straight into the mine water discharge dam. This water was physically trucked to the surface wash-pad, which is equipped with an efficient oil/water separation system. The solvent degreasers previously used on site were replaced with quick-break water-based detergents and a general spill prevention and clean up campaign commenced.

The spill prevention campaign involved a number of components including routine maintenance checking and change-out of hydraulic hoses at the first sign of cracking to minimize failures, equipping underground machinery and work-shops with peat-absorb oil absorbent to immediately clean-up spills and, delivering hydrocarbon awareness training to all operators. The reporting of underground oil spills as environmental incidents also commenced for the first time. A hydrocarbon accounting system was introduced to track oil usage and recovery. The impact of the above measures was remarkably successful with the average level of oil and grease contamination in the discharge water decreasing from 10 ppm in 2000 to 2 ppm in 2002 with 7 of the 12 months having undetectable levels of contamination. Oil spills are reported by operators and there is a general high awareness and diligence in managing this issue in the work-place. In early 2002 an ecological risk assessment on the impact of oils and greases in mine discharge water on Lake Wownaminya was completed. This assessment was based largely on indicator organisms such as dragon and damsel flies. It indicated that the current level of oil and grease contamination is having no detectable impact on the ecological health of the Lake and that the ecology could accommodate intermittent contamination levels of 10 ppm which would occur from time to time following hydraulic hose failures on loaders. The Department of Environmental Protection subsequently approved a license amendment to increase the upper allowable level of oil and grease from 5 ppm to 10 ppm. The internal target remains 5 ppm but the risk of non-compliance for intermittent events not creating significant or lasting harm has reduced greatly. The integrated focus on managing hydrocarbons underground has delivered environmental benefit for the wetland system of Lake Wownaminya and financial benefit through the provision of clean water to the Mill. The business benefits of an increased awareness of mining people of their potential downstream impacts on the environmental and other parts of the business cannot be overstated.

#### 4.1.5.3.4    *Land contamination*

The old adage says an ounce of prevention is better than a pound of cure and this is particularly true of land contamination. Golden Grove currently has spillage and emission controls to prevent contamination from acidic soil contaminated with metal sulfide.

A large volume of sulfide-contaminated material (i.e. thousands of tons) from severely contaminated areas has been excavated and placed in the tailings dam and replaced with clean soil. Areas of lesser contamination are being remediated in-situ with significant financial, energy and land disturbance savings. In the past two years Golden Grove has applied over 1,000 tons of lime to sulfide-contaminated soil and the response has been significant. Bare trees, thought long dead, have sprouted foliage and seedlings are emerging in areas that were recently bare acidic deserts. The process of liming has been assisted by the strategic placement of topsoil windrows directly planted out with trees and irrigated. In the past two years Golden Grove has planted over 10,000 seedlings. The use of lime to reverse soil acidity is well known in agriculture but is not a technique commonly used in the mining industry. The in-situ liming of contaminated soils, has already reduced the final closure liability of the mine by approximately $AUD500,000.

#### 4.1.5.3.5    *Solid waste management*

Golden Grove has effective waste management procedures based on the three-tiered structure of minimizing waste at the source, maximizing recycling and appropriately disposing of the remainder.

The success of this approach is reliant on work-force commitment, which in turn is developed through a combination of training, rewarding and, making people accountable for noncompliance with standards. Monthly work area inspections have been a particularly effective tool in this process. Each operational area is jointly inspected each month by a representative from the environmental department and the area supervisor and scored for performance on a range of aspects including waste management. The best performer on site receives and excellence certificate. Poor performance over the course of the year results in a poor performance appraisal, which is linked to a bonus payment or, in the case of contractors, to contract renewal and competition for the excellence certificate is intense.

#### 4.1.5.3.6   *Energy management*
Newmont Australia is a signatory to the Australian Greenhouse Challenge and Golden Grove has been participating in the program since 1999. The success of Golden Grove's energy efficiency programs is a good example of site performance enabling the achievement of corporate targets. The Newmont Australia co-efficiency target for energy set in 2001 was for a 10.9% improvement by 2004. Golden Grove's reduction in kg of $CO_2$ per ton of ore milled from 78.49 kg/ton to 68.25 kg/ton from 2001 to 2002 was 13%. The main reason for this improvement was improvement in mining efficiency. Waste rock previously brought to the surface now being immediately deposited into available stopes underground. This has had an added advantage of limiting the expansion of the waste rock stockpile, which will have to be rehabilitated at the end of Mining at great financial and, energy expense.

The above examples of cleaner production initiatives at Golden Grove all had significant financial benefits for the company. This has facilitated management support for environmental programs on site with a realization that environmental expenditure gives a high financial return on investment. The cleaner production achievements at Golden Grove mirror similar achievements at other Newmont Australia sites and have ensured that Newmont Australia meets it's commitment and obligations as a signatory to the Western Australian Cleaner Production Statement.

### 4.1.6   WASTEWATER MANAGEMENT

#### 4.1.6.1   *OneSteel/Whyalla Steelworks*

The scarcity of water is an intrinsic part of life in most of Australia. Better use and conservation of this vital resource will improve the quality of life for many Australians, especially those who live and work in the semi-arid region. One such area is Whyalla in South Australia, where OneSteel's Whyalla Steelworks has been operating since 1964.

#### 4.1.6.1.1   *The process*
Water is essential to the operations of a steelworks, being used for cooling, cleaning, lubrication and numerous other purposes. In Whyalla it is a scarce and expensive resource. OneSteel is planning to reduce water consumption and is also aware of the need to efficiently treat waste effluent before it is discharged. Studies over a number of years have looked into the steelworks' wastewater discharges into the Spencer Gulf. The studies have identified the effluent from coke ovens as a significant source of organic matter and ammonia. OneSteel needed to reduce and/or eliminate these materials from its wastewater prior to discharging it.

#### 4.1.6.1.2   *Cleaner production programs*
In Australia, reed beds have been used for stormwater run-off and sewage treatment, but little is known about other uses. The Llanwern plant represented the first reed bed technology trial on coke ovens effluent. In soil-based reed bed systems, the effluent to be treated percolates through

the biologically active soil and roots of a large bed of reeds and then drains through a pipe at the base of the bed. The function of the reeds is to pump oxygen into the soil through the roots. Near the roots, there is an aerobic (oxygen-containing) zone and further away, there is an anaerobic (oxygen-free) zone. Thus, within the soil, a range of processes exist that allow the transformation of environmentally undesirable components of wastewater.

Construction of the trial beds commenced in February 1993. Surveys of reeds in the surrounding areas were undertaken and information on reeds best suited to waste water treatment was reviewed. Accordingly, five reed varieties, all of Australian native species, were selected to make up the trial. Once the system was established the process of adapting the reeds to the effluent was started. The trial lasted 18 months and paved the way for the planning of the full-scale system.

After the trials, a large scale (2 ha) trial system was constructed. Commissioning commenced in 1997. This involved the adaptation of the plants and biological life within the system to pollutants in the wastewater. Ongoing work is occurring to increase the effluent load removed by the reed beds. Currently, in excess of 70% of the ammonia is removed from the treated coke ovens effluent, with removal of other organic and inorganic materials running at or above 90%.

### 4.1.6.1.3   *Advantages of the process*
The benefits of the process were:

- improving the quality of OneSteel's waste water discharges into the Spencer Gulf
- future recovery of a valuable resource of fresh water for recycling on the plant
- improving the quality of reclaimed land that previously had no value in the coke ovens area, while improving the visual appearance of that part of the plant
- reducing the impact of wind blown dust in an area with no vegetation
- providing a shield against the searing hot north winds during the summer months.

### 4.1.6.1.4   *Barriers*
Strong winds, dusty conditions, hot summers with high evaporation rates, salts within the soil and water, have all posed problems at various stages. These setbacks have gradually been overcome through modifications to the successful European design to suit Whyalla's environment. The results of the trial to date, to a large extent, reflect the effort which has been put into monitoring and managing the system. Identification of suitable soil for the trial reed bed at Whyalla had to be identified through extensive soil sampling because of the high salinity in the area. Due to the shallow water table in the Whyalla works area, the trial beds had to be built up from ground level.

## 4.1.7   WASTE MINIMIZATION

### 4.1.7.1   *Alcoa – Portland Aluminum*

The Portland Smelter in south-western Victoria, Australia, produced its first aluminum in 1986 upon commissioning of the first of two smelting potlines. The second potline, commissioned in 1988, raised the intended production capacity of the plant's 408 smelting pots to approximately 320,000 tons of primary aluminum per year. Portland Aluminium is currently a joint venture between Alcoa of Australia, Eastern Aluminum, the Chinese Government and the Japanese trading company Marubeni. The smelter was designed and built using Alcoa technology. On commissioning, the smelter was placed on a full production setting. However, as might be expected of the plant of the size of Portland, considerable adjustment to design and processes was required in the early stages to achieve satisfactory performance. During this start-up phase, many of the raw materials used in the aluminum production process ended up in the waste stream, becoming part of over $1,000\,m^3$ of waste going to landfill each month during 1989.

### 4.1.7.1.1 *The process*

Portland Aluminum uses the "Hall/Heroult" process to convert alumina to aluminums. Alumina ore (containing aluminum oxide) is subjected to high electric current in smelting pots. The current passes from an anode through a molten bath of cryolite to a cathode which removes the oxide, leaving a cake of molten aluminum. Substantial infrastructure surrounds the complex chemical and physical aluminum production process. Alumina, the major component in the process, is produced at the Alcoa refineries in Western Australia and delivered to Portland by sea. Carbon products that are processed to produce 170,000 carbon anodes per year are also delivered by sea. Electricity is transmitted 700 km from brown coal fired power stations in the Latrobe Valley via a 500 KV tower system. Natural gas is supplied by underground pipeline primarily to fuel the anode baking furnaces.

Raw materials are vacuum unloaded at the wharf and conveyed to the plant 4.2 km away, via an underground and overland enclosed belt conveyor. The final product from the process is near-pure aluminum in the form of 22.5 kg ingots.

### 4.1.7.1.2 *Cleaner production program*

In 1990, Portland set itself two basic and at the time unique goals. The first objective was to have no process materials going to landfill; the second being to have zero general waste going to landfill by the end of 1995. At the time that these goals were set, the overall financial loss to landfill was estimated at around $AUD1.3 million per year. Cleaner production initiatives were subsequently undertaken in the four key stages of the aluminum production process – raw materials, electrode, smelting, and casting.

### 4.1.7.1.3 *Raw materials*

The raw material ship unloader at the port uses a vacuum system. During certain weather conditions and 24-hour unloading operations, this system was capable of creating nuisance noise in residential areas around the port. Silencing equipment has now been fitted to alleviate this problem.

Considerable amounts of raw material, mainly carbon and alumina, were lost at the transfer points along the 4.2 km belt conveyor from the wharf to the site storage silos. The spills resulted in cross-contamination to the extent that recovery was not possible and landfill the only option. Installation of vacuum duct systems at the conveyor transfer stations and the implementation of a product recovery procedure after each product shipment has allowed raw materials to be returned to the process. A further benefit of the initiative is that system maintenance has been reduced.

Poor handling practices for both anthracite and fluoride resulted in significant material spillage. Ground contamination was a major concern. Handling methods have now been revised using designated bulk handling and storage silos, reducing material loss and minimizing environmental risk.

Pencil pitch was formerly used in the aluminum production process as a binding agent in the making of carbon anodes. In its solid form, the material required complicated and stringent materials handling and management practices. It had a further disadvantage in that it needed to be melted before it could be used. Pencil pitch has now been replaced by liquid pitch, resulting in substantial gains in efficiency and improved occupational health and environmental practices.

Installation of a natural gas pipeline to Portland Aluminum has improved process efficiencies, optimized process control and eliminated the need for on-site storage of gas. An assessment of the use of natural gas as a fuel option for vehicles has also been undertaken.

### 4.1.7.1.4 *Electrode*

The anode baking furnaces require major maintenance after approximately 100 baking cycles. Maintenance was traditionally done in-situ. A method has been devised at Portland to allow full units of wall sections to be pre-built off-site, transported and installed with minimal need for people to enter the hot baking furnace. Bricks from carbon bake furnace re-builds are returned to the manufacturer for re-processing. New products are manufactured from the recycled bricks and sold back to Portland Aluminium or on the open market. The aim of the furnace maintenance initiative is to totally recycle all furnace components to the specification standards for new materials. Anodes consumed in the smelting process are removed from the pots after about four weeks. The spent anodes are transported from the pot rooms in purpose built vehicles to reduce the escape of emissions. Anodes are then placed in a cooling tunnel, which allows fluoride-rich gases to be passed through a filter system and recycled.

### 4.1.7.1.5   *Smelting*
After smelting pots have been relined, placed in the potline and made ready for commissioning, they are pre-baked using a natural gas fired baking system. This reduces start-up stress, extends pot-life and subsequently reduces the amount of spent potlining material generated. A specially designed dust and soundproof facility has been constructed where smelting pot shell demolition and maintenance is conducted. Specialized equipment has been developed to allow processing of large aluminum metal pads left in the pots, and to recover for recycling, steel cathode bars, steel vapor barrier, and pot repair steel.

The smelting pot reline facility allows several pot shells to be re-built at once. Part of this operation is to prepare new cathodes for pot installation. The cathodes have a steel electrode glued into them. In heating up the pots, the glue gives off solvent gases into the working area of the pot rooms. Now, prior to pot start-up, the cathodes are baked in a special purpose, EPA-licensed furnace. Volatile gases are collected and scrubbed within the furnace prior to atmospheric discharge.

A process has been developed in conjunction with Ausmelt, an Australian furnace development company, which is unique to the world aluminum smelting industry. It is projected that the process, once in full operation, will allow the old linings from smelting pots to be heat treated to remove residual cyanide and to recover the valuable fluorides. These will then be recycled back into the smelting process. Preliminary trials show that by-products from the process have the potential to be used in various civil industries applications. In the meantime, spent potlining material destined to be recycled is being stored in special containers and buildings.

Major up-grades of the bath handling facility have reduced both occupational health and environmental risks. Bath products are recovered during the recycling of spent anodes from the smelting pots. Traditionally, this caused an extremely dusty work environment. Dust has now been greatly reduced and contained.

### 4.1.7.1.6   *Casting*
The quality and presentation of aluminum ingots is important to market acceptance. A continual effort is being made to reduce the level of impurities in metal during smelting. The installation of robots to skim ingots during casting has complemented these efforts. Skimming improves surface quality, reduces the risk of moisture ingress during storage and transport, and the subsequent dangers during ingot re-melting. This task was previously performed manually with operators exposed to risk of both hot metal splashes and fumes.

### 4.1.7.1.7   *Advantages of the process*
All the above process improvements are aimed at increasing commercial returns to the plant. These improvements are highlighted below:

- *Raw materials*: The cost of material loss and waste disposal has been reduced from over $AUD1.3 million in 1990 to less than $AUD200,000 in 1997, increasing revenue by more than $AUD1 million per year. Minimizing raw material loss and substituting pencil pitch for liquid pitch have both led to significant energy savings and have improved process control and anode performance. The use of natural gas has improved baking efficiencies and reduced gaseous emissions from scrubbers. The goals that Portland Aluminium set itself in 1990 in relation to landfill have largely been achieved. Through the pursuit of efficiencies and implementation of waste reduction initiatives, waste going to landfill in 1992 had fallen to 1,100 cubic meters. In 1997, the amount of landfill waste was only 21 cubic meters.

- *Electrode*: Off-site maintenance of the anode baking furnaces minimizes health and safety risks, increases furnace life, and reduces the generation of refactory waste. Furthermore, by improving the quality of anode, smelting pot processing performance is enhanced. The initiative cost $AUD9 million and has an estimated payback period of 6–8 years. Emissions have been substantially reduced as a result of the initiative.
- *Smelting*: The cost of refurbishing each smelting pot is between $AUD80,000 and $AUD100,000. In the process, around 100 tons of spent potlining is generated, which requires treatment at a cost of $AUD200 to $AUD250 per ton. Considerable savings are derived by extending pot life beyond the normal production period of 5,000 days. The smelting pot maintenance processes developed by Portland Aluminium were designed to minimize downtime and increase productivity. Portland Aluminium's methods are now used by other smelters to obtain improved performance, reduce emissions, and lessen health and safety risks.

### 4.1.7.1.8   *Cleaner production drivers*

With so much raw material being wasted and going to landfill in the early years of smelter operations, there was strong incentive to recover as much raw material as possible to increase production revenue. Waste minimization has thus become, since the early 1990s, an integral part of Portland Aluminium's operations. The company now actively embraces cleaner production practices as a means of achieving improved process performance in all aspects of its operations. Process changes reflect the company's commitment to its environmental, health and safety policies. The Portland smelter has set many process benchmarks, strengthening its competitive position in the world aluminum market. Waste management and cleaner production tools have provided the vehicle for competitive change, impacting strongly on the economic viability of the plant while reducing environmental risk and creating a healthier and safer working environment.

## 4.1.8   WATER EFFICIENCY

### 4.1.8.1   *Western Mining Company (WMC)*

WMC is one of Australia's largest mining companies. Its main business is the discovery, development and processing of mineral resources. The company owns huge underground reserves of copper, uranium, silver and gold at Olympic Dam in South Australia, and has invested approximately $AUD1.1 billion in an integrated underground mine and processing plant to extract these minerals. The company produces refined copper, gold and silver, and uranium oxide concentrate.

The mine is close to the rim of the Great Artesian Basin, which covers $AUD1.7 million km² in Central Australia. The Basin is one of the largest sedimentary basins in the world. It is estimated that 425 ML of water flow into the South Australian section of the Basin each day. The overall flow is so large that water under pressure seeps to the surface in mound springs throughout several regions of arid South Australia. These springs are of high conservation significance, and WMC has a comprehensive monitoring program for the springs and also the bores in the south-west region of the Basin.

### 4.1.8.1.1   *The process*

The mining operations data use 13.4 ML of water a day (1998), while the nearby town of Roxby Downs, which was built to support the operations, uses another 1.6 ML. Roxby Downs has developed to be an important regional centre. In the 1980s, water was extracted from a single borefield located 110 km from the mine and town. The water requires desalinization before it is suitable for human consumption or can be used for plant processes at Olympic Dam. An extensive monitoring program ensures that impacts on mound springs on the southern rim of the Great Artesian Basin are minimized. In 1996, the company was granted permission to begin drawing water from a second borefield a further 90 km into the Basin. The goal was to reduce withdrawal pressures on the original borefield and to ensure an additional water supply for expansion of mine operations. One of the conditions for approval was that WMC should monitor carefully the water flows to nearby water mounds. Further, the company is committed to minimize its withdrawals, to conserve water and to recycle whenever possible.

### 4.1.8.1.2    *Cleaner production program*

Since 1997, the overall approach to reducing water consumption at Olympic Dam has been to: develop more efficient work practices; substitute lower quality recycled water where practicable; and modify metallurgical processes to reduce water consumption or increase water recovery. Various processes were being modified so that less water is used in flotation and separation of the minerals from the ore. This has included:

- use of high density thickeners to reduce water passing to the tailings system
- recycling the acidic liquids from mine tailings that historically had been evaporated
- using highly saline water which seeps from the mine for drilling and dust control (water quality being of lower concern in such instances)
- implementing various other minor water conservation programs, including re-use of wash waters.

Since 1998, WMC has also worked closely with the 3,000 residents of Roxby Downs in water conservation activities, such as:

- providing trees, plants and drip irrigation systems to households
- encouraging use of and providing advice on arid zone gardens
- fostering mulching to retain garden moisture
- introducing low water consumption trees and shrubs
- recycling treated town effluent to the local golf course and sports ground
- introducing synthetic turf for recreational use
- capturing and reusing storm water.

### 4.1.8.1.3    *Cleaner production drivers*

For individuals and companies living and operating in an arid climate such as South Australia, water conservation and management is an important consideration. WMC recognizes its responsibilities to assist with management of the Great Artesian Basin and is strongly committed to minimizing its withdrawals, to conserve water and recycle whenever possible. Because of the relatively high cost of water, WMC has considerable incentive to minimize its own use, and to encourage residents also to be water conscious. As at 1998, the unit cost of water for the project was $AUD1.61 per kL for general process water, and $AUD2.40 per kL for potable quality water. This compares with the cost, i.e. $AUD0.88 per kL, charged by the South Australian Government to other users. It is a company decision that water be sold to Roxby Downs residents at the same rate as paid elsewhere in the State.

### 4.1.8.1.4    *Barriers*

The main barriers to further water conservation are process constraints and the capital and operating costs of recycling equipment. Owing to the high cost of supply, the processing facilities at Olympic Dam were designed and built, and are being operated, to be efficient in the use of water. However, the continuous reuse of process water results in the build-up of salts (particularly chlorides) and other contaminants, which originate either from the initial water source or from the ore or process chemicals used. Eventually the concentrations of some salts and other contaminants become so high that they have detrimental effects on process efficiency.

WMC will continue to research and develop methods to reduce consumption and to encourage water conservation through educational programs for employees and other Roxby Downs residents.

The costs of water conservation initiatives within the plant are many times higher than the cost of equivalent potential water savings in the pastoral industry. WMC has therefore offered to assist pastoralists in the borefields region to more efficiently utilize their water by providing assistance for the closure of boredrains and their replacement with piping, tank and trough systems.

The response to this initiative has been well received by pastoralists in the region. The potential water savings identified are between 14.6 and 23.8 ML per day. This is significantly greater than any potential water savings available at the mine and town. However, the company is continuing to investigate possible further cleaner production initiatives available at the mine and plant.

This collection of case studies covers a wide range of the environmental issues facing modern mining operations. These case studies demonstrate a need for an integrated response by mining companies to the various environmental, social and economic impacts at their operations. It also highlights that different operations face different challenges, depending on the location of their operations and indicate the need for mining operations to work with the natural environment in their area. For example, selected Alcoa World Alumina operations are relatively close to areas of higher population densities and therefore blasting noise and vibration are an important consideration for management. The WMC case study illustrated measures undertaken to improve water efficiency in remote regions, which suffer water shortages.

A number of the reported case studies emphasized the importance of working with and providing employment for indigenous communities. This is becoming an increasingly important expectation for many mining companies striving for an improved triple bottom line. At times, the financial and environmental benefits are difficult to quantify although organization were committed to the ongoing effort to improve eco-efficiency and adopt a triple bottom line approach to accounting.

In summary, the Australian minerals industry case studies have shown the following:

- There are numerous companies in the minerals in Australia that are putting sustainable development and cleaner production into operation at their sites.
- Environmental and social benefits can often be realized economically at mining operations and will not always incur a cost.
- It is critical that mining companies work closely with businesses and suppliers to identify new processes that realize the benefits of cleaner production approaches.
- Local communities often provide the key to a mining operation realizing its full potential in contributing to a region's economic and social well-being.
- Improvements to waste management practices and waste prevention, more often than not, lead to cost reductions and often increased revenues.
- Energy and water efficiency improvements will be needed by any mining company planning to remain viable in the future.

The Mining, Metals and Sustainable Development (MMSD) project is embracing the concept of sustainability and examining how the mining industry can contribute to the global transition to sustainability. MMSD has put forward an agenda for change. These case studies, and particularly the integrated case studies, are well aligned with this agenda. They illustrate that Australia provides a good example of the mining industry identifying how it can contribute locally, to the global transition to sustainability.

REFERENCES

1. Van Berkel, R., Application of Cleaner Production Principles and Tools for Eco-Efficient Minerals Processing, Paper presented at the Green Processing Conference, May 2002, Cairns, Queensland.
2. Center of Excellence in Cleaner Production (CECP), Cleaner Production – Comalco Aluminum (Bell Bay) Limited – Dry Scrubbing Project Case Study, 2001, http://www.ea.gov.au/industry/eecp/case-studies/comalco.html.
3. Australian Center for Cleaner Production (ACCP), Cleaner Production – Limestone Mine Sells Dust as By-Product – Blue Circle Southern Cement Case Study, 1997, http://www.ea.gov.au/industry/eecp/case-studies/bcsc.html.
4. CECP, Cleaner Production – Improve Process Efficiency – Tiwest: Kwinana Pigment Plant Case Study, 2001, http://www.ea.gov.au/industry/case-studies/tiwest.html.
5. CECP, Iluka Resources Limited – Synthetic Rutile Plant Waste Heat Recovery Power Generation Plant Case Study, 2001, http://cleaner production.curtin.edu.au/pub/case_studies/iluka_wacasestudy.pdf.

6. CECP, Cleaner Production – Utilization of Methane from Collieries – BHP Coal Illawarra Case Study, 1997, http://www.ea.gov.au/industry/eecp/case-studies/bhp.html.
7. Environmental Management Industry Association of Australia (EMIAA), Cleaner Production – Reduction in Energy Consumption – Mt. Isa Mines Case Study, 1998, http://www.ea.gov.au/industry/eecp/case-studies/mim.html.
8. CECP, Cleaner Production Initiatives – Alcoa World Alumina Case Study, 2001, http://www.ea.gov.au/industry/eecp/case-studies/alcoa.html.
9. Miles, N., More Sustainable Approaches In Mining: Encouraging Progress at the Granny Smith Gold Mine, State Sustainability Unit of the Western Australian Government, 2001, http://www.sustainability.dpc.wa.gov.au/casestudies/grannysmith.html.
10. Stanton-Hicks, E., Argyle: "Creating a Future", State Sustainability Unit of the Western Australian Government, 2001, http://www.sustainability.dpc.wa.gov.au/casestudies/argyle.html.
11. Tyler, J., Cleaner Production at the Golden Grove, Paper presented at the Waste and Recycle 2002 Conference, October 2002, Perth, Western Australia.
12. Business Council of Australia, Cleaner Production – Artificial Reef Beds for Treatment of Industrial Wastewater – OneSteel Whyalla Steel Works Case Study, 1997, http://www.ea.gov.au/industry/eecp/case-studies/onesteel.html.
13. EMIAA, Cleaner Production – Waste Minimization – Alcoa – Portland Aluminum Case Study, 2001, http://www.ea.gov.au/industry/eecp/case-studies/portland.html.
14. EMIAA, Cleaner Production – Reduction of Water Consumption In Sensitive Arid Zones – WMC Limited, 1997, http://www.ea.gov.au/industry/eecp/case-studies/wmc.html.

## 4.2 Blasting impacts and their control

Paul Worsey

*University of Missouri-Rolla, Rolla, MO, USA*

### 4.2.1 INTRODUCTION

Blasting is extremely important both to mining and the world economy. The saying is often used, "If it can't be grown it has to be mined," however if the ground is too hard to be mechanically mined economically, it has to be blasted. Certainly many materials, such as iron, copper and concrete to name but a few would be significantly more expensive if it weren't for explosives and our ability to easily drill holes to use these explosives efficiently. Between two and three megatons (Mt) of explosives are used in the United States each year, 2.57 Mt in the year 2000 according to the USGS mineral industry survey (Kramer [1]), which is probably under-reported. The vast majority (90%) of these explosives are used in mining. Taking a conservative estimate of 2.5 tons of rock blasted for every kilogram of explosives used means that well in excess of 5,000 Mt of rock are blasted in the U.S. each year in mining alone.

Common effects of blasting are airblast, ground vibration, flyrock, fumes and dust. To make things worse, blasting is a sudden event and is therefore more noticeable, and sometimes frightening to the uninformed public. In addition, many mining operations are being encroached by development. Because of the vast amount of explosives used in mining and the violent nature of blasting, it is therefore evident that the consideration of blasting is important for sustainable mining.

### 4.2.2 HISTORICAL BACKGROUND

The question is, "Why is so much explosive used in modern mining?" and to answer this we have to look back in history. Originally man hoed ore from the ground to make metal. When the easy surface material was taken, primitive mining began. When the rock was hard it was broken by using fire to heat the rock, causing thermal cracking. The first recorded use of an explosive in mining was black powder in 1627 in the Hungarian mines (Konya and Walter [2]). Holes were hand drilled using hammer and chisel and then filled with black powder. The use of black powder in blast holes provided a tremendous increase in productivity and in the availability of metals. However, the drilling rate in hard rock was still insignificant in comparison with today's productivity.

In 1846 Sobrero invented nitroglycerin, which was soon used in blasting because of its potency, specifically its ability as a high-powered detonating explosive to break hard rock many times more efficiently than black powder. However, the liquid was extremely dangerous and not until the invention of dynamite (nitroglycerine based), "the first safe explosive," by Nobel in 1866 did nitroglycerine based products rapidly replace black powder as the explosive of choice. In fact dynamite was such a step forward in the economics of mining that it was snapped up at the expensive price of $1 and above per pound when it was first introduced into North America over 120 years ago (Hopler [3]). Dynamite was used in the first tunnel through the Alps between France and Italy under Mont-Cenis. The tunnel was started in the 1850's, and the builders thought that it would take 40 years to build. However, in 1867 dynamite became available, which more than doubled the speed of construction and the tunnel was opened in 1871 (Grady [4]).

At this point drilling was the expensive and time-consuming part of the blasting operation. With the advent of power equipment drilling costs rapidly fell, and with the introduction of ammonium nitrate based explosives in the 1960's mining costs dropped significantly. The costs of mined materials in general have continued to drop ever since. Because of the continued economy of drilling and blasting over mechanical grinding and cutting, drilling and blasting continues to be the major technique for breaking insitu rock in mining. In fact drilling and blasting continues to be extremely cheap (at less than \$1 per ton in the United States, down to as cheap as 25 cents per ton of rock blasted at many mining operations). Economics is why mining relies so heavily on the use of explosives.

## 4.2.3   ENCROACHMENT OF MINING OPERATIONS

With increasing prosperity has come the encroachment of development on mining operations. Some may say that encroachment is inevitable if careful planning is not done. The reasons for encroachment may be quite varied, depending on the type of mining carried out and the geographic location. Some of these are discussed below.

Quarries typically produce crushed stone for construction. Crushed stone is used for a variety of purposes, including concrete aggregate, road sub base and black top (road surface). The majority of these quarries are located in the vicinity of major population concentrations, i.e. their market. Because of the low cost of the product versus high transportation costs, the quarries are initially often sited on the edge of cities to give a competitive edge. Inevitably the cities prosper and grow outwards towards the quarries, in many cases eventually surrounding them. The standard of living is always rising, which equates to the construction of larger, newer houses and the migration of people away from town centers to new suburbs. This in one sense is advantageous for existing quarries, as new markets are opened up on their door steps, and haulage distances are far less than for new operations opening up that are now much further from the population centers. This makes existing quarries more competitive. On the other hand it produces problems with complaints about blasting.

In contrast to quarries, mines for coal and metals are generally in more remote locations. However the establishment of a new mining district inevitably encourages the migration of people and the establishment of towns and other economic activities, along with the necessary infrastructure for the mine or mines. As the mining district matures and old deposits are mined out, either the mines expand geographically or new mining operations are started. This also inevitably leads to conflict with existing property owners. The solution is to buy up enough land to provide a buffer, and to mine from the direction of possible encroachment first. Another solution is for the State and local governments to preclude residential development in areas of mineable reserves. However, these solutions need a lot of proactive effort by the mining company and the local community, which does not happen many times.

To add to these problems the standard of living in a broad sense is rising on each populated continent of the world. As people acquire more wealth they feel empowered and move further away from city centers to where mining operations are already established. Often neighbors are irritated by a combination of aspects of the mining operations such as noise, dust and heavy truck traffic as well as blasting. However, blasting is often used as the reason in an attempt to close down the operation, because it is a more dramatic and dynamic event and therefore people tend to deal with it on a more emotional basis, whereas the other aspects are constant background aggravations.

Basting complaints are not just confined to developed nations such as the U.S., but occur wherever there is a high population density around or next to a mine site. For example in Kampala, Uganda, blasting at the Muyenga quarry has recently resulted in strong complaints from residents near the site (Olanyo [5]), so much so that the National Environment Management Authority is conducting an Environment Audit to evaluate the operational impacts of quarrying on the environment. In developing countries business as usual using the old techniques is rapidly becoming a thing of the past, as evidenced by the recent conference "New Paradigm Shift in Drilling and Blasting" held in Johannesburg. According to The Uganda Quarry Operators' Association Chairman, George Kyaligonza, quarries can co-exist where human settlement has encroached; it is a matter of using the right and most appropriate technologies. Most countries are keen to seek these technologies.

## 4.2.4 REGULATIONS

For the present, and probably for the foreseeable future the economic fragmentation of hard rock is focused on blasting, which requires the use of explosives. Explosives are hazardous materials and therefore their use, storage and transportation are highly regulated. When one talks about blasting related regulations, blasting vibrations usually come to mind and these are discussed later. Accuracy in numbers plays a key part in complying with federal, state and local regulations, such as scaled distance requirements, shot size limitations, or storage and inventory maintenance. Pressures to abide by regulations while trying to fulfill all of the demands of a blasting professional can, at times, entice workers to compromise on safety. Such pressure can often find employees working too fast, skipping safety precautions, or not checking for numerical accuracy because time is valuable. However, laws and regulations are meant to protect everyone involved, including the individual, the company and the general public, as well as the explosives industry. In addition to the obvious safety factors involved, failure to abide by regulations and maintain accurate records can result in penalties relative to the seriousness of the infraction. Such penalties can range from monetary fines and revocation of blasting permits and licenses to increased liability for civil claims or even criminal charges (Nobel Insurance [6]).

The country that has done the most research work on blasting vibrations is the United States, having been the major player for many years in world mining and having taken the lead in environmental issues in the 1970's and 1980's. Because of this the U.S. has some of the earliest regulations on blasting vibration limits, and strangely enough some of the least conservative. Even more strangely, although the federal government regulates the sale, transportation and storage of explosives, there are no uniform blasting regulations throughout the United States. For example in 2002, the state of Missouri had no blasting regulations apart from coal mining and blasting on highway projects, whereas certain cities, municipalities and counties within the state had extensive and widely varying regulations and vibration limits, with the majority of rural counties having no regulations whatsoever. In fact, even different types of mining can have different limits in the same state. Another way of looking at this is that as the U.S. is a conglomeration of 50 different states, local pressures and interests determine the extent of the regulations.

The old US Bureau of Mines (USBM) recommended limit of 50 mm/s (2 in/s) is still used by many municipalities and is good for smaller scale blasting using up to 100 mm holes. However, the work was based on small to medium diameter holes, up to 150 mm. With time and advances in drilling technology, the trend in the 1970's and 1980's was towards large diameter holes in large-scale mining. For instance, the Phelps Dodge Morenci mine in Arizona moved from 150 mm to 245 mm and was contemplating increasing to 310 mm diameter holes as late as the mid 1990's. Doubling the diameter of a hole usually results in quadrupling the weight of explosives in each hole, and if the same geometry is used, the associated doubling of bench height can result in an eight-fold increase in explosives per hole. The economy of this is that less drill holes are required, and drilling costs can be significantly reduced. The use of large diameter holes in stripping operations at coal mines, especially with the introduction of blast casting has resulted in significant increases in hole loads, with two tons of explosives per hole no longer uncommon, and with stripping blasts anywhere between 50,000 and 4,000,000 kg of explosives per blast (500,000 kg being an average blast in the Powder River Basin coal mines in Wyoming). Obviously this type of blasting cannot be done next to other people's dwellings, which have to be a long, long way away. One of the attributes of blasting vibrations is that their frequency decreases with distance, with the effects of blasts at the kiloton size similar to those of small earthquakes. This effect is clearly shown in Figure 4.1 where the relative frequencies of coal mine, quarry and construction blasting were compared by the USBM.

Because of this problem the USBM (Siskind et al. [8]) did further studies on the effect of low frequency blasting vibrations from coal mine blasting and the results of this study were subsequently adopted to a large extent by the U.S. Office of Surface Mining (OSM), which regulates coal mining in the U.S. They found that low frequency ground vibrations had a more detrimental effect on structures than high frequency vibrations and that the two-inch per second safety criteria had to be modified for frequencies below 40 Hz, as illustrated in Figure 4.2.

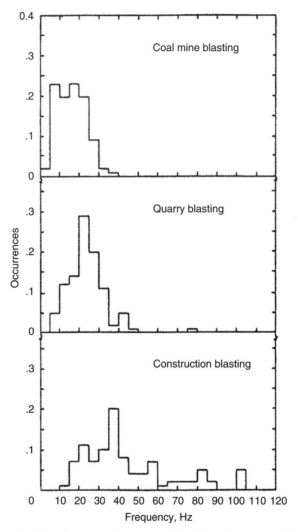

Figure 4.1.   USBM predominant frequency chart comparison for coal, quarry and construction blasts, reprinted from RI 8506 (Stagg and Engler [7]).

Residential construction in the U.S. is predominantly timber frame. This form of construction, being more flexible, is very resistant to blast vibrations. This is shown in Figure 4.2 where older construction plaster lath has a lower vibration limit. This is an important point because many countries have different building techniques, due to a combination of climate, natural building resources, labor and historical traditions and practices. Also, the building standards of many countries have to be taken into consideration. As such the majority of countries have a lower ground vibration limit than the U.S., Brazil for example being 15 mm/s.

## 4.2.5   THE ENVIRONMENTAL EFFECTS OF BLASTING

To head off any potential problems with blasting and/or to be able to mitigate their effects, it is important to understand what these potential problems are likely to be. The problems that can arise

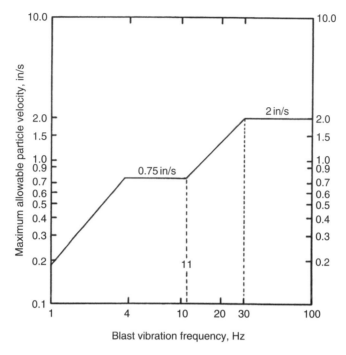

Figure 4.2.   Safe levels of blasting vibration for houses using a combination of velocity and displacement, from USBM RI 8507 (Siskind et al. [8]).

are grouped into the following categories; airblast, ground vibrations, flyrock, fumes, dust and other excavations problems, and are briefly discussed below.

### 4.2.5.1  *Airblast*

Airblast is the pressure pulse from a blast that is transmitted through the air. All production blasts in mining cause the displacement of air and thus produce noise. Airblast is generated in three ways; first through the displacement of the ground causing the displacement of air similar to a subwoofer speaker in an audio system, second by the premature release of gas pressure from the blast hole, and third by the detonation of unconfined explosives such as the surface detonating cord. The displacement of the ground generally results in low frequency noise, which is difficult to hear but rattles dwellings, whereas the high velocity release of explosion gases, usually at supersonic velocities, causes intense noise in the audible range, often startling neighbors. Because the pressure waves in the air cause residential structures to flex and the air to move within them, they are often confused with the effect of ground vibrations by occupants. In fact, in many cases complaints of excessive blasting vibrations can be attributed to airblast rather than ground vibrations.

In the U.S. the generally accepted limit for airblast is 133 decibels (dB) using modern equipment. The decibel range has traditionally been used because it mimics human hearing and perception. The scale is logarithmic when compared to pressure. In this scale, an increase of 6 dB is equivalent to a doubling of pressure, therefore an airblast of 109 dB may look near the limit on paper but in fact only represents 1/16th of the legal limit for pressure. Airblast rarely damages structures, yet is very startling and can be the cause of much anxiety for residents. The first type of structural damage to occur is the breaking of windows, which starts at about 143 dB for large or poorly seated glass of low quality. The 133 dB limit was chosen by OSM because higher levels, although non-damaging would be intolerable.

The number one cause of excessive airblast levels is the premature venting of explosion gases and this represents the loss/waste of energy in a blast. Unusually high airblast levels are always linked to inefficiencies in blasting, which are realized as higher mining operating costs. The four causes of premature venting are: weak seams, through the blast hole collar (as illustrated in Figure 4.3) due to inadequate and inappropriate stemming material, overburdening of blast holes, and poor delay timing. Unnecessarily annoying neighbors with excessive airblast costs the mining operator more for every ton that is mined, and this is a key fact that is not understood by many mining operators.

### 4.2.5.2   *Ground vibrations*

Ground vibrations are the result of energy lost from the blast which is distributed in the form of vibration waves through the ground away from a blast. Ground vibrations are caused by the detonation of the explosives, the breaking of the rock and displacement of the rock (for example, rock falling onto the bench from a highwall). The second two causes are difficult to control because they are usually the effects that we want in the majority of mining operations. The efficient use of energy liberated from the detonation of our explosives is something that we can control to a greater degree. Ideally, all the blast energy should be used in rock fragmentation, and none wasted by loss into the surrounding ground to annoy neighbors. The general correlation is good fragmentation with lower ground vibrations, and poor fragmentation with high ground vibrations. Improved fragmentation has been shown time and again by many researchers to reduce overall mining costs, and minimize ground vibrations. Factors which result in poorer fragmentation, increased mining costs and increased blasting vibrations are: low powder factor, improper delay between holes, improper delay between rows, the improper functioning of explosives within a hole, and the use of inadequate stemming.

First, a low powder factor results in reduced fragmentation. Fragmentation is composed of both fracturing and the disassociation of rock fragments, and their movement away from the standing faces into a muckpile on the bench or pit floor. An excessive mass of rock to break and move in front of a hole results in overburdening, and slower movement of the rock. This results in a longer

Figure 4.3.   Picture of excessive stemming ejection during a cast blast, resulting in unnecessarily high airblast and reduced throw.

pressure profile in the hole and more energy transmitted into the rock mass behind the hole rather than into the muckpile. If we increase the burden in front of the hole we eventually approach pre-splitting conditions where no significant movement of the rock occurs. Pre-splitting typically produces four times the ground vibration compared to similar weight explosive charges used in less than optimal bench blasting (Worsey et al. [9]). Reducing burden increases fragmentation but there is a point at which the increase in blasting costs reverses the economic gains. However, for most mining operations, the blasting patterns are much larger than for optimum economics. It is important to take the whole cost of rock breakage and transportation into consideration for the determination of optimum fragmentation, which not only includes drilling and blasting costs but loading, hauling and crushing costs. Depending on the type of blasting and material blasted, the powder factor for bench shots should be about 0.6 kg/cubic meter, and for large-scale underground stopes and headings, double that number.

Second, an improper delay between holes causes choking of a blast and reduced fragmentation. It has been known for a long time that the use of delays along a row of blast holes not only reduces ground vibrations but also dramatically improves fragmentation. Work by the USBM (Stagg and Engler [7]) and data compiled by Grant [10] shows that the optimum fragmentation occurs when the delay between neighboring holes is on the order of three milliseconds per meter of burden. Figure 4.4 clearly shows the effect of hole to hole delay. The effect can be explained by reduced relief for each hole when no delay is used, and improved fragmentation from the interaction of holes.

At excessive delays, this interaction is eliminated, but this is outweighed by the increase in relief, giving far better fragmentation compared with no delays at all. Improper delay between rows is often

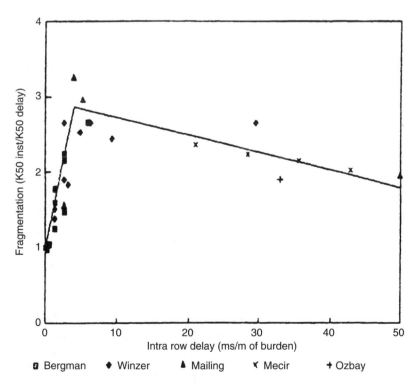

Figure 4.4.   Effect of hole to hole (intra-row) delay timing on fragmentation, after Grant [10]. Note: increasing fragmentation on the vertical axis and increasing delay (ms/m burden) on the horizontal axis.

evidenced as high vertical movement of the blast and excessive material thrown behind the blast rather than into the muckpile, as illustrated in Figure 4.5. Too short a row to row delay results in the material from preceding rows not having moved sufficiently out of the way before the next row fires and tries to move. The resulting overburdening causes vertical movement and increased ground vibrations, often along with shearing of the ground, resulting in the partial cut off of individual holes. In contrast, excessive delays between rows often result in the cut off of surface initiating systems. The minimum delay between rows is 10 millisecond per meter (ms/m) of burden, 10–30 ms/m of burden being an acceptable range. For large numbers of rows, this delay should be increased moving back into the blast to provide extra time for material to move out of the way for the back rows. Too small a row to row delay on the last row is always indicated by excessive back splatter from the blast.

Improper functioning of an explosive charge or charges within a blast decreases the overall effective powder factor, which results in decreased fragmentation and increased ground vibrations, and often an increase in airblast as well. The misfire of a hole in a multi-row blast results in the overburdening of holes, and results in excessive vertical movement of the rock from those holes. The most common causes of misfires are poor design, most usually not having a blast fully activated, poor choice of explosives and initiating systems, and incomplete hook up.

Figure 4.5.    Overburdened and under delayed shot in Minnesota iron range resulting in excessive vertical movement (top) and back splatter on bench behind (bottom). Note: Final face beyond ridge on left and back of the bench to the right.

### 4.2.5.3   *Flyrock*

Excessive flyrock can be devastating to a mining operation. Being hit by rock from a blast when it is intentionally initiated is the leading cause of blasting fatalities (D'Andrea et al. [11]). Not only are financial penalties assessed by government mine safety authorities but also expensive and time-consuming lawsuits generally ensue, and the mining operations may be temporarily or permanently halted. In addition, flyrock can cause extensive and expensive damage to equipment and plant and stoppages in production. Rock thrown an excessive distance from a blast is a hazard to equipment tires and can cause unnecessary delays related to cleanup of the flyrock.

   There are three major causes of flyrock; insufficient burden, insufficient stemming and weak layers or seams. Other contributing factors include poor blast design, and insufficient delays between rows. Insufficient burden is the primary cause of flyrock out from the face, and there are several instances where rocks up to one ton in size have been projected as far as half a kilometer. Insufficient burden is most commonly caused by a combination of poor drilling, excessive face height and not paying attention to the face profile. A certain percentage of blast holes in the front row come too close to the free face, resulting in an explosive charge with only a scant amount of rock in front of it. This results in excessive rock velocities and projection distances. Weak seams afford the easy release of energy from the blast hole and areas for premature venting, so that the explosion gases escape out of this easier route rather than propelling the rock forward. The rapid escape of the gases rips material with it, projecting it at high velocity. Insufficient or poor stemming is the major cause of flyrock from the top of a surface shot. A properly stemmed shot that has no stemming ejection will very rarely have flyrock from the top of the blast.

### 4.2.5.4   *Fumes*

Fumes are always generated during blasting. The majority of fumes come from the improper combustion of the explosives due to a variety of reasons including; poor loading conditions, ground water, improper priming, improper selection of explosives and explosives that are incorrectly formulated or in deteriorated condition. The amount of fume generation in ideal conditions is dependant on the oxygen balance of the formulation. Some explosives have excessive fumes and require substantial ventilation when used underground; others are formulated to produce very low fumes. The more explosives you use and the more confined the conditions, the worse fume problems will be. Not only are fumes a serious problem underground but also in other confined conditions such as trenching, where fumes are often trapped under the blasted material, and sometimes migrate into neighboring voids such as sewers and basements. In April 2000, a family of three received carbon monoxide poisoning in Armstrong County, Pennsylvania (Eltschlager et al. [12]). The gases entered the basement of their home from production blasting at the Milliren mine, a surface coal mine 400 ft away. Where vast amount of explosives are used (hundreds of thousand of kilograms) extremely large fume clouds may be generated and take tens of minutes to dissipate, even in windy conditions. Such an example occurs in coal mining in the Powder River Basin in Wyoming where this is a particular problem (Turcotte et al. [13]). In these types of operations special precautions should be taken to protect both employees and the public.

### 4.2.5.5   *Dust*

Blasting can be a substantial source of dust for short time periods. Where this is a problem, sites have been successfully hosed down immediately prior to and during blasting in a number of operations. The key is to use the right amount of water distributed in the optimum fashion such that any explosive with poor resistance loaded in the shot is not compromised.

### 4.2.5.6   *Excavation*

The excavation process itself can cause problems, which unfortunately are often blamed on the blasting. Such problems include instability and the draw down of water tables. Every face has a stable angle of repose. In certain geological conditions this may be low, 30 degrees or even less. When mining in such conditions care should be taken that any instability or even minor movement is confined to the mining permit area and that mine plant is not affected. There are several examples of damage to commercial and residential property due to instability, including a condominium complex in Michigan where substantial monetary claims were made. Digging a hole in the ground usually results in the influx of water, which has to be pumped out of the pit. Ground water drainage into mining excavations often results in the draining of perched aquifers and the drying up of neighboring springs above the elevation of the pit floor. An example of this occurred in Alabama, where a spring fed stream that contained endangered species in a state park near a sizeable rock quarry dried up. This problem has caused particular uproar from environmentalists and state agencies.

## 4.2.6   EFFECTIVELY MANAGING BLASTING LIABILITY

Many mining operations run into problems with complaints, lawsuits and regulatory fines. In most of these cases, they may have annoyed neighbors, but the neighboring property is rarely damaged by the blasting. However, the mining the company is usually guilty until it can prove itself innocent. Fortunately, this can be avoided if the company is proactive in its approach to its blasting liability. Proactive items which are extremely effective include; blast monitoring, keeping good blast records, active programs to decrease ground vibrations and airblast, public relations, the use of blasting specialists or consultants and pre-blast surveys.

### 4.2.6.1   *Blast monitoring*

Blast monitoring is performed using blasting seismographs. A modern, four-channel, digital seismograph should be used, as illustrated in Figure 4.6. Modern seismographs are relatively inexpensive, easy to use and have plenty of useful technical innovations. The instrument is used to measure ground vibrations and airblast. It should be placed on the blast side of the nearest off-site structure of interest, three to six feet from the foundation where possible, with the arrow pointing towards the blast. The geophone should be buried or spiked in compacted dirt. The regulated limits are for ground vibration rather than structural response and therefore the geophone should not be placed inside or on any portion of the structure. The geophone has three sensors mounted perpendicular to one another and measures the ground vibrations in the vertical, radial and transverse directions. A low frequency microphone is used to measure the airblast. The unit not only records but also processes the data to give vibration values in the field. The data can also be downloaded for processing using a cable or automatically in some units by cell phone. The two most important items in blast monitoring are that the person setting the instrument in the field and interpreting the data is properly trained (and certified) in blast vibration monitoring and use of the equipment, and that the instrument is well maintained and sent for calibration by the manufacturer on a regular basis, usually once per year.

### 4.2.6.2   *Keeping good blast records*

Keeping good blast records is of paramount importance to demonstrate that you blasted in an appropriate and safe manner. Attention to detail in the blasting records, with all the lines filled in an appropriate manner shows attention to detail, pride in the blaster's work and that the mining operation cares about meeting regulated or reasonable limits. Seismograph and good blasting records are a mining operation's best defense when dealing with regulatory agencies and court cases.

### 4.2.6.3   *Active programs to decrease ground vibrations and airblast*

Work by the U.S. Navy Medical Research Institute (Goldman [14]) and the U.S. Army Corps of Engineers [15] shows that the structural response of a building to blasting vibrations can be

Figure 4.6.   Modern digital blasting seismograph. Geophone spiked to ground (center), miniseis digital recording device and processor (left), and microphone sitting on carrying case (right).

extremely unpleasant and worrying, even at ground vibration levels far below damaging levels. This is illustrated in Figure 4.7, which shows that although lower frequency vibrations are the most damaging for buildings, higher frequency vibrations are perceived by people as being of the most concern. Thus, ground vibrations and airblast should be minimized where possible, rather than taking the attitude that "we have always stayed below the limit."

### 4.2.6.4   *Public relations*

The public's perception of blasting varies greatly. Generally, people who live in noisy environments such as those having teenage children or living next to busy roads or railroads are not likely to complain, neither are people who are at work when blasting occurs. The people who are most likely to complain are those who are present during blasting and live in a quiet environment, i.e. retired individuals. Humans perceive the intensity of an event relative to their surroundings, for example, hearing a pin drop in a quiet room compared to someone screaming right in your face in a noisy machine shop. This is why retired people rather than people with active young families are the most likely to complain.

Your worst enemy is silence on your part because it breeds suspicion that you are hiding something or you do not care. Being proactive using public relations is an important step you have to take. Education is one of the best tools that can be used. Both regulatory and public officials and local residents should be informed of what you are doing, that you care about public protection, that there are regulated or acceptable vibration limits that you have to meet, that you use seismographs to stay well below those limits, and that you document your blasts. In addition, you should inform the public that blasting may rattle residences for a second or two and that both studies in the U.S. and internationally have conclusively shown that weekly changes in temperature and humidity put far more stress on their houses for long durations of time in comparison. Be open with people and offer to show them what you do. Blasting can be an extremely interesting subject for the public and knowledge goes a long way to dispel fear. Always try to work on a one to one basis, or with small groups on your property.

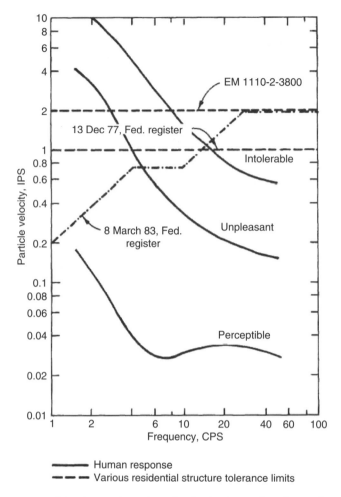

Figure 4.7.   The response of structures and people to vibration levels at various frequencies with OSM limits, from U.S. Army Corps of Engineers [15]. Note: Human and dwelling frequency responses are opposite.

#### 4.2.6.5   *Use of blasting specialists or consultants*

Bringing in outside blasting specialists or consultants can greatly help in maximizing blasting efficiency, reduce costs and reduce environmental effects. Hiring a consultant is also useful for public relations and as proof that you are doing your job properly and also are always looking to improve. Blasting expertise is almost always available from the major explosive suppliers, who usually have technical representatives available. If you are bringing in a consultant, have him/her conduct a full blasting audit every few years. Items to be looked at include blast design, drilling and loading practice, safety, records, and environmental effects. For large operations, the use of third parties adds credibility to your blasting program. Vibration monitoring is an ideal task for consultants. Where permanent vibration monitoring stations are set up, it is prudent for the mining company to purchase appropriate seismographs and have the vibration consultant set up, service and operate the seismographs.

For medium size operations, shot service has become the trend, where the explosives distributor not only delivers the explosives but also loads and to a large part shoots the blasts. For smaller operations, the trend is to subcontract drilling and blasting entirely. This can be very cost effective since contractors generally have a wealth of extra experience, operating at many different locations.

These contractors generally provide modern, high-productivity drilling equipment, which a single smaller mine would find hard to justify, due to low utilization. These drills are very mobile and are moved from mine to mine by the contractor.

### 4.2.6.6   *Training and certification*

Proper third party training and certification of drilling and blasting supervisors and front line personnel is a great asset to a mining operation, and is one of the best ways to ensure blasting compliance, minimize blasting effects both off and on the property, increase safety and minimize operational costs. A study on the effectiveness of 10 years of blasters certification in the state of Missouri was presented in 2002 by Henderson [16]. His numbers dramatically show a decrease in blasting costs which is correlated with the number of blasting personnel obtaining specialist training as part of the Missouri Blasters Certification Program. Many mining operations are remote; however, distance learning opportunities for technical personnel are available, including for-credit courses and post graduate degree programs. For example, the internet-based distance learning Masters of Engineering degree program in Mining at the University of Missouri-Rolla includes three blasting courses. Keeping abreast of new technology gives a company a competitive edge.

### 4.2.6.7   *Pre-blast surveys*

Pre-blast surveys were first mandated by OSM in 1983 [17] in the U.S. for new permitted coal mines. They are an excellent method of determining the condition of an existing structure before blasting commences. The rules were designed to protect both the property owner when legitimate blasting damage occurs, and to protect the mining operator from unjustified claims. Pre-blast surveys have subsequently been adopted by many municipalities, counties and states in the U.S. A pre-blast survey is made before blasting at a new operation or permitted area is started, and comprises the thorough inspection of buildings concerned, identifying every flaw including cracks and nail pops for the outside, inside, foundations and any adjoining concrete work such as footpaths and driveways. The purpose is to document and date these defects so that they can be conclusively proven to pre-exist before the start of blasting, and are therefore not related to blasting vibrations. Having the defects documented quickly resolves any complaints from neighbors. These surveys should be performed by a third party, i.e., a company that specializes in this type of work and whose personnel have the training, experience and qualifications to do this work. The pre-blast surveys should be kept for the life of the mining operation because buildings often change hands and you may have to deal with new owners. Pre-blast surveys more than pay for themselves in the long run.

### 4.2.7   IMPORTANCE OF PROFESSIONAL ASSOCIATIONS IN THE REGULATION OF BLASTING

The preemptive self-regulation of our industry where regulations do not exist tends to ward off or delay regulation. Codes of good practice that the industry can easily live with are an excellent alternative to unsound, restrictive legislation. The technical authority of active associations assures a place on committees that draft or review blasting legislation. Blasting certification programs run by associations are the ideal candidates for adoption by states and large municipalities, with only minor changes. This has been the case in the state of Missouri, where neighboring states have adopted its association run certification programs.

An example of where problems can arise if the associations are not proactive is Brazil where in 1986 a blasting limit of 15 mm/s was mandated (Brazilian Law 18.923 NBR 9653 [18]). Unfortunately, the calculation of this value is prescribed as the vector summation of the peaks of the vertical, radial and transverse vibrations, which more often than not do not occur at the same point in time during the vibration trace. Thus, this regulation is both unscientific and excessively conservative for probably over 75% of blasts. The Brazilian mining engineering associations are currently attempting to get this regulation changed to the maximum of the real time vector sum.

4.2.8   NEW TECHNOLOGY

There have been dramatic changes in the mining industry in the past 25 years, and drilling and blasting has been a significant part of these changes. The old stand by, dynamite, has been replaced by emulsion and slurry explosives, with only one dynamite factory remaining in the U.S., and in countries such as Mexico, dynamite is no longer available. We have seen a transition from stick explosive to bulk explosive, which is augered, slurried, blown or pumped into holes, and shot service has become commonplace. In the U.S. we have seen the virtual replacement of electric blasting and detonating cord by nonel, and the world is now experiencing the introduction of electronic blasting caps. Electronic blasting caps, with their ability to produce precise firing times, to shoot substantially larger blasting patterns, reduce blast vibrations, and to be fully activated to reduce cut offs and misfires gives the potential for further leaps forward in improving the way the industry blasts. Also, the potential for substantial improvement in fragmentation through the precise control of the sequencing of a blast will ultimately prove this technology cost effective for the majority of mining operations.

REFERENCES

1. Kramer, D.A., *Explosives, 2001 Annual review*. USGS Mineral Industry Surveys, 2002.
2. Konya, C.J. and Walter, E.J., *Surface Blast Design*. Prentice Hall, Englewood Cliffs, 1990.
3. Hopler, R.B., *Explosives 100 years Ago More or Less The Book*. International Society of Explosives Engineers, Cleveland, 2001.
4. Grady, S.M., *Explosives: Devices of Controlled Destruction*. Lucent, San Diego, 1995.
5. Olanyo, J., Quarry body tells residents to wait for NEMA report. *Associated Press,* December 09, 2002.
6. Nobel Insurance., Every number tells a story. *Safety Update*. 21(3) (2002), p. 3.
7. Stagg, M.S. and Engler, A.S., *Measurement of Blast-Induced Ground Vibrations and Seismograph Calibration*. United States Department of the Interior, Bureau of Mines Report of Investigations, RI 8506, 1980.
8. Siskind, D.E., Stagg, M.S., Kopp, J.W. and Dowding, C.H., *Structure Response and Damage Produced By Ground Vibration From Surface Mine Blasting*. United States Department of the Interior, Bureau of Mines Report of Investigations, RI 8507, 1980.
9. Worsey, P.N., Giltner, S.G., Drechsler, T., Ecklecamp, R. and Inman, R., Vibration Control During the Construction of In-Pit Lime Kiln. *Proceedings*: Twenty-second Annual Conference on Explosives and Blasting Technique. The International Society of Explosives Engineers, Orlando, 1996, Vol. 2, pp. 180–190.
10. Grant, J.R., Initiation Systems – What does the Future Hold? *Proceedings*: Third International Symposium on Rock Fragmentation by Blasting. The Australian Institute of Mining and Metallurgy, Parkville, 1990, pp. 369–372.
11. D'Andrea, D., Fletcher, L. and Peltier, M., Blasting Accidents in Mining. *Proceedings*: Thirteenth Annual Conference on Explosives and Blasting Technique. The International Society of Explosives Engineers, Miami, 1987, pp. 291–305.
12. Eltschlager, K.K., Shuss, W. and Kovalchuk, T.E., Carbon Monoxide Poisoning at a Surface Coal Mine ... A Case Study. *Proceedings*: Twenty-Seventh Annual Conference on Explosives and Blasting Technique. The International Society of Explosives Engineers, Orlando, 2001. Vol. II, pp. 121–132.
13. Turcotte, R., Yang, R., Lee, M.C., Short, B., and Shomaker, R., Factors affecting Fume Production in Surface Coal Blasting Operations. *Proceedings*: Twenty-Eighth Annual Conference on Explosives and Blasting Technique. The International Society of Explosives Engineers, Las Vegas, 2002, pp. 307–316.
14. Goldman, D.E., *A Review of Subjective Responses of Vibrating Motion of the Human Body in the Frequency Range, 1 to 70 Cycles per Second*. Naval Medical Research Institute, Project NM 004001, Report 1, Washington D.C., 1948.

15. U.S. Army Corps of Engineers, *Engineering and Design: Blasting Vibration Damage and Noise Prediction and Control*. Department of the Army, U.S. Army Corps of Engineers ETL 1110-1-142, Washington D. C., 1989.
16. Henderson, K., Blaster's Certificates Past, Present and Future. Seventh Mid-America Blasters Conference. The International Society of Explosives Engineers Heartland, Mississippi Valley, Ozark and UMR Chapters, Lake of the Ozarks, 2002.
17. Office of Surface Mining, *Surface Mining Law Revision January 1983*. OSM 30CFR816,8179850, 1983.
18. Brazilian Law 18.923 NBR 9653.: *Guia Para Avaliacao Dos Efeitos Provocados Pelo Uso de Explosivos nas Mineracoes em Areas Urbanos*. November 1986, Part 4.2–4.3, p. 3.

# 4.3    Minimizing surface water impacts

Raj Rajaram
*Complete Environmental Solutions, Oak Brook, IL, USA*

## 4.3.1    INTRODUCTION

A large amount of water is used by mining companies for dust control, process water, and other uses. With proper use of recycling and water conservation techniques, the amount used can be minimized. The use of runoff controls, proper tailing management design and operation, and monitoring the discharges from the mine can minimize impact to surface water quality. Implementing the EMS during mine operations results in the following benefits:

- reduced siltation of ponds and other surface water bodies
- prevention of oil and material spills that could degrade surface water quality
- minimizing water use for dust control
- water conservation through recycling of process water, and other techniques
- prevention of surface water quality degradation through the use of treatment methods, and monitoring the mine discharges.

State Environmental Protection Agencies (EPA) promulgate regulations for controlling mine discharges and non-point source mine discharges into State water bodies. In Illinois, United States, compliance with the numerical standards are based on the 24-hour composite samples averaged over any calendar month (IEPA [1]). In addition, no single 24-hour composite may exceed two times the numerical standards prescribed nor shall any grab samples taken individually or as an aliquot of any composite sample exceed five times the numerical standard. The numerical standards for pH and settleable solids are not subject to averaging. The location of sampling shall be before entry into or mixture with waters of the state.

## 4.3.2    SURFACE WATER MANAGEMENT PRACTICE

The duty of the mine operator is to provide the best degree of treatment of wastewater consistent with technological feasibility, economic reasonableness, and sound engineering judgment. In deciding the best degree of treatment, the following should be considered:

- the degree of waste reduction that can be achieved by process change, improved housekeeping and recovery of individual waste components for reuse
- segregation of individual wastewater streams to the extent possible to reduce the cost of treatment.

Effluent standards for mine discharges in Illinois are as follows (IEPA [1]):

- 5 mg/L ammonia nitrogen (as N)
- 3.5 mg/L iron
- 15.0 mg/L fluoride (total)
- 6 to 9 pH
- 35.0 mg/L total suspended solids
- 5.0 mg/L zinc (total).

In addition, no mine discharge shall contain settleable solids, floating debris, visible oil, grease, scum or sludge solids. Colour, odour and turbidity shall be reduced to below obvious levels.

Illinois also regulates non-point source mine discharges from land areas affected by a mine, including disturbed areas which have been graded, seeded or planted (IEPA [1]). Effluent from a reclamation area shall not have settleable solids exceeding 0.5 ml/L. This requires that water leaving land areas affected by a mine should have a sedimentation pond or a series of sedimentation ponds to meet the settleable solids requirement. Discharges caused by precipitation within any 24-hour period greater than the 10-year, 24-hour precipitation event shall be subject only to the pH limitation mentioned above.

Illinois also states that mine discharges should not violate the state water quality standards. Therefore, a mine permit issued under the water quality based permit conditions can state that the total dissolved solids, chloride, sulphate, manganese and iron be under specified limits (IEPA [1]). The water quality based permit can be obtained if the following conditions are met:

- demonstrating that the discharge is not causing an adverse effect on the environment in and around the receiving stream
- monitoring the discharge and proving that it has a concentration of less than or equal to 3,500 mg/L sulphate and 1,000 mg/L chloride;
- proving that the discharge will not adversely affect any public water supply
- proving that the mine operator is using good mining practices.

Good mining practices include the following:

- practices that stop or minimize water from coming into contact with disturbed areas
- retention and control within the site of waters exposed to disturbed materials
- control and treatment of waters discharged from the site
- unconventional practices.

The regulatory agency may approve any of the following methods to minimize contact with disturbed areas:

- bypass diversions to collect or divert water to a receiving stream that would otherwise flow through disturbed areas
- on-site diversions to convey water around disturbed areas
- minimizing the disturbed surface area that is open at one time
- keeping gradients and inclines to the active pit as short as possible in order to minimize the amount of drainage going to the active pit
- soil stabilization measures such as revegetation and mulching that will reduce the potential for exposing materials producing dissolved solids
- sealing of boreholes or water-yielding fractures zones which allow uncontrolled flow of water to underground mines or to active areas of surface mines
- leaving sufficient barriers whenever mining adjacent to abandoned underground workings that may be inundated with water
- disposal of potential contaminant producing materials as soon as possible in areas that will minimize contact with surface and groundwater
- covering or treating potential contamination producing materials so as to minimize adverse effects on water quality.

Retention and control of water exposed to disturbed materials can be achieved as follows:

- sedimentation controls, routing and segregation or combination of wastewater and mine runoff water to minimize the effect on the quality of the receiving stream
- reuse of discharges containing high concentration of total dissolved solids through recirculation ponds or holding ponds to provide water for mine dust control, irrigation of reclaimed land or adjacent crop land

- application of water management practices, either continuously or at frequent intervals, in order to minimize water contact with disturbed materials
- prevention of accumulation of waters in active pits, roads, processing areas, surface depressions and underground mine workings where dissolution of contaminants can take place
- removal of water to diversions and appropriate impoundments as soon as possible to minimize additional loadings of total dissolved solids.

Control and treatment of discharges can be accomplished as follows:

- regulation of discharges can be implemented when other control measures are insufficient and chemical treatment is uneconomical, through suitable discharge control structures so that the discharges remain within the established water quality standards
- rerouting over economically feasible distances, involving collecting discharges and conveying them to more suitable discharge points where dilution and/or water quality is better.

Unconventional practices that could be used as part of good mining practice include:

- diversion of groundwater by intercepting the flow path prior to entering a surface or underground mine, if treatment is uneconomic
- dewatering practices that remove clean groundwater before contacting dissolved solids-producing materials
- any other method that the operator demonstrates to be effective in reducing levels of total dissolved solids, chloride, sulfate, iron and manganese in the discharges.

Case studies of water management in mines are discussed in Section 4.1. A case study of surface water management in a proposed underground mine in Wisconsin, United States, is presented below.

Nicolet Mining Company proposes to mine the Crandon deposit consisting of copper, zinc, and minor amounts of precious metals. Figure 4.8 shows the deposit location near Crandon, Wisconsin. The Wisconsin Department of Natural Resources (WDNR) has specified strict limits for mine discharges to protect the surface waters of the state, which include several creeks and lakes located in the vicinity of the proposed mine. The surface water bodies have been classified as follows (Foth and Van Dyke [2]):

- Level I to include water bodies for which substantial impacts are predicted from the mining operation
- Level II to include water bodies which could possibly experience substantial impacts
- Level III to include water bodies which are not likely to experience impacts, but could be impacted given the proximity of the water body to the proposed mine
- Level IV to include water bodies which are highly unlikely to experience impacts, but could be impacted given their proximity to the proposed mine.

The mitigation plan for Level I water bodies will be constructed prior to start of the mining operation. The mitigation plan for Level II water bodies will be designed but will be implemented only if the data collected during the operation indicate that the system is needed. A conceptual mitigation plan is to be developed for Level III water bodies, and a generic outline for the development of a mitigation plan will be prepared for Level IV water bodies. The trigger criteria to upgrade or downgrade mitigation planning levels are based on water level, water chemistry monitoring data such as dissolved oxygen, climatological data, and fish data obtained from the creeks and lakes in the vicinity of the proposed mine. The water levels in the creeks and lakes could be affected by the mine dewatering requirements, and hence, the drawdown in the aquifer measured at key wells between the surface water bodies and the mine will be carefully monitored.

Surface water management measures incorporated include minimization of erosion and the release of sediment from disturbed areas throughout the life of the mine.

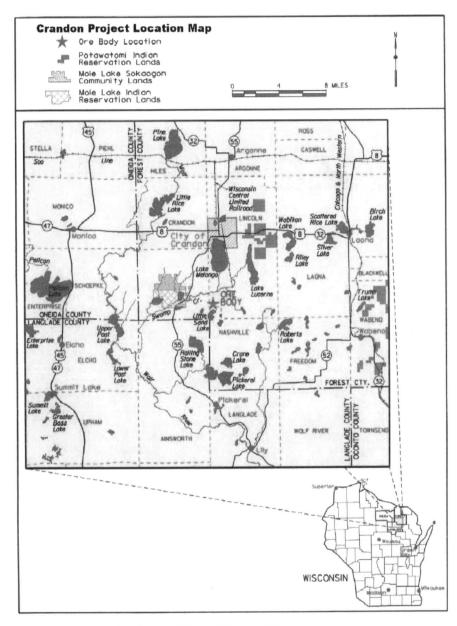

Figure 4.8.   Crandon mine location map (Source: Wisconsin DNR).

The measures planned for maintaining the water quantity and quality in surface water bodies include:

- recirculation and reuse of the process water from the concentrator facility;
- contact water intended for discharge will be treated to groundwater discharge standards by the on-site wastewater treatment facility – the wastewater treatment facility will include wastewater storage basins, lime and sulfide precipitation, filtration and reverse osmosis – reject water from

the reverse osmosis system will be treated in an evaporator system. Condensate from the evaporator will be treated for ammonia removal in an air stripper;
- runoff basins to provide retention, treatment, and discharge of non-contact runoff water; and
- discharge of the treated water to groundwater using a soil absorption system.

The soil absorption system at Crandon is unique, and mitigates the groundwater impacts at the site.

REFERENCES

1. Illinois Administrative Code, Title 35, Subtitle C, Section 406, Nov. 2002.
2. Foth & Van Dyke, Environmental Impact Statement for the Crandon Mine, Wisconsin, Submitted to the Wisconsin Department of Natural Resources, Madison, WI, USA, 1998.

## 4.4   Minimizing ground water impacts

Raj Rajaram
*Complete Environmental Solutions, Oak Brook, IL, USA*

### 4.4.1   INTRODUCTION

Many mining sites have impacted the ground water through an improper understanding of the impacts of waste disposal in unlined ponds, and inadequate characterization of the site hydrogeology. A large lead and zinc mine in India was forced by the government to implement lined ponds for waste disposal after it was discovered through monitoring that down gradient wells were adversely impacted by the tailings disposal facility.

Adverse ground water impacts are very expensive to remediate, as demonstrated in the United States over the last two decades (EPA [1]). Prevention of ground water impacts can be accomplished through adequate characterization of the subsurface geology, including the aquifers and aquitards, and proper planning of waste disposal and water management issues. A mining permit is not issued by the regulatory authorities in many jurisdictions unless the ground water resources in the mining area are discussed in detail, along with impacts to the ground water during operation and closure/post-closure periods. An environmental impact statement (EIS) report issued by a mining company for obtaining the mining permit discusses the ground water issues in detail.

### 4.4.2   GROUND WATER MANAGEMENT PRACTICE

A large proposed underground mining operation in the United States was found to impact the water levels in surrounding areas due to the large amount of dewatering required for conducting the underground mining. The mining company is constructing a 25-foot thick grout blanket at the base of the crown pillar of the water bearing formation to minimize water inflow into the mine. It also plans to remove pyrite from the tailings and backfill the tailings in a cemented paste form in underground openings. This backfilling will lead to underground mine stability, and reduce the water flow into the mine. However, the most unique aspect of groundwater management is the recharge of the aquifer through a soil absorption system to raise the groundwater table and minimize the impact to surface water bodies (Foth and Van Dyke [2]).

A study of the regional groundwater quality and potentiometric surface was completed. In addition, wetland delineation and mapping, subsurface characterization, topographic and archaeological surveys, and groundwater quality sampling were completed in potential areas that could be used for recharge of the aquifer. An important parameter that was measured was the infiltration rate and hydraulic conductivity of the near surface soils. Treated wastewater is stored in a holding pond for testing prior to being pumped to the soil absorption system site. The soil absorption system is designed to handle discharge rates of 205 gallons per minute (gpm) to 386 gpm. The system is laid out in six individual cells, as shown in Figure 4.9. Each cell is designed to apply treated effluent through the use of distribution laterals which are laid out at a uniform elevation over the entire cell. Beneath the laterals, 6 inches of gravel will be placed to further help distribute water evenly over the cell surface. Each lateral will have orifices spaced at 10-foot centers, with the laterals being 20 ft apart. The laterals are then covered with sand, a non-woven geotextile, and soil and vegetation. The vegetation will insulate the soil absorption cells so that they can function even in cold weather without freezing. Each cell is designed to operate independently of the others, and this is accomplished by a central distribution chamber which feeds each cell through designated controls (Foth and Van Dyke [2]).

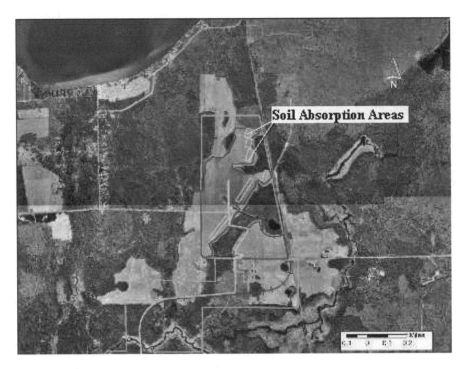

Figure 4.9.   Soil absorption cells (Source: Wisconsin DNR).

This is the kind of planning required to obtain permits for mining projects in the twenty first century, with heightened awareness of environmental impacts by the public and governmental agencies.

A proper ground water management plan for a mining operation has to include the following:

- fractures and faults controlling ground water entry into the mine
- dewatering required to keep mining operations dry
- mine sealing plan to minimize ground water entry into the mine
- pump system layout, and layout of sumps
- power requirements for pumping, and methods to minimize power costs
- ground water handling, including aquifer recharge
- segregating dirty and clean water in the pumping system
- emergency plan to manage large quantities of unexpected ground water, for example, from unmapped karst conduits and fault zones
- ground water use and recycling within the mine.

Ground water impacts can be minimized through implementation of an EMS system at the mine. The EMS system looks at the ground water withdrawal rate, impacts of waste management systems on the ground water, ground water usage and recycling issues, and impacts of the ground water withdrawal on other water users in the area. The EMS system also considers various ways of minimizing water usage, and details measures to prevent adverse impacts to groundwater during mine operations. A typical EMS system used for a copper mining operation in the U.S. is presented below.

REFERENCES

1. U.S. Environmental Protection Agency (EPA), Effectiveness of Ground Water Pump and Treat Systems, Technology Innovation Office, Washington, D.C., 1977.
2. Foth and Van Dyke, Soil Absorption System, Environmental Impact Statement Report, Crandon Mine, Forest County, Wisconsin, Submitted to the Wisconsin Department of Natural Resources, Madison, Wisconsin, USA, 1998.

# 4.5   Subsidence considerations

M.M. Singh

*Engineers International, Inc., Chandler, Arizona, USA*

## 4.5.1   INTRODUCTION

Many persons insist that mining should be discontinued because of its adverse environmental impacts. This is not realistic since mining provides an essential foundation for our civilization. Without mining there would be no electric power (even windmills, solar cells, and hydro-generators are built with mined materials), roads, or even houses. Impacts of mining can be minimized, however. Kennecott Minerals Company (Kennecott [1]) has a sustainable development policy, and in December 2002, Rio Tinto was named the highest ranking mining company in the Dow Jones Sustainability Index (DJSI). Studies have been undertaken to achieve sustainable mining by the Bureau of Land Management (BLM [2]) and for environmentally acceptable energy development by the World Bank (World Bank [3]). In the United States, successful sustainable mining projects have been conducted at the Flambeau and Ridgeway mines (Fox [4]). In Italy, natural aggregate is being mined using sustainable development principles (Langer [5]). In Peru, the Antamina Project is similarly planned (Botts [6]). Generally, most minerals are mined from open pit operations – including sand and gravel, clay, limestone, even coal and copper. However, many are still obtained from underground workings (such as coal, lead, zinc, fluorspar) or by liquid extraction (for example petroleum and sulfur). Extraction of underground minerals results in surface subsidence. This may create harmful effects to the land and its use. For sustainable mining we need to mitigate its effects as much as possible. Hence we need to understand the nature of mine subsidence and examine the factors that affect it. Understanding the underlying principles not only helps the mining community, but all stakeholders in the land surrounding the mine.

## 4.5.2   NATURE OF MINE SUBSIDENCE

Whenever a void is created in the rock strata below the surface the surrounding material tends to move into the cavity. This is the case when a mineral is mined. Over a period of time as the cavity enlarges, or the supporting structures deteriorate, instability results. This induces the superjacent strata to occupy the void (Figures 4.10 and 4.11). Gradually these movements work up to the surface, manifesting them as a depression. Thus, mine subsidence may be defined as ground movements that occur due to the collapse of the overlying strata into the mine openings. In petroleum extraction the removal of fluids from the strata have a similar effect. Subsidence from petroleum extraction has subsided areas in the Los Angeles (United States) area, among others. Even pumping of water can instigate subsidence, as experienced in the vicinity of New Orleans, United States. Surface subsidence entails both vertical and lateral movements. This section addresses solid mineral removal.

All strata achieve a balance of forces over periods of geologic time by processes of deposition and tectonic change. The natural stress field in the strata is mainly due to the weight of the overlying rocks though some areas of the world have reported active geologic horizontal stresses. The average density of bedded rocks is approximately 2.4 g/cc (150 lb/cu ft). Therefore, the vertical pressure gradient due to gravity is approximately 22.6 kPa/m (1 psi/ft) of depth. The creation of a void in the strata by underground mining produces a disturbance in the natural stress field. The effect of the stress change results in strata deformation that, depending on the dimensions of the

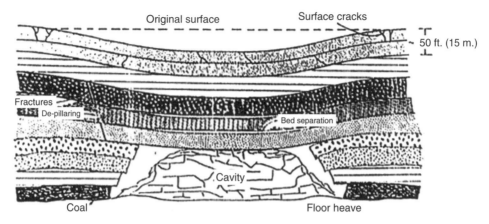

Figure 4.10.   Strata disturbance and subsidence caused by underground mining.

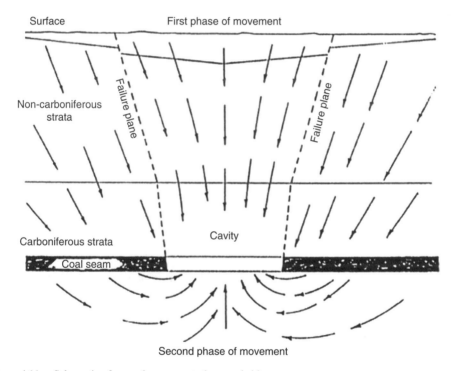

Figure 4.11.   Schematic of ground movements due to subsidence.

excavation, can extend to the surface. Surface depressions, cracks and pits, with or without damage to the surface structures, are the common effects of subsidence induced by mining.

The term "subsidence" as used in this section, implies the total phenomenon of surface effects associated with the mining of minerals, and not just the vertical displacement of the surface as sometimes used in the technical literature.

The mechanics of subsidence are widely discussed in the technical literature (Fejes [7] and Singh [8]) and are not presented here. There are four types of subsidence: cracks (tensile or shear), buckling (due to compression or shear), pit (also locally termed pothole, sinkhole, chimney,

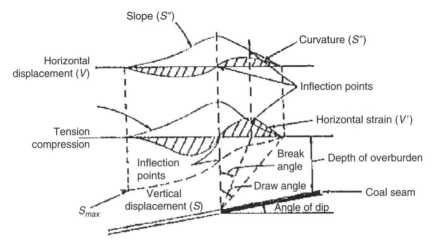

Figure 4.12.   Schematic representation of the components of mine subsidence.

crownhole, or pipe), and trough (or sag). In bedded deposits, pits are generally associated with room-and-pillar mining or partial extraction methods, and troughs with longwall mining or total extraction (de-pillaring). Trough type subsidence, in essentially horizontal strata, has been investigated more thoroughly and is more predictable.

Subsidence entails five major components that may cause damage to the land surface and structures (Figure 4.12):

- vertical displacement (settlement, sinking or lowering)
- horizontal displacement (lateral movement)
- slope or tilt (derivative of the vertical displacement with respect to the horizontal)
- horizontal strain, tension or compression (derivative of the horizontal displacement with respect to the horizontal)
- vertical curvature or flexure (approximated by the derivative of the slope, or the second derivative of the vertical displacement with respect to the horizontal).

Horizontal strains are the most damaging to surface structures (such as buildings, impoundments and utility lines), although curvature may lead to distortion or shear strains and flexural bending. In bedded strata, as the mine workings progress the subsidence trough enlarges and the horizontal tensile and compressive strains move along in tandem (Figure 4.13). Hence, these are often referred to as traveling strains.

### 4.5.3   FACTORS GOVERNING SUBSIDENCE

Several geological and mining parameters affect the magnitude and extent of subsidence that occurs due to coal mining. The nature of the structure also plays a role:

- *Effective Bed Thickness*: Obviously the greater the thickness of the bed extracted, the larger the amount of surface subsidence that is possible. In many cases, however, the entire bed may not be mined or conversely more than one adjoining bed may be mined together. Also, some pillars or other non-mineable minerals may be left in place. Hence, the effective bed thickness should be considered. In thick beds, the slenderness (height-to-width ratio) of the pillars is higher for a given extraction ratio. Slender pillars are more prone to failure.

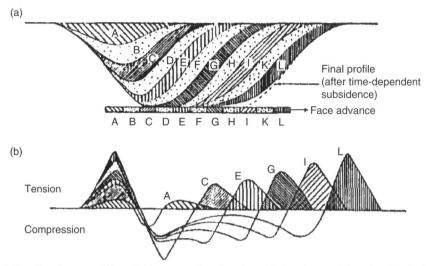

Figure 4.13. Development of the subsidence trough and strains with the advance of the mine face ("a" indicating trough development and "b" indicating traveling strains).

- *Depth of Deposit*: A school of thought exists that at greater depths an arch is formed over the mine cavity, preventing surface subsidence. However, this has been refuted. Perhaps the time period which elapses before subsidence effects are observed on the surface is prolonged, but the total amount of subsidence does not appear to be changed, that is, subsidence is independent of depth (Gray [9]).
- *Dip of the Bed*: When the mineral stratum being mined is inclined, an asymmetric subsidence trough is formed, which is skewed towards the rise, that is, the angle of draw is greater on the dip side of the workings. The strains are also smaller towards the dip direction. Pillars in steeply dipping beds tend to be less stable.
- *Competence of Mine Roof and Floor*: Since subsidence propagates from the mine level, the characteristics of the mine roof and floor are vital in the initiation of subsidence movements. For example, soft fireclay floors below coal seams, especially if susceptible to further weakening due to moisture, induce pillar punching or heave. Weak roofs composed of shales, siltstones, and limestones, permit falls that are accentuated if punching also occurs. Competent roof beds tend to support the overlying strata longer and hence delay the subsidence. Also, when these fracture, they occupy a greater bulk volume than weaker strata (which compact more). When both the roof and floor are strong, the mine pillars tend to spall and crush.
- *Nature of Overburden*: Strong massive beds above the mine level tend to prop the overburden for a prolonged period and defer (but not eliminate) the occurrence of subsidence.
- *Near-surface Geology*: The soils and unconsolidated rocks near the surface tend to accentuate subsidence effects. Near-Surface geologic materials are less homogeneous and isotropic than the underlying strata, and often behave in an inconsistent manner. Cracks and fissures may initially form in a 15-m (50 ft) thick layer from the surface (Singh and Kendorski [10]). Later, these may be filled by plastic deformation or material transportation by water. Occasionally, however, water flow may accentuate these fissures and form gullies. Structures and renewable resource lands are thereby adversely affected.

  The composition of the rock/soil cover is important; if the material is of a fine, sandy nature containing large amounts of water, it may flow to a rock fracture and drop into the underground workings. Besides, water accumulating in an abandoned mine may seep upwards into the unconsolidated strata above through natural fissures and cracks in the rock and increase the potential for soil collapse.

- *Geologic Discontinuities*: The existence of faults, folds, and the like may increase subsidence potential. Mining disturbs the equilibrium of forces in the strata and may trigger movement along a fault plane, due to ease of slippage, causing either settlement or upthrust at the surface, which may appear as a series of step fractures. Lateral movements concentrate near the fault, but the strains may become very small on either side. Structures that straddle fault planes tend to be severely damaged, but nearby buildings remain relatively intact. Joints and fissures in the strata affect subsidence behavior in a manner similar to faults but on a smaller scale.
- *Fractures and Lineaments*: Natural fractures and lineaments affect surface subsidence, but a strong correlation has not been established.
- *In-Situ Stresses*: High horizontal stresses tend to suspend surface subsidence by forming a ground arch in the immediate mine roof (Lee and Abel [11]). The arch height and stability are sensitive to the ratio of vertical to horizontal stresses. Highly stressed arches may fail violently. Roof instability and floor heave, resulting from high horizontal stresses and their orientation, need to be taken into account when laying out coal mines.
- *Degree of Extraction*: Lower extraction ratios tend to delay subsidence. It is less prevalent in areas superjacent to first mining, since sufficient pillar support is generally available without crushing of pillars. In second mining, the cross-section of the pillars is reduced by splitting and slicing. Localized stress buildups promote crushing, and excessively wide roof areas exposed between pillars stimulate roof failure. Third mining is almost invariably followed by roof collapse in the workings. Surface manifestations are a function of time, dependent on the rate of upward propagation of settlement.
- *Surface Topography*: As may be anticipated, sloping ground tends to emphasize downward movements because of gravity. Tensile strains may become more marked on hilltops and decrease in valleys. Surface effects are influenced accordingly.
- *Groundwater*: Deformation of the strata around mined areas may alter drainage gradients, resulting in the formation of surface or underground reservoirs (in aquifers). Low-lying areas, such as in central Illinois (USA), may become flooded. Rocks may be weakened by saturation. Erosion patterns could change, and in limestone areas, caverns or karst areas may be created over a period of time.

  Where surface runoffs from precipitation or water from leaky mains are allowed to accumulate, water may percolate down through the soil to the fractures and fissures in the bedrock, and finally into the mine openings. The erosion and lubrication effects induce failure.
- *Water Level Elevation and Fluctuations*: Water reduces the strength and stiffness of pillars, the roof and the floor markedly. Periodic changes in mine humidity promote deterioration of all these members. Instability and subsidence also results from floor softening since this permits punching. Flow through fissures cause seepage pressures, endangering the stability of the rock mass. Cleavage and bedding planes are lubricated by water, inducing movements.
- *Mined Area*: The critical width of the mined area must be exceeded along both the lateral and longitudinal axes, to achieve maximum subsidence. This is especially important if competent strata present in the overburden tend to bridge across the panel and defer total subsidence when the panel width is less than the critical width, even though the length of the panel is greater.
- *Method of Working*: In bedded strata, the type of initial subsidence experienced, namely pit or trough, depends on whether room-and-pillar or longwall mining is being practiced. With room-and-pillar mining, the eventual collapse of pillars may lead to trenching or sagging of the surface. The displacements and strains over short distances, when they start appearing on the surface, are significant. Nearly immediate but predictable subsidence occurs with longwall mining. Harmonic mining (discussed later in this section), either by working adjacent longwalls in the same bed or superposed panels in different beds, can be effectively utilized to neutralize compressive and tensile strains and thereby protect surface structures. However, the method is not readily applied.
- *Rate of Face Advance*: Surface subsidence follows the face as it progresses in the panel. If the coal extraction rate varies markedly, the traveling strains also fluctuate. This results in large differential settlements. A fairly rapid, even rate of advance is best.

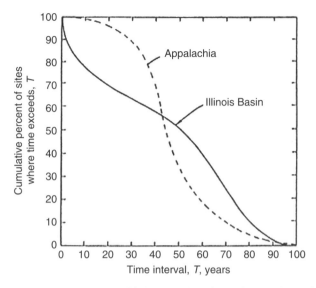

Figure 4.14. Curves depicting the relationship between time elapse since mining and subsidence.

- *Backfilling of the Gob*: Partial or complete backfilling of the gob (mined out area) reduces, but does not eliminate, subsidence. The amount of subsidence that occurs depends upon the type and extent of backfilling adopted. Thus, for example, hand packing is not as good as pneumatic stowing or hydraulic backfilling.
- *Time Elapse*: The amount of subsidence observed is a function of time. In room-and-pillar operations, no surface effects may be noted for some time after the mining is complete until the pillars deteriorate or punch into the floor. In longwall mines, the surface may start sagging almost immediately after the face passes below an area. However, the occurrence of massive beds in the overburden could delay this. With longwalls, surface movements are complete within a few years, but when pillars are left intact for support, this may take decades. Room-and-pillar mining with removal of pillars may produce surface effects similar to longwall mining, with the degree of similarity dependent upon the amount of coal left as fenders or stumps. With room-and-pillar mining, however, subsidence may take over 100 years to complete (see Figure 4.14).
- *Structural Characteristics*: The extent of damage to a structure is dependent on the type of structure and its size, shape, age, foundation design, construction materials and techniques used, standards of maintenance and repair, and purpose. The surcharge due to building loads may induce soil compaction, generating instability. Tall structures cannot tolerate much tilt, poorly constructed or older buildings are readily damaged, and a large edifice is more liable to crack because of the strains and curvature induced by subsidence.

### 4.5.4 IMPACT OF GEOLOGICAL DISCONTINUITIES ON MINE SUBSIDENCE AND THEIR EFFECTS ON SURFACE WATER

If faults or lineaments are known to exist in the strata where major structures are to be located, such sites are avoided. Displacements along these can be sudden and severe, with disastrous consequences. However, joints and cracks are much more common and cannot be avoided. These merit some discussion.

Sedimentary strata, besides being bedded, have joints whose spacing and orientation vary depending on the type of rock, thickness of individual beds, geological age, depth, and various geological stresses to which they have been subjected during deposition and subsequently. Joints can be defined as surfaces across which rock has lost cohesion without displacement. Jointing systems are mainly characteristic of brittle or nearly brittle rocks. They are generally not associated with ductile or plastic materials. Joints are universally present and have a highly variable geometry and spatial distribution. Few new cracks are formed by the disturbances induced by the mining of bedded deposits, and movements along pre-existing joints and cracks induce most of the subsidence.

### 4.5.4.1   *Types of joints*

Sedimentary rocks have generally 2 or 3 sets of joints, which may be classified into the following four categories:

1. Sedimentary rocks have generally 2 or 3 sets of joints, which may be classified into the following four categories.
2. Systematic joints, which intersect other joints and bedding surfaces.
3. Non-systematic joints, which terminate at other joints and bedding surfaces, and may be strongly curved in plan.
4. Cross-joints, which is a special type of non-systematic joints.
5. Bedding joints, which are parallel to the bedding surfaces and may be continuous or discontinuous.

Some of the common characteristics of joints in sedimentary rocks are as follows:

- Spacing often depends on lithology and the thickness of the bed. Denser rocks have wider spacing. Spacing also increases with depth of the bed and ranges from a few inches to several hundred feet. However, 0.3 to 3 m (1 to 10 ft) spacing is common in coal measures strata.
- In alternating sandstone-shale sequences (such as found in coal-bearing strata), joints are generally better developed in the sandstone and may be confined to it.
- In limestone, joints are generally widely spaced.
- Joints in shale are inconspicuous unless composition makes them rigid. However, shales may develop closely spaced joints near the surface.
- A minimum 35-micron opening is needed to transmit water.
- Joints may be smooth, rough, filled, "healed", or open. Openings tend to close with great depth due to weight of the superincumbent strata, or high horizontal stresses.
- Joint openings may have rough, interlocking, or polished (slickensided) surfaces or they may be coated with thin films of clay.

As a result of total extraction of a bedded mineral either by the room-and-pillar method or the longwall method, the immediate strata above the extracted area caves into the void. This induces disturbance in the strata within the zone of influence of the extracted area. The disturbance on the surface manifests itself as mining subsidence and surface cracks depending on the depth-to-seam thickness ratio, besides other factors. As regards subsurface disturbance, it is known that the strata constituting the immediate roof layers caves into the void and the layers above the caved zone are subjected to compressive and tensile stresses. The tensile strain in the vertical direction generally gives rise to bed separation, whereas in the horizontal direction it tends to open up joints in the rock formations. An understanding of these phenomena can be a great help in planning mine layouts for working under bodies of water. However, the available information regarding the following factors is insufficient at the present and needs to be investigated further:

- The manner in which the strain tends to distribute itself over the entire length of the strata which is under tensile stress, that is, whether the strain distributes over the entire length or it tends to concentrate at a few points.
- How the constraining depth effects the distribution of the horizontal strain among different joints in different strata.
- The fashion in which the spatial distribution of joints and their geometry influence development of cracks wide enough to provide water pathways.

- Whether the thickness of individual beds affect the development of cracks cutting across the bedding planes.
- Whether alternating multiple beds of shale, sandstone, and limestone are more important for development of cracks, or individual bed thickness of these materials is more important.

### 4.5.4.2 *Closure of cracks*

Some factors tend to close the cracks soon after they are induced as a result of mining operations. For working under water bodies, the closure of cracks in the constrained strata above the caved zone is of particular interest:

- *Induced Lateral Strain*: Several continuum mechanics theories have been developed for different rock behavior. These theories deal mainly with the mode of strata deformation. It may be sufficient to assume for closure of cracks, that the constrained strata, above the caved zone and below some depth from the surface, behave more or less elastically. If this is true, any fractures induced by mining tend to close up due to lateral expansion of the rock beds as a consequence of the vertical load of the overlying strata. Poisson's ratio for clays and shale is more than for sandstones and limestone, and hence their lateral expansion is greater. Plastic deformation may be more significant.
- *Natural Horizontal Stress Field*: High horizontal stress fields have been detected in some mining areas. Whenever high horizontal stresses are encountered, they help in reclosing cracks that may be induced in the strata on account of mining activity, and need to be investigated further. Excessively high horizontal stresses may aid in the disturbance of rock above caving areas by promoting breakage and deformation of stressed blocks of rock.
- *Effect of Creep*: Creep pertains to the time-dependent behavior of rock. It may be defined as an inelastic deformation of rock at some constant load below the yield point or fracture strength, considered as a function of time. The dependence or the creep rate on stress is rather complex. It has been reported that the creep deformation for a shale specimen is of the order of 20 microstrains at a stress of nearly 8,300 kPa (1,200 psi). This shows that at shallow depths, creep may not be of much significance in the closure of fractures.
- *Filling of Cracks*: Freshly exposed fractured surfaces are subjected to weathering and cracks may fill up with weathered material. The cracks also can get filled up with sand, clay, or calcite in due course of time. When working below surface bodies of water, it is important that the cracks get filled up soon after they are formed. It is presumed that materials such as ocean ooze and estuarine silt act as effective filling materials for cracks. It is, however, not desirable to depend solely on such materials to provide the safety for working below surface water bodies, such as streams and lakes, owing to their low strength and consequent lack of resistance to higher water pressures.

Subsidence causes damage to buildings, other structures (dams, bridges, impoundments, water reservoirs), utilities (pipelines, sewers), and renewable resources (agricultural and silvicultural lands, grazing pastures, aquifers and their recharge areas, aquatic life supporting regions). Most of the discussion in this section is directed towards buildings, but the same principles apply to other facilities and resources.

### 4.5.5 COMPREHENSIVE PLANNING

Comprehensive planning is critical to plan for subsidence and thus achieve sustainable mining. Planning principles are the same as for any other type of sustainable development, that is, to adopt a holistic and integrated approach, involving all stakeholders. Although simple in concept, this may be quite complex and arduous to achieve. Singh [12] has described this in depth, and highlights are provided here.

Clearly it is desirable to plan both the surface land use and the mine, with full knowledge of the requirements of each. Deep cuts for highways, railroads, or other structures, or excavations for utility tunnels or basements, may reduce the competent overburden thickness above abandoned workings to induce subsidence. This type of situation could be prevented with planning.

For planning, or any other measure to be successfully implemented, it is paramount that everyone affected by subsidence fully comprehend what is being done and why. Hence, an intensive effort of public education about the subject is in order. This should not only involve the general community, but also mine personnel, builders and developers, government officials at all levels, and civic groups.

Four situations may be identified: existing and future development with subsidence potential currently present and existing and future development in an area where forthcoming mining is planned. Each requires a slightly different approach to planning. Essentially all these approaches entail either coordination or control of both the surface and subsurface development.

### 4.5.5.1    *Coordination of surface/underground development*

Although not a comprehensive list, principles that may be followed are:

- Avoid construction near outcrops or faults.
- Build only specially designed structures over shallow workings. Surface effects are magnified as the depth decreases.
- Locate buildings above steeply dipping seams, since the strains induced are reduced.
- Erect communications or other significant structures in unmined or completely subsided areas.
- Alter routes of highways, railroads, canals, and other structures to suit mineral conditions.
- Site linear structures (canals, railroads) so that they can be uniformly lowered along their entire length.
- Avoid building important structures near mine boundaries, since coordination with several surface landowners and mine operators is onerous. Also, boundary pillars may introduce higher stresses.

Collaboration between mining companies and surface owners and developers is essential in regional or zonal planning; otherwise, problems will arise.

### 4.5.5.2    *Land use and development control*

Currently, development of land areas overlying mines must be economically justifiable, as well as socially and culturally acceptable. This implies that regional plans should not only be discussed with surface owners and mine developers, but should also be open to public comment before adoption. Changes in these plans also deserve an equally protracted treatment.

Federal, state, regional, county, and local governmental authorities exert considerable control over development of land that is potentially liable to damage due to subsidence. In the United States, this is accomplished in a number of ways: through the Surface Mining Reclamation and Control Act (SMCRA) of 1977 (Public Law 95–87); environmental impact requirements; zoning and subdivision regulations; building provisions (issuance of permits); mining regulations; safety requirements; insurance needs; investigative requirements for public buildings (for example, Act 17 of 1972 in Pennsylvania); special local ordinances; and interagency coordination. Unfortunately, this fragments the individual actions and the overall plan may not get the attention it deserves. It is also difficult for citizen stakeholders to understand this complex set of rules and requirements during public hearings.

Perhaps it can be mandated that mine operators prepare plans that depict predicted subsidence locations, extent, trough centers, maximum subsidence, values and direction of tilt, compression and extension zones, and other pertinent data. These could then be circulated to building authorities, highway commissions, railroads, water supply and other utility agencies, pipeline operators, and others who may be affected, for comments and suggestions. On the other hand, these groups as well as builders/developers should be required to incorporate proper precautions in the design of their respective structures. In extreme cases, construction may be barred from particularly risky areas. These lands could be used for parks, forest preserves, and open spaces.

An approach that could be readily implemented would be one that would be particularly applicable to flat prime-farmlands (for example, in the Illinois basin in the U.S.) is that more of the coal be mined by longwall procedures, so as to produce large troughs. Then, as the mining operators move away, the structural integrity of the barrier pillars could be destroyed. The surface drainage could then be modified, to remove water that may collect because of the high water table. The productivity of this land should not be affected significantly, since no disturbance of the various layers occurs, except for an elevation drop. Current room-and-pillar methods develop "dips" and "humps" over a period of time, and serve as troughs or gullies to collect water, impairing productivity. Besides, considerable amounts of valuable coal are sterilized by the method. After complete subsidence with a full extraction method, there will also be little opportunity for acid water drainage. Illinois mines are generally not gassy, so removing the pillars should not pose any major problems. Structures that are built in the post-subsidence period will remain stable.

If the full economic impacts of mining were taken into account, there will be more incentive for the various stakeholders in mining, mining personnel, shareholders, insurance companies, landowners in the area, developers, government agencies, tax payers, and the citizenry at-large, to present creative methods of mining without permanent damage to the surroundings.

In developing sustainable mining procedures and make comprehensive planning possible, it is essential to understand the techniques available for minimizing the effects of mine subsidence.

## 4.5.6  SUBSIDENCE CONTROL METHODS

There are three types of measures that may be adopted to control subsidence damage: alterations in mining techniques, post-mining stabilization, and architectural and structural design. These techniques encompass several methods and are discussed below.

### 4.5.6.1  *Alterations in mining techniques*

#### 4.5.6.1.1  *Partial mining*
This may be accomplished in a number of ways. First, leaving protective zones is the most common procedure in use (Figure 4.15). The zones may entail:

- leaving the entire pillar unmined beneath structures, such as factories, railroads, major highways, and bodies of water
- partially extracting the pillar and then backfilling
- room-and-pillar mining, with up to 50% extraction – a practice common in Pennsylvania coal mines and recommended in the regulations – this method does not account for pillars deteriorating with time, especially if the mine is flooded.

Any structure supported by a protective zone is liable to become perched at a higher level than the surrounding ground, after it subsides. This may not affect railroads or highways, but could disturb the utilities to a building. An island may form if the water table is high.

Second, use of sized pillars – pillar width between longwall panels should be adjusted so as to uniformly lower the ground surface (Figure 4.16).

Third, mining subcritical widths in order to reduce the maximum subsidence.

### 4.5.6.1.2  *Backfilling*
This may be done using hydraulic or pneumatic techniques. It reduces the amount of subsidence, but does not eliminate it entirely. It is a very effective method of mitigating subsidence effects, since it not only minimizes the ground deformation forces, but also conserves the hydrologic regime. Cost-effectiveness studies should consider the beneficial effect on the environment, such as reduced acid-water drainage, savings in waste disposal and reforestation, prevention of refuse fires, reduced ground fissuring, and escape of mine gases, as well as the advantages of long-term

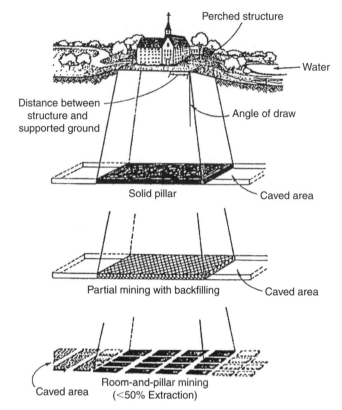

Figure 4.15.   Methods for protecting surface structures.

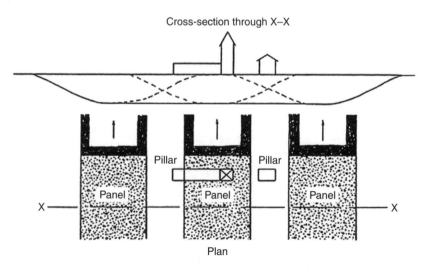

Figure 4.16.   Uniform lowering of the surface with sized pillars.

strata stability and decreased roof support. Backfilling may become essential in flat regions with a high water table to prevent flooding, and in areas reclaimed from water bodies.

### 4.5.6.1.3   *Harmonious mining*
The technique entails superimposing compressive surface strains on the tensile strains induced by another longwall face, in a manner that they move along together. This may be accomplished by staggering two simultaneously-worked faces that advance at the same rate, with multiple mineral beds where one face is superjacent over another and in single seams, where the panels adjoin.

It is evident that total cancellation of the traveling strains can only occur if the displacement curves are congruent and symmetrical – the bed thickness, influence factors, width of compressive and tensile zones, and stowage density (if backfilling is adopted) are identical. Time factors for the mining sequence must also be available from prior experience.

### 4.5.6.1.4   *Mine layout or configuration*
Layout controls the strains experienced by the structure. It may be possible to locate the panel with respect to the building in such a manner as to expose it to deformations that it can withstand (Figure 4.17). In some cases, it may be best to leave material to support the building.

### 4.5.6.1.5   *Extraction rate*
Face advance cannot be readily altered in mining, and its range is generally limited with the available equipment. A faster rate is desirable in unfractured, viscoelastic strata, because it lowers the tensile peak and moves it closer toward the working face. However, in fractured, clastic rocks (such as over previously mined beds), rapid face advance may accentuate displacements and strains and, thus, induce greater damage.

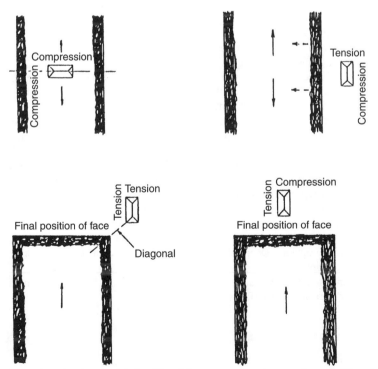

Figure 4.17.   Sketches showing the relationship of the nature of strains experienced with a structure with respect to face layout.

#### 4.5.6.2   *Post-mining stabilization*

These techniques are most commonly used in the U.S. They may extend over large areas, or be restricted to support a specific structure.

##### 4.5.6.2.1   *Stabilization of large areas*
The four main methods used include:

- *Backfilling*: This is done by mechanical, hydraulic or pneumatic means. In each case, the procedure may be controlled, when the mine is accessible and barricades can be manually built, or be remote (blind), through boreholes when the openings cannot be entered, such as in abandoned mines.

   With hydraulic stowing, the water level may rise temporarily in dry mines, acid water may be flushed out into the hydrologic system, and surface drainage may be affected by siltation, pollution, or flooding (especially for shallow mines). The technique is usable in water-filled mines.

   The Dowel process (GAI [13]) is a special blind hydraulic flushing technique where the slurry is pumped at a high velocity. The mixture deposits its load when the velocity drops on entering the mine cavity, forming a doughnut-shaped pile. As the pile height nears the mine roof, the slurry velocity in the gap increases. This keeps the solids in suspension longer so that the pile grows outward.

   Pneumatic stowing causes considerable sparking and may pose a hazard because of the potential for gas ignition.

   Often mine tailings may be flushed back into the mine to rid the surface of unsightly piles. This should be regarded more of a waste disposal method than backfilling, since little attention is paid to the degree of compaction. In coal areas, a commonly used material for backfilling, both hydraulic and pneumatic, is fly ash. This is because of its abundance at coal-fueled power plants.
- *Grouting*: This method entails the use of cementitious mixtures and provides stronger support. Additives used include Portland cement, pozzolanic mixtures (including fly ash), or organic compounds.

   Gravity grouting is used to simply fill the mine voids to whatever extent possible. There is little control, although a perimeter wall may first be built with a thick grout, which is then filled with an expansive grout to achieve good roof contact.

   Pressure grouting is needed if a number of joints need to be filled or roof caving has occurred.

   Bag grouting entails lowering a bag through a 150-mm-diameter (6-in.-diameter) borehole, and filling that with grout until roof contact is obtained.
- *Excavation and Fill Placement*: It is only feasible in shallow abandoned mines with no surface obstructions to excavation. The entire overburden and coal are removed and replaced with compacted fill. Flooded mines may yield large quantities of acid water.
- *Blasting of the Roof and Floor to Fill the Cavity*: This a patented technique (U.S. Patent No. 1 004 419) that has not been used often. Over time, the broken rock would compress, but the movements may be expected to be gradual and evenly distributed.
- *Site-Specific*: These techniques are mostly used to support isolated structures. Grout columns may be built remotely, but floor and column strengths are variable. Water may impede construction. Piers or cribs may be constructed in mine openings that are accessible, if the mine floor and roof are competent. Deep foundations may be used with shallow workings. They are, however, liable to damage by lateral shear forces that may be experienced.

   Grout column entail placing casing between the mine roof and floor, and filling it with grout. These supplement existing coal pillars.

#### 4.5.6.3   *Architectural and structural design*

##### 4.5.6.3.1   *Orientation*
It is preferable to have the long axis of the building parallel to the subsidence contours. If a fault exists nearby, the shorter axis should be oriented perpendicular to the fault.

#### 4.5.6.3.2   *Location*

Faults tend to concentrate ground strains. Hence, structures should be located at least 15 m (50 ft) away.

A single building should not be constructed on dissimilar soils, owing to the possibility of differential deformations or settlements.

#### 4.5.6.3.3   *Subsidence-resistant construction*

The basic principles are presented below. There are four major construction categories under which these may be considered.

- *Rigid*: In this case, both the foundation and superstructure are rigid in design. Often the foundations are highly reinforced concrete rafts or beams, capable of withstanding ground displacement and curvature. The structures generally span or cantilever over a subsidence wave. Foundations are of small plan area. Elevator shafts and the like are designed with extra clearances.
- *Flexible*: Often slab foundations are adequate for small buildings, such as houses, in this type of design. The slab should preferably be less than 20 m (60 ft) along the side, poured in a single operation, without joints, and finished close to ground level. It is generally underlain by granular material. Reinforcement should be near both the top and bottom so as to accommodate tensile and compressive strains.

  If the building has a basement, there should be an open gap around it or filled with a compressible or granular material. Larger buildings may have rollers or slip joints between the superstructure and foundation. Trenches around structures absorb some of the strains.

  Flexible structures are designed to track the traveling subsidence wave without cantilevering, permit free ground movement below the foundation, provide sufficient superstructure support in spite of the ground flexing, and accommodate subsidence deformations that are larger than anticipated without jeopardizing structural stability. An often-quoted design is that developed under the Consortium of Local Authorities Special Program (CLASP) in the UK, but several others exist.
- *Semi-flexible*: In instances where the structures can tolerate minor damage, it maybe possible to use designs and construction techniques that do not strictly adhere to the rigid or flexible criteria outlined above. It may be more cost-effective to perform minor repairs as required.
- *Use of Releveling Devices*: Many buildings can be provided with jacks to prevent tilting. Excessive tilt may cause the gap between them to be reduced to the extent that they touch.

  Gaps need to be provided between all buildings to allow for both compression and tilt. Other precautions that are helpful, depending on design philosophy, include:
  - Provide expansion joints, to accommodate ground movements and thermal expansion.
  - Minimize the number of door and window openings. Use flexible frames. Their location should not significantly weaken the structure. Do not position front and back doors opposite each other.
  - Avoid weak skin materials within rooms. Partitions between building segments should be strong. Instead of plaster on ceiling and walls, use plasterboard.
  - Floors and roofs should be secured to walls.
  - Allow for tensile strains at all structural connections. Movements should be possible for staircases.
  - Exclude masonry arches.
  - Do not have corner or bay windows, or porches.
  - Detach outbuildings from main building.
  - Provide excessive falls for gutters.
  - Do not pave immediately adjoining buildings. Have planting areas or gaps instead. Use bituminous type materials for paving where necessary (for example driveways and parking lots).
  - Use flexible damp-proof courses (such as bitumen).
  - Use light fences around properties, rather than walls.
  - Replace rigid retaining walls with earth banks.

- *Modification of Existing Structures*: Total repair expenses may sometimes be reduced if a building is suitably modified before experiencing ground movements. Possible alterations include:
  - Cutting out a part of a house or removing an entire house from a row of buildings. Unit lengths should be about 20 m (60 ft) with cuts extending into trenches and gaps bridged with flexible materials. Preferably locate cuts in connection corridors or unit divisions.
  - Digging trenches around building (and filling with compressive material weaker than the surrounding soil) to below foundation level, without disturbing the foundation. Trenches may be covered, if desired, with concrete slabs that do not butt.
  - Slotting rigid pavements or floors, and even superstructures (generally wood, brick, or stone do not present difficulties; concrete may).
  - Introducing artificial slip planes, especially in new buildings.
  - Providing temporary supports or strengthening to parts susceptible to damage. Support screens, partitions, and ornaments independently of the walls and floor.
  - Using tie rods, if it is anticipated that the roof trusses will be pulled out from their seats. However, indiscriminate use of tie rods may needlessly disfigure the building. Stress concentrations at tie-rod bearing plates may pull these through the walls. Often temporary corbels provide adequate support for trusses.
  - Installing pre-tensioned steel mesh around the exterior walls (this could be dismantled and re-used later).
  - Taping windows (especially with metal frames) to avoid flying glass.
  - Removing and storing stained glass windows until subsidence is complete.
- *Remedial and restorative measures*: Constructing structures that are easily repaired after subsidence damage is becoming more common. Since a tension wave is usually followed by a compression wave, cracks should not be patched until all movements have stopped. Debris in the fractures should, however, be removed before compression.

  In low-lying areas, the water table may create difficulties, necessitating the installation of drains and pumps.

## 4.5.7   SUBSIDENCE FROM ABANDONED MINES

The physical principles that govern subsidence in abandoned mines are the same as for active mines. There are a few restrictions that come into play. The mine workings are generally not accessible. Hence remote methods of stabilization have to be adopted. Mostly this implies some method of backfilling (Bureau of Mines [14]). In the U.S. this procedure has been successfully used in Streator, Illinois, for the foundation of a water treatment facility, and for several residences in Indiana. Another major aspect that needs to be considered in the case of abandoned mines is the source of funding for rehabilitation of the property. Often, most of the mining has been done in the past and the mine operators are no longer available to fund the restoration. This leaves the State as the source of funding. In the U.S. a tax on currently mined coal has been used to mitigate such problem areas. There are also examples of surface mines that have been rehabilitated for productive use. One abandoned mountaintop operation in Kentucky has been turned into a landing area for small planes. Another old strip mine in central Illinois has been turned into a lake and stocked for fishing. In the U.S. regulations require that currently operating mines follow closure procedures, and do not result in future problems.

Another aspect of concern with abandoned mines is pollution of the hydrologic regime. If the mines are full of water, it generally becomes acid, generally because of sulfur containing minerals. This then gradually seeps into nearby waterways. Some attempts have been made to seal off the mines, but this is difficult and may not be successful. Invariably the water finds alternate routes. A method of treating the water as it exits through a controlled outlet needs to be installed. Dewatering of the mine voids would cause surface settlements and subsidence.

If large areas have been completely mined without any intermediate support, the surface may already be completely settled. Then surface structures would remain relatively stable. However, if pillars or other supports have be left in the mines, these will continue to deteriorate and sudden impacts may be experienced on the surface.

## REFERENCES

1. Kennecott Minerals Company (Kennecott), 2001 Social and Environment Report, Kennecott Minerals Company Sustainable Development Policy, April 2002. www.kennecottminerals.com.
2. Bureau of Land Management (BLM), Sustainable Development and its Influence on Mining Operations on Federal Lands – A Conversation in Plain Language, U.S. Department of Interior, Washington, D.C., 2002, 4 p.
3. World Bank, Fuel for Thought: An Environmental Strategy for the Energy Sector, Report No. 20740, World Bank Group, Washington, D.C., 2000, 138 p.
4. Fox, F.D., Flambeau and Ridgeway Mines – Lessons Learned, SME Annual Meeting Preprint, 03-141, Cincinnati, OH, Feb. 2003
5. Langer, W.H., Giusti, C. and Barelli, G., Sustainable Development of Natural Aggregates with Examples from Modena Province, Italy, SME Annual Meeting Preprint 03-045, Cincinnati, OH, Feb. 2003
6. Botts, S.D., The Antamina Project – The Challenge of Sustainable Development in Peru, SME Annual Meeting Preprint 03-033, Cincinnati, OH, Feb. 2003.
7. Fejes, A.J., Subsidence Information for Underground Mines – Literature Assessment and Annotated Bibliography, U.S. Bureau of Mines Information Circular 9007, Pittsburgh, PA, 1985, 86 p.
8. Singh, M.M., Mine Subsidence, Section 10.6, SME Mining Engineering Handbook, 2nd Ed., Vol. 1, SME, Littleton, CO, 1992, pp. 967–971.
9. Gray, R.E., Bruhn, R.W. and Knott, D.L., Subsidence Misconceptions and Myths, Proc. 15th Conference on Ground Control in Mining, Colorado School of Mines, Golden, CO, Aug. 1996, 10 p.
10. Singh, M.M., and Kendorski, F.S., Strata Disturbance Prediction for Mining Beneath Surface Water and Waste Impoundments, Proc. First Conference on Ground Control in Mining, West Virginia University, Morgantown, WV, 1981, pp. 76–89.
11. Lee, F.T. and Abel, J.F., Subsidence from Underground Mining: Environmental Analysis and Planning Considerations, U.S. Geological Survey Circular 876, 1983.
12. Singh, M.M., Review of Coal Mine Subsidence Control Measures, Trans. SME of AIME, Vol. 276, SME, Littleton, CO, 1985, pp. 1988–1992.
13. GAI, State of the Art of Subsidence Control, Appalachian Regional Commission Report ARC-73-111-2550, Dec. 1974, 256 p.
14. Bureau of Mines, State of the Art Techniques for Backfilling Abandoned Mine Voids, Information Circular 9359, U.S. Department of Interior, Washington, D.C., 1993, p. 17.

## 4.5.8    Use of Time Domain Reflectometry in monitoring subsidence

Kevin M. O'Connor
*GeoTDR, Inc, Westerville, Ohio*

### 4.5.8.1    INTRODUCTION

This discussion is focused on applications of Time Domain Reflectometry (TDR) in the mining industry particularly monitoring of subsidence. This technology has been used to monitor strata movements over active and abandoned mines for purposes of determining subsurface changes that occur prior to surface subsidence, monitoring slope movement, O'Connor 1999 [15], O'Connor 1995 [16], and monitoring mine roof strata separation, O'Connor 1991 [17]. Automated remote operation is particularly applicable to post-closure monitoring where savings can be made in personnel costs and alarm systems can be implemented.

TDR was developed by the power and telecommunications industries to locate faults in cables. It is a form of RADAR in which voltage pulses are transmitted along coaxial cable, and reflections are created at every location where the cable is deformed such as the crimp shown in Figure 4.18. The basic hardware consists of a coaxial cable embedded in the ground being monitored and a TDR cable tester to interrogate the cable. The cable generally used for monitoring subsidence is a solid aluminum 22.2 mm (7/8 in.) diameter coaxial cable, CommScope General Instrument [18].

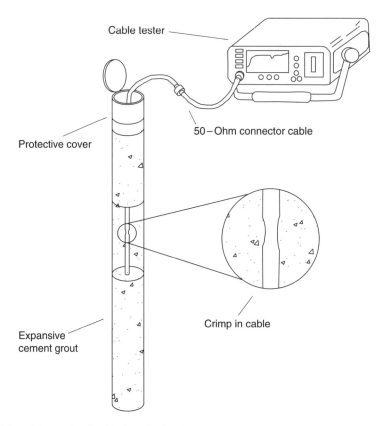

Figure 4.18.    Schematic of cable installation in a vertical borehole.

Prior to installation, the cable is crimped as shown in Figure 18, to provide reference reflections at known physical locations. In vertical and inclined borehole installations, the crimped cable is lowered into the hole and then bonded to the surrounding rock with an expansive cement grout that is tremied into the hole. In trench installations, the cable is placed along the bottom of the trench (typically 1 m (3 ft) deep) and bonded to the surrounding soil with a sand/cement backfill suffi-cient to cover the cable, and then the trench is backfilled with crushed aggregate. When ground movement is sufficient to fracture the grout, cable deformation occurs that can be monitored with a TDR cable tester, O'Connor 1999 [15].

The time for a pulse to reach a cable fault and be reflected back to the TDR unit is converted to distance by knowing the propagation velocity of the cable. Consequently, it is possible to display all reflections and identify the type and location of cable damage. Cable faults can be located with an accuracy of about 2% of the distance from cable tester to fault. For example, a fault at a distance of 150 m can be located with an accuracy of approximately 3 m. This accuracy can be improved by crimping a cable at some desired spacing. Since the reflections caused by crimps are at known physical locations, these reflections provide control markers in the TDR records and cable damage can be located to within about 2% of the distance between reference crimps.

Not only is it possible to locate movement but also to distinguish shearing from tensile defor-mation and to quantify the magnitude of deformation, Dowding 1998 [19]. If the cable is subjected to localized shearing, the TDR reflection is a characteristic spike. The magnitude of each TDR reflection spike will increase as progressive ground movement continues to deform the cable. Based on calibrations that have been developed in the laboratory and at field sites, the magnitude of TDR reflections can be converted to magnitude of cable deformation. By acquiring TDR records for a cable over time, it is possible to track the location, magnitude, and rate of cable defor-mation caused by the progressive ground movement.

### 4.5.8.2   REMOTE MONITORING

A remote modem, datalogger and TDR cable tester, Campbell Scientific [20], are used to acquire and store data (Figure 4.19). Data can be processed by the datalogger and also downloaded to any

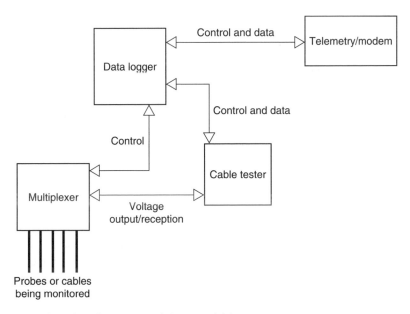

Figure 4.19.   Flow chart for automated data acquisition system.

convenient location over a phone line. This capability allows for analysis and interpretation of TDR waveforms. As a consequence of this remote monitoring capability, TDR technology has made it possible to implement real time monitoring systems with a call back alarm capability.

A site is divided into sectors and each cable within a sector extends to a data acquisition system (DAS) for that sector. Each cable is connected to a coaxial multiplexer installed within an enclosure. The multiplexer and TDR cable tester are controlled by a datalogger. The datalogger is also attached to a storage module and modem so each DAS is accessible remotely via a phone connection. The datalogger interrogates each cable once during a preset time interval (e.g., once every three hours) and stores the measurements. The datalogger is programmed to compare the current waveform point-by-point against the stored baseline waveform for each cable. If the datalogger detects a change from the baseline value that exceeds a preset alarm threshold, the datalogger initiates a phone call to assigned responsible personnel.

A default value for the alarm threshold is established by experience. This value is set independently for each cable and it was adjusted periodically as responsible personnel gain experience with the system and changing site conditions. For example, it would be temporarily increased for a cable during periods when noise levels on that cable are high and then lowered as conditions stabilized. The objective is to keep the threshold as low as possible to maximize sensitivity to changes from the baseline waveform.

In response to alarm calls from the TDR monitoring system, it is vital to isolate the cause. It is necessary to download and analyze TDR data files to determine if the alarm condition is associated with ground movement and the rate at which movement is occurring. Based on this information, it is possible to make an informed decision when talking with personnel on site about the appropriate plan of action. Once it is determined that the alarm call was triggered by a condition that is persistent, it is typical to validate the condition using on-site manual interrogation of the coaxial cable, identify the physical location, conduct a visual assessment, and implement the action plan developed by project personnel.

### 4.5.8.3    EXAMPLES OF MINE SUBSIDENCE MONITORING

The pits, sags, and troughs that develop on the surface over underground mines are the ultimate result of subsurface strata movements. In the case of important structures located over mines, it is not enough to say that subsidence is imminent. In some critical areas, a means must be provided to detect strata movements and quantify the rate at which they are occurring so that appropriate measures can be taken to mitigate damage. TDR monitoring of cables embedded in strata overlying mines provides a means of locating and quantifying subsurface deformation which is a precursor to surface subsidence, Dowding 1994 [21], O'Connor 1997 [22].

Note that this approach is tailored to monitoring localized occurrence of subsidence beneath important structures. In order to assess the risk of subsidence over an entire mine, GIS techniques must be used to integrate the data acquired by TDR and data acquired by surface survey monitoring. However, for purposes of intensive local monitoring, TDR is listed in the Ohio Department of Transportation risk assessment manual, Ruegsegger 1997 [23], as one of the available state-of-the-art technologies that is acceptable.

#### 4.5.8.3.1    *Interstate Highway I-70, Cambridge, Ohio*

A need for such monitoring occurred when the eastbound driving lane of Interstate Route 70 in Guernsey County, Ohio collapsed in March 1995. Subsidence was initiated in response to dewatering of an underlying abandoned coal mine. In spite of periodic visual monitoring, a 5 m diameter hole, 3 m deep suddenly opened and three cars and a truck were damaged, Ruegsegger 1999 [24]. The remediation effort by the Ohio Department of Transportation involved air rotary drilling of approximately 1,800 boreholes down to mine level at a depth of approximately 20 m to tremie approximately 18,000 m$^3$ of fly ash grout into the voids. Two land bridges, having lengths of 215 m and 34 m, were constructed over areas where the drilling and grouting program encountered high concentrations of caved and broken material in the mine interval. It was decided to monitor, rather than grout, remaining voids. In January 1996, two coaxial cables were grouted into holes drilled down to mine level. Data collection is automated and data plots can be viewed on the project internet web page.

### 4.5.8.3.2    *Interstate Highway I-70, Washington, PA*

Two longwall coal mine panels were mined at a depth of approximately 156 m ( 520 ft) beneath I-70 east of Washington, Pennsylvania, such that the highway crossed the width of one panel at two locations, O'Connor 2001 [25]. The Pennsylvania Department of Transportation assumed responsibility for real-time monitoring of both ground deformation and changes in highway conditions. Innovative monitoring of ground deformation was accomplished with TDR to interrogate coaxial cables installed in seven deep holes and an array of 32 tiltmeters along the highway shoulder. Tiltmeters were connected to a central remote data acquisition system that automatically recorded and stored measurements. When specified tilt values were detected, the system initiated a phone call to key Pennsylvania DOT personnel, who then monitored tiltmeter measurements in real time via a phone-line connection. Based on the TDR measurements, the lateral extent of subsurface movement was over 300 m from the active longwall mine face well outside the limits of the conventionally accepted angle of draw.

### 4.5.8.3.3    *Interstate Highway I-77, Akron, OH*

Ohio DOT encountered abandoned underground coal mine subsidence during an investigation conducted as part of a project to widen I-77 in Summit County, Ohio, O'Connor 2002 [26]. A decision was made to stabilize 2000 feet of four-lane highway by backfilling the mine voids with cement grout while traffic remained on the interstate. Due to the potential for additional mine subsidence during remediation work, Ohio DOT required the installation of a real time monitoring system to activate an alarm when movement or settlement of the road base exceeded threshold values. TDR monitoring cables were installed by a directional drilling contractor in horizontal holes drilled 5 feet or more beneath each lane of the entire section of highway. The cables were connected to a central remote data acquisition system that automatically recorded and stored measurements. The initial record for each cable was stored as the baseline measurement. When the magnitude of any TDR reflection differed from the baseline value by a specified amount, the system initiated a phone call to project personnel who could then analyze TDR measurements in real time via a phone line connection to verify that the alarm was associated with cable deformation. If it was determined that actual movement of soil had occurred, Ohio DOT personnel were alerted to intensify visual reconnaissance and determine if lane closures were necessary. This project demonstrated the capability of real time monitoring of ground movement over a wide area utilizing a horizontal application of TDR technology.

### 4.5.8.3.4    *State Route 91, Saltville, VA*

Closure activities at an inactive mine in Saltville, Virginia include realignment of an existing highway outside the predicted limits of long term subsidence, O'Connor 2003 [27]. Concerns about the possibility of subsidence along the highway being induced by construction activities, or in the former plant area where excavated rock is being placed, motivated the installation of a real time monitoring system. Rapid development of sinkhole features is of particular concern because they can pose nearly non-predictable and sudden safety hazards. The goal is to provide an early warning system for catastrophic subsidence.

Manual monitoring of three TDR cables began in June 2001. Since then, the system has been expanded to automated real time monitoring of 21 cables with lengths that vary from 10 m (30 ft) to 270 m (886 ft) for a total of over 2500 m (8200 ft) of cable. They are installed in angled holes beneath the existing highway, in trenches along the highway, and in trenches over the former plant area where rock excavated for the new alignment is being placed. The sensitivity of TDR to rock mass movement is related to the proximity of cable installations to subsiding ground. The lateral extent of monitoring was enhanced by installation of cables in trenches over a wide area, and the depth of monitoring was enhanced by installation of cables in deep vertical or angled holes beneath critical structures.

The automated monitoring incorporates a call back alarm capability. Whenever the difference between the baseline profile and the current profile at any location along a cable exceeds a preset alarm threshold, the datalogger will initiate a call to responsible personnel.

### 4.5.8.4   SUMMARY

The principle of TDR has proven to be economical, efficient, and resistant to sabotage for monitoring rock mass response at active and abandoned mines. Significant improvements have been made since this technology was first used for geotechnical applications. Cable types that reduce costs have been identified; the database of TDR records and experience with interpretation of TDR reflections has increased; digital TDR records can be obtained; telemetry for remote data acquisition is available; and automated monitoring with call back alarm capabilities is in operation at several inactive mines.

### REFERENCES

1. Kennecott Minerals Company (Kennecott), 2001 Social and Environment Report, Kennecott Minerals Company Sustainable Development Policy, April 2002. www.kennecottminerals.com
2. Bureau of Land Management (BLM), Sustainable Development and its Influence on Mining Operations on Federal Lands – A Conversation in Plain Language, U.S. Department of Interior, Washington, D.C., 2002, 4 p.
3. World Bank, Fuel for Thought: An Environmental Strategy for the Energy Sector, Report No. 20740, World Bank Group, Washington, D.C., 2000, 138 p.
4. Fox, F.D., Flambeau and Ridgeway Mines – Lessons Learned, SME Annual Meeting Preprint, 03–141, Cincinnati, OH, Feb. 2003.
5. Langer, W.H., Giusti, C., and Barelli, G., Sustainable Development of Natural Aggregates with Examples from Modena Province, Italy, SME Annual Meeting Preprint 03–045, Cincinnati, OH, Feb. 2003.
6. Botts, S.D., The Antamina Project – The Challenge of Sustainable Development in Peru, SME Annual Meeting Preprint 03–033, Cincinnati, OH, Feb. 2003.
7. Fejes, A.J., Subsidence Information for Underground Mines – Literature Assessment and Annotated Bibliography, U.S. Bureau of Mines Information Circular 9007, Pittsburgh, PA, 1985, 86 p.
8. Singh, M.M., Mine Subsidence, Section 10.6, SME Mining Engineering Handbook, 2nd Ed., Vol. 1, SME, Littleton, CO, 1992, pp. 967–971.
9. Gray, R.E., Bruhn, R.W., and Knott, D.L., Subsidence Misconceptions and Myths, Proc. 15th Conference on Ground Control in Mining, Colorado School of Mines, Golden, CO, Aug. 1996, 10 p.
10. Singh, M.M., and Kendorski, F.S., Strata Disturbance Prediction for Mining Beneath Surface Water and Waste Impoundments, Proc. First Conference on Ground Control in Mining, West Virginia University, Morgantown, WV, 1981, pp. 76–89.
11. Lee, F.T., and Abel, J.F., Subsidence from Underground Mining: Environmental Analysis and Planning Considerations, U.S. Geological Survey Circular 876, 1983.
12. Singh, M.M., Review of Coal Mine Subsidence Control Measures, Trans. SME of AIME, Vol. 276, SME, Littleton, CO, 1985, pp. 1988–1992.
13. GAI, State of the Art of Subsidence Control, Appalachian Regional Commission Report ARC-73-111-2550, Dec. 1974, 256 p.
14. Bureau of Mines, State of the Art Techniques for Backfilling Abandoned Mine Voids, Information Circular 9359, U.S. Department of Interior, Washington, D.C., 1993, 17 p.
15. O'Connor, K.M. and Dowding, C.H. *GeoMeasurements by Pulsing TDR Cables and Probes*. CRC Press, Boca Raton, ISBN 0849305861, 1999.
16. O'Connor, K.M., Peterson, D.E. and Lord, E.R. Development of a Highwall Monitoring System using Time Domain Reflectometry, Proc., 35th U.S. Sym. Rock Mech. Reno, Nevada, June, pp. 79–84, 1995.

17. O'Connor, K.M. and Zimmerly, T. Application of Time Domain Reflectometry to Ground Control. Proceedings of the 10th Int. Conf. on ground Control in Mining, Morgantown, WV, June 10–12, pp. 115–121, 1991.

18. CommScope General Instrument, Trunk & Distribution Cable Catalog, Hickory, North Carolina, 1995 http://www.commscope.com/html/db_trunk.shtml

19. Dowding, C.H., Su, M.B. and O'Connor, K.M. Principles of Time Domain Reflectometry Applied to Measurement of Rock Mass Deformation, Int. J. Rock Mech. and Min. Sci., v. 25, No. 5, pp. 287–297, 1988.

20. Campbell Scientific, Inc., Time Domain Reflectometry System, Logan, Utah, 2002. http://www.campbellsci.com/tdr.html

21. Dowding, C.H. and Huang, F.-C. Telemetric Monitoring for Early Detection of Rock Movement With Time Domain Reflectometry. J. Geotech. Eng., Am. Soc. Civ. Eng., v. 120, No. 8, pp. 1413–1427. 1994.

22. O'Connor, K.M. and Murphy, E.W. TDR Monitoring as a Component of Subsidence Risk Assessment Over Abandoned Mines, Int. Journal of Rock Mechanics & Mining Science, Vol. 34, Nos. 3–4, Paper 230, 1997.

23. Ruegsegger, L.R., Abandoned Underground Mine Inventory and Risk Assessment Manual, Ohio Department of Transportation, Office of Materials Management, Geotechnical Design Section, Columbus, August, 150 p., 1997. http://www.dot.state.oh.us/mines/MineManual/HTML/Main.htm

24. Ruegsegger, L.R. and Lefchik, T.E. Managing Car-Crunching Sinkholes. Public Roads, July/August Vol. 63, No.1, 1999 http://www.tfhrc.gov/pubrds/julaug99/minehole.htm

25. O'Connor, K.M., Clark, R.J. Whitlatch, D.J. and Dowding, C.H. Real Time Monitoring of Subsidence along I–70 in Washington, Pennsylvania. Transportation Research Record No. 1772, Soil Mechanics 2001. http://64.118.69.9/acb1/showdetl.cfm?&DID = 92&Product_ID = 6093&CATID = 1&series = 1.

26. O'Connor, K., Reugsegger, R. and Beach, K. Real Time Monitoring of Subsidence Along Interstate I–77, Summit County, Ohio. Fourth Biennial Interstate Technical Group on Abandoned Underground Mines, Davenport, Iowa May 1–3, 2002, http://www.fhwa.dot.gov///mine/oconnor.htm

27. O'Connor, K.M., Crawford, J. Price, K. and Sharpe, R. Real Time Monitoring of Subsidence Over Abandoned Mines in Virginia using TDR. 25th Annual Meeting, National Association of Abandoned Mine Land Programs, Louisville, KY, September, 2003, http://www.surfacemining.ky.gov/NR/rdonlyres/3122CEAB-6256-4BF9-A40C-8747415C49A7/0/MonitoringSubsidenceVA.pdf

# 4.6 Role of environmental indicators in mining operations

Subijoy Dutta

*S&M Engineering Services, Crofton, MD, USA*

## 4.6.1 INTRODUCTION

Watersheds and airsheds are the major focus of various environmental programs in United States federal, state, and local agencies today. A significant part of mining regulations are designed for protection of the environment. Environmental management thus takes up a significant fraction of the resources. Regulatory requirements of sampling and analysis of mine-drainage spoil piles, and emissions are on the rise with the advancement of technology. These requirements are often hard to comply with, especially for the small miners and quarry operators. A healthy environment is often reflected by a few common indicators, which could be easily observed and monitored at a much lower cost and in a more definitive manner.

As the urban sprawl continues its stride towards suburban areas, health of local watersheds is becoming increasingly important to the resident population because of the dearth of clean and pollution-free environment. In response to this and other water quality concerns, the U.S. Environmental Protection Agency (EPA) has stepped forward with a revision to the regulatory requirement of the Total Maximum Daily Loads (TMDL) under section 303(d) of Clean Water Act (CWA). Amendments to regulations 40 CFR part 122.4(i) and 40 CFR part 130 were made to provide states, territories, and authorized tribes with the necessary information to identify impaired water bodies and to establish TMDLs to restore water quality. States, territories, and authorized tribes establish the section 303(d) lists of impaired waters and submit to EPA for approval of the lists or to add waters to the submitted lists, if EPA determines that the list is not complete. TMDLs have been applied to several mining sites because of the causal link between the impaired water body and the source of pollution. The most common examples of mining sites where TMDLs have been applied include the following sites from EPA Regions 8 and 10 (USEPA [1]):

- Animis River
- Coeur d'Alene site
- South Dakota Black Hills area (e.g. Whitewood Creek)
- Colorado's Cripple Creek
- West Fork of Clear Creek.

The top three causes of impairment of water bodies nationally include sediments, nutrients, and pathogens (USEPA, 2000). Mine runoffs are oftentimes linked with high sediment concentrations. As a result, the surface water sampling and monitoring requirements in mining operations have been increased in certain areas. For example, in case of greater water quality concerns, the State of Minnesota sometimes requires the sand and gravel operators, and rock quarries to submit water quality analyses including pH, maximum turbidity, total suspended solids (TSS), total phosphorus, and other potential pollutants, at certain times during their operation, in addition to their initial permit application and renewal. The cost of these additional sampling and monitoring could be formidable for small operators. Also, some operators are unable to perceive the broad picture involving the local watershed and fail to employ the most effective control for silts and sediments in the mine runoff.

This paper examines the use of various environmental indicators to provide a more comprehensive picture to the mine operators and communities in a watershed concerning the health of their watershed and possible signs of ecological "brown spots". By using these indicators, all of the stakeholders from the watershed can develop an integrated approach for protecting the health of their watershed. This should lead to a more sustainable growth and development in that watershed as compared to the standard regulatory mandates.

## 4.6.2   USE OF INDICATORS BY THE LOUDOUN COUNTY ENVIRONMENTAL INDICATOR PROJECT (LEIP)

To study the environmental impact of proliferating growth and development in Loudoun County, Virginia, various indicators of change are studied by LEIP, under the auspices of the Geography Department at the George Washington University. There are fifteen different indicators used by the LEIP. They are Roadside imagery, Aerial imagery, Digitized imagery, Conventional map imagery, Forest areas, Agricultural lands, Wetlands, Riparian areas, Impervious surfaces, Urbanized areas, Listed plant species, Key soil types, Water quality, Air quality and Historic and cultural sites. These indicators are monitored at various sites in the county and the changes studied and analyzed. Seventeen sites throughout the county were selected for long term monitoring. These sites represent a wide variety of environmental conditions ranging from forest areas and riparian zones to shopping centers and residential areas of varying densities. Amongst the findings on various indicators as (LEIP [2]) the following few sample observations from a few indicators are summarized here.

### 4.6.2.1   *Roadside imagery*

Continuous changes to familiar landmark or its setting to a new form have been routinely observed in LEIP's roadside image logs. Almost daily changes were apparent along Route 50. The concentration and extension of commercial facilities along Route 28 have placed a deeply engraved rubber stamp of urbanism in this suburban setting by placement of several multistory buildings along the Route 28 corridor. The inventory of images captured by LEIP in 2000 has been marked with both significant and subtle changes when compared to the previous years.

### 4.6.2.2   *Remote sensing technologies for mapping changes in forest cover*

A series of landscape metrics was calculated and presented by LEIP to show how patterns of forest fragmentation have changed in the county during the past three decades. The results revealed that there were increases in forest fragmentation in most watersheds of the county, especially between 1987 and 1999. Later, LEIP acquired a new digital land cover map of the Washington-Baltimore metropolitan region which was used as a basis for analyzing the relationship between the size of the forest fragments and the percent of urbanized area per major watershed. For the eleven major watersheds the calculated mean forest fragment size is plotted against the percent urbanized land as shown in the Figure 4.20. Each of the data points in the figure represents one major watershed in Loudoun County.

With the progress of urbanization the forest patches are getting smaller and smaller. These patches are less likely to support an ecologically diverse wildlife and plant species. A diverse wildlife population normally requires a large continuum of forest cover without fragmentation. For example, many songbird species found in the mid-Atlantic region usually occur in forest patches of 1 km$^2$ or greater in area. Using the relationship shown in the Figure 4.20, the amount of urbanized area likely to result in a mean forest patch size of 1 km$^2$ for maintaining viable population of songbirds is 52.58% (Fuller [3]). This could be an approximate value of maximum urbanization potential in a watershed while maintaining the minimum forest patch size requirement for a viable songbird population. Similarly, the forest fragmentation could be tied to other ecological factors identified in a watershed.

### 4.6.2.3   *Environmental indicators for monitoring impacts of mining operations*

Impacts of mining operations on the watershed and sensitive water bodies are generally monitored by regulatory agencies a little more carefully by requiring the operators to submit water quality

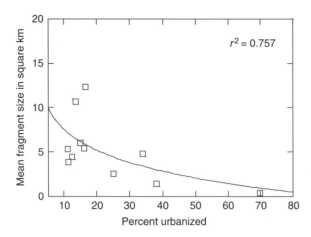

Figure 4.20.   Mean forest fragment size (in km²) versus percent urbanized land in major watersheds, Loudoun County, VA (Source: Fuller, 2000).

data at certain times during the operation. For small scale mining operations this requirement becomes cost intensive. The validity of such data is also questionable because of the possibility of bias exercised in the sampling time and location. In addition, the reported results are subject to sampling and analytical errors and oftentimes do not provide a clear and comprehensive picture of the actual damage done to the local environment due to a mining operation. Effective use of environmental indicators in such operations could possibly resolve these problems. Amongst many different indicators, such as aquatic life, vegetational stress, satellite imagery, aerial imagery, roadside imagery, forest areas, water quality, air quality, wildlife species, and genetic biomarkers, only the following three different indicators will be covered briefly in this paper due to the high relevance in mining applications.

### 4.6.2.4   *Aquatic life*

Aquatic environments generally reflect the effects of acid mine drainage (AMD) quite clearly. Most lakes and streams have a pH between 6 and 8 (USDI [4]). A pH between 6.5 and 9.0 is harmless to most aquatic species. However, near an acid mine spoil site, AMD water flows to streams, lakes, and ponds which may greatly acidify the water. Lakes and streams become acidic when the water itself and its surrounding soil cannot buffer the acidic water enough to neutralize it. The pH of many AMD waters fall below 5.0. Generally, the young of most species are more sensitive to changes in acidity than adults. When the acidity of a lake or stream drops the pH to less than 6.0, there are decreases in the reproductive success in many aquatic species. As the acidity increases to a pH level below 5.0, the number of aquatic species that live in lakes and streams decreases. Some species like frogs are able to tolerate acidic waters but eventually disappear due to low prey (mayfly) populations. At a pH of less than 4.5, most fish species can not survive. Figure 4.21 shows that not all adult fish, shellfish, or their food insects can tolerate the same amount of acid (top wide bars). It also shows the levels of acidity that are harmful to reproduction for each species listed in the chart (narrow bars below the top wide bars).

Biomonitoring methods using macrophytes, phytoplankton, and periphyton also have the potential to be useful tools for monitoring impacts of mine effluents on the aquatic environment. All of these methods are undergoing further testing and evaluation by various agencies. A good example is EPA's Mid-Atlantic Integrated Assessment (MAIA) project, where the evaluation of anthropogenic activities on aquatic ecosystems and integration of watershed studies at different scales are being conducted to develop a landscape assessment on the environmental condition of the Mid-Atlantic region.

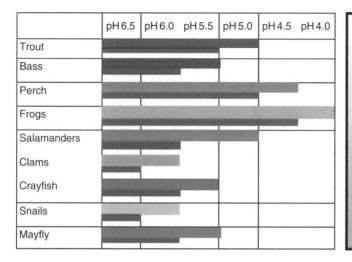

|  | pH 6.5 | pH 6.0 | pH 5.5 | pH 5.0 | pH 4.5 | pH 4.0 |
|---|---|---|---|---|---|---|
| Trout | | | | | | |
| Bass | | | | | | |
| Perch | | | | | | |
| Frogs | | | | | | |
| Salamanders | | | | | | |
| Clams | | | | | | |
| Crayfish | | | | | | |
| Snails | | | | | | |
| Mayfly | | | | | | |

**Monitoring Baffle**

A small waterbody (part of a stream or a small baffle part of a stream or lake), downstream of the active mine boundary, could be stocked with some of the sensitive aquatic species and used as a monitoring area. This waterbody could be periodically checked to evaluate the impact of mining operations.

Figure 4.21.   Degree of acidity endurable by different aquatic species.

### 4.6.2.5   *Vegetational stress*

Some terrestrial plant species have better survivability under acidic condition in the mine area. Examples are Pines (*Pinus spp.*), Black Locust (*Robinia pseudoacacia*), Broomsedge (*Andropogon virginicus*), Common cattail (*Typha latifolia*), Common reed (*Phragmites australis*), Flowering dogwood (*Cornus florida*), River birch (*Betula nigra*), Sericea lespedeza (*Lespedeza cuneata*), Sugar maple (*Acer saccharum*), and American sycamore (*Platanus occidentalis*) (USDI, 2000 [4]) (http://plants.usda.gov).

However, there are a few sensitive plant species that will exhibit stress and show decline in population and growth when impacted by acidic drainage or high metals or other contaminants from the mining operation. Amongst a wide variety of species from the plant database provided by Natural Resources Conservation Service of the US Department of Agriculture (USDA [5]), the following few species are listed as samples of sensitive vegetation which could reveal adverse impacts of mining operations because of their high sensitivity to low pH and low tolerance to other adverse impacts.

1. Purple crowberry (*empetrum nigrum*, pH 4.3–7.8)
2. Kura clover (*trifolium ambiguum*, pH 6.0–7.4)
3. Redosier dogwood (*cornus sericea* L., pH 4.8–7.5)
4. Dotted blazingstar (*liatris punctata*, pH 6.0–7.8).

Similar to the aquatic indicator, a baffle area planted with some of the sensitive species could be monitored using photographic imagery and reported to the regulatory agency. Proper selection of the monitoring area will be a critical factor for its effective use as an indicator.

### 4.6.2.6   *Remote sensing methodologies*

Aerial photography and Satellite imagery could also be used to monitor the periodic changes at a mine site. The data from a remote sensor is digitally stored as a matrix of numbers. A picture element (*pixel*) is the smallest element of a digital image representing the reflectance from one specific location (*i, j*) of row and column out of a matrix comprising of the whole picture. A variety of multispectral and hyperspectral remote sensing systems are being used today. A few commonly used remote sensing technologies categorized by the type of technology are furnished below (Jensen [6]):

- multispectral imaging using discrete detectors and scanning mirrors
  - Landsat Multispectral Scanner (MSS)
  - Landsat 7 Enhanced Thematic Mapper Plus (ETM+)

Figure 4.22.    Typical vegetative indicator species for mining operations – (a) flowering dogwood – (b) Kura clover.

- – NOAA Geostationary Operational Environmental Satellite (GOES)
- – NOAA Advanced Very High Resolution Radiometer (AVHRR)
- • multispectral imaging using linear arrays
  - – SPOT 4 High Resolution Visible Infrared (HRVIR) and Vegetation Sensor
  - – Indian Remote Sensing System (IRS)
  - – Space Imaging Inc. (IKONOS)
  - – Earthwatch, Inc. (Quickbird)
- • imaging spectrometry using linear and area arrays
  - – Airborne Visible Infrared Imaging Spectrometer (AVIRIS)
  - – NASA Terra Moderate Resolution Imaging Spectrometer (MODIS)
- • digital frame cameras
  - – Litton Emerge Spatial, Inc. and Positive Systems, Inc.
- • satellite photographic systems
  - – Russian KVR-1000 and NASA Space Shuttle Photography.

Changes in disturbed area, spoil pile accumulation, and watercourse diversions at a mine site can be easily mapped by using remote sensing technologies. By reviewing temporal changes in the satellite imagery of an area, the forest canopy cover, the forest types (deciduous vs. evergreen), and the vegetative cover (forest vs. savannah) of the area could be determined (Brakken [7]). Determining water quality using hyperspectral data is also possible with new high resolution imagery provided by such satellites as IKONOS and Quickbird. Using high resolution and clear distinction between the reflected wavelengths in the blue ($0.45\,\mu$m), and yellow ($0.8\,\mu$m) region, the sediment load in a surface water near a mine site can be estimated. Remote sensing can be used to periodically monitor the changes in permit boundary, forest cover, site runoff, drainage ditches, slope erosion, and spoil piles by the regulatory agencies and other stakeholders. One similar program is being used for enforcement by the Technology Information Processing System (TIPS) program of the Office of Surface Mining (OSM). The remote sensing technique would allow an early warning and identification of environmental problems at a mine site. By proper use of the remote sensing imagery, a timely alert could be relayed to the mine operator to resolve the problem before it grows too large, difficult and expensive to handle.

REFERENCES

1. US EPA 2000, *Total Maximum Daily Load (TMDL) FAQs*, Office of Water, Mining Waste Scientist to Scientist Meeting, EPA/ORD National Exposure Research Laboratory, Las Vegas, NV, June.MMSD, The

Mining Mineral and Sustainable Development Project, a joint effort of a multidisciplinary team from 10 countries headquartered in London, 2001.
2. LEIP 2000, *Annual Report*, Published by the George Washington University, Washington, DC.
3. Fuller, D.O. 2000, *The Effect of Urbanization on Forest Fragmentation in Loudoun County*, LEIP Annual Report 2000, The George Washington University, Washington, DC.
4. USDI 2000, *Old Ben Scout Reservation Natural Resources and Mining Handbook*, Office of Surface Mining, Alton, IL, May.
5. USDA 2002, *Online Plants Database*, http://plants.usda.gov, Natural Resources Conservation Service. April.
6. Jensen, J.R. 2000, *Remote Sensing of the Environment: an earth resource perspective*, Prentice-Hall, Inc., New Jersey.
7. Brakken, K.T., 2002, *Personal Communication*, U.S. DOE, Rocky Flats Environmental Technology Site, Golden, CO, April.

# 4.7   Emerging mining technologies

Subijoy Dutta[1] and Bonnie Robinson[2]
[1]S&M Engineering Services, Crofton, MD, USA
[2]Geologist, Yamuna Foundation for Blue Water, Inc, Crofton, MD, USA

## 4.7.1   INTRODUCTION

To cope up with the pace of modernization and technological advancement the need to explore and develop new and emerging technologies has never been greater. A wide variety of techniques are used to extract minerals from the earth. Generally, mining consists of removing ore and associated rock or matrix in bulk form from a deposit and transporting it away from the mined site. In the interests of economic efficiency, the extraction process is designed to remove ore of a predetermined grade or higher, leaving behind lower-grade material and barren rock, if at all practicable. In practice, an ideal separation is not often possible, so that some lower-grade rock is mined and some higher-grade ore is left behind. Although mining processes may be classified according to the numerous techniques that are employed in removing ore, they can be broken down into two broad categories associated with the general setting of a mining operation: (1) surface mining or open-pit processes; and (2) underground mining processes (USEPA [1]). Different mining methodologies that are currently used in various parts of the world are summarized in the following paragraphs.

### 4.7.1.1   *Surface mining*

Surface mining techniques are used for most of the major metallic ores in most parts of the world. This method is generally used when the ore deposit is near the surface, or is of sufficient size to justify removing overburden.

Major trends are emerging in surface mining technologies which are designed to improve productivity and safety. Large shovels, draglines, and bucketwheel excavators have increased the stripping ratios that can be economically mined. Increased mechanization and automation in the mining industry is increasing productivity by orders of magnitude compared to the productivity that was achieved only a few decades ago. The machine operator can accurately achieve the limits of excavation through the use of Global Positioning System (GPS).

The improvements in safety have been achieved mainly through improved rollover protection systems, and by increasing the range of visibility of the machine operator. The operator cab is air conditioned, and a large amount of information is available to the operator to increase the precision of drilling, excavating and loading. Truck sizes have increased, and the improvements in technology have made them safer.

### 4.7.1.2   *Underground mining*

Historically, underground mining was the major method used for the extraction of many metal ores. However, it is now much less commonly used in hard rock mining in the United States. There are many variations and combinations of underground mining techniques which have been developed in response to specific or unusual characteristics of the ore body.

In situ mining is another method of underground mining that is applicable to certain ores under some specific hydrogeologic condition. In situ leaching is currently used in the uranium and copper industries. In situ techniques have also been used experimentally in the gold industry. Use of in situ methods has obvious geochemical restrictions based upon the amenability of the ore minerals to being solubilized and the cost and practicality of solvents, and based on concerns related to groundwater quality. Hydrologic requisites include: (1) the host rock must be permeable to circulating fluids; and (2) the host rock must be overlain and underlain by impermeable formations or rock units that restrict the vertical flow of fluids. Wastes associated with in situ mining include spent leaching solutions.

Many different types of leaching and extraction of ores are used by the mining industry. A few common types of leaching and extraction include heap leaching, tank leaching, solvent extraction, floatation, and electrowinning.

Although surface and underground mining of coal is practiced in various mining operations throughout the world, the economic considerations and physical recoverability of coal are the key factors in determining the specific mining methodology to be adopted at a particular operation. The gas associated with the coal seam, known as coal bed methane (CBM), is also extracted commercially by various methodologies.

Coalbed methane extraction was on the rise in late 1980s, but has taken off in recent years. In the 15 years between 1987 and 2001, annual U.S. natural gas consumption grew by about 30%, from 17 trillion cubic feet (Tcf) to 22.2 Tcf, according to the U.S. Department of Energy (DOE). Looking ahead to 2015, DOE and Gas Technology Institute (GTI), and others project that U.S. gas demand could reach 32 Tcf per year – an additional 44% increase (GTI [2]).

It will be an insurmountable task for the natural gas industry to cope up with this surging demand. As traditional gas sources are depleted, producers will turn to so-called unconventional resources, i.e. reservoirs that are difficult to find and tap. From 1990 to 2000 annual production from unconventional resources roughly doubled from 2 Tcf to 4 Tcf. The projections of the DOE and others call for another doubling of production by 2015 from unconventional resources, to 8 Tcf. Continued technology development will be more important than ever in the coming decade.

Similarly, conventional coal resources will soon fall short in meeting the increasing demand for coal. Unconventional resources and techniques have to be applied to meet that demand. One unconventional technique, known as named Hydro-evacuation mining system, used on an exploratory basis in Oklahoma, is a good example of an emerging technology which can recover steeply dipping coal, or other minerals, which are difficult to mine by conventional mining methods. Hydro-evacuation also has the potential of extracting CBM or coal mine methane (CMM) effectively.

GTI is pursuing research and development towards new technologies for exploration and production of unconventional natural gas. GTI has led the development of CBM technologies which transformed coal bed methane from a safety hazard in the early 1980s into a resource that provides about 7% of current gas supply. Their research has provided improved formation evaluation, core sampling, well-casing design, and fracturing. Today, more than 37,000 U.S. wells produce shale gas, providing about 3% of total U.S. gas supply. A GTI study completed in 2002 evaluated shale-gas resources in the Western Canada Sedimentary Basin (WCSB). Researchers found that shales are the most common type of sedimentary rock in the WCSB, that the potential volume of gas in five formations studied was significant, and that stimulation techniques likely would be needed to produce gas at economically attractive levels. Building on that legacy of research, several current GTI programs are addressing the needs identified in the 2002 roadmap project, as well as related issues. Two new technologies (InSpect$^{SM}$ and FluidPro$^{SM}$ ) have been recently identified by GTI for more-accurate identification of the most promising productive zones (the "sweet spots") in unconventional-gas formations. Improved accuracy cuts the risk of drilling a non-productive well. InSpect analyzes seismic data using a new kind of spectral decomposition to reveal "indicators" that show the presence of hydrocarbons in a formation – indicators not obvious using conventional seismic analysis. It can distinguish between gas and brine, can analyze subsurface layers as thin as 15 ft, and can spot discontinuities in reservoir structure not readily visible on broadband seismic charts. FluidPro quantifies the probability of finding commercially viable gas deposits at a particular subsurface location. It integrates geologic, geophysical, and seismic data to characterize the most meaningful properties (e.g. modulus and density) of subsurface fluids such as gas, brine, and oil. Fusion Geophysical developed InSpect and FluidPro with support from GTI and the Oklahoma University Geophysical Reservoir Characterization Consortium. Both of these high-resolution tools are expected to be of significant value to gas producers in coming years as they work to better characterize various unconventional gas reservoirs.

Details on all of these emerging technologies are beyond the scope of this book. However, some details on two different unconventional technologies – coal bed methane (CBM) extraction and Hydro-evacuation Mining Systems will be covered in this chapter.

### 4.7.2   COAL BED METHANE (CBM) EXTRACTION

Methane is the chief component of natural gas, and extraction of this resource can occur in association with coal mining, but is increasingly considered a resource in and of itself and worthy of extraction even in areas where the amount of coal is insubstantial. Environmentalists are becoming increasingly concerned with water quality and aquifer depletion as production increases due to high natural gas prices and demand.

According to an October 2000 issue of Hart's magazine, an independent publisher serving the energy exploration and production industry, the lower 48 states contain 400 trillion cubic feet (TCF) of coal bed natural gas, also known as coal bed methane. Alaska contains another 1,000 Tcf, but only 57 is recoverable.

The areas with high reserves include the San Juan Basin in New Mexico and Colorado with 88 Tcf, North Appalachia with 61, Powder River Basin in Wyoming and Montana with 39, Illinois River Basin with 21, Black Warrior Basin in Mississippi and Alabama with 20, Raton Basin in New Mexico and Colorado with 18, and Central Appalachia with 5. In comparison, the United States now consumes about 23 Tcf of natural gas annually and the Energy Information Administration expects that to grow to 34.7 by 2020 (Henry [3]).

There are two basic ways to extract methane from coal beds. One is through existing coal mines whose main function is to produce coal. This is called coal mine methane and the practice began as a way to reduce methane emissions, since methane is a greenhouse gas 10 times more effective than carbon dioxide at causing global warming, according to the U.S. Geological Survey.

With the increased demand and prices commanded by natural gas, capturing methane emitted from coal mines not only reduces greenhouse gas emissions, but has also become a local source of natural gas for the coal mine operators and for local consumers. According to Karl Schultz at the Environmental Protection Agency's voluntary coal bed methane outreach program, coal mines emit 170 to 199 billion cubic feet (BCF) of methane, "a fairly significant" source.

But the majority of methane from coal is not associated with mining. Schultz says the United States produces 1.3 Tcf of coal bed methane every year, accounting for 7% of the nation's gas production (1999 figures), whereas captured coalmine methane accounts for less than 50 Bcf, which is up from 14 Bcf in 1990.

In areas where the conventional coal resource is not significant enough to warrant a mine, energy exploration companies are now working to only extract the methane that lies inside the coal seams. The coal bed usually ranges in thickness from inches to feet, which is fairly thin, and can be near the surface or several thousand feet below ground.

First, operators perform a process called hydraulic fracturing, which is a form of well stimulation that accesses the methane. Gas producers drill a well, put casing inside, and pump extremely high-pressured water inside, aiming for a particular area where the coal seam is. This "literally fractures the ground," and fracturing takes one to two hours.

Then producers use material, such as sand or ceramic beads, to prop the fracture open. Once the fracture is opened and secure, gas operators pump the water out again and with it comes the methane, which they can capture. At first, mostly water is released, about 20 to 25 gallons per minute. Eventually, it becomes mostly methane and as little as one gallon of water per minute.

On the positive side, methane can be liquefied into natural gas, which is clean-burning and emits less greenhouse gases compared to other fossil fuels. However, environmentalists are becoming increasingly concerned about water quality issues, especially as the industry seems to be taking off. In Wyoming alone, the Bureau of Land Management estimates there are 14,000 wells and forecasts 70,000 by 2060.

The water used for hydraulic fracturing and pulling out the methane comes out of the ground with high levels of sodium. The water is potable for humans and animals, but it can be deadly to plants.

There are two basic ways to dispose of the water. The most common way is to discharge it as surface water. But in areas that depend on agriculture, this high-sodium water can kill valuable crops, unless the gas companies treat it beforehand. If the water is discharged as surface water or used for irrigation, it is treated first. Coal bed methane extraction in Montana is under moratorium until state and federal agencies perform an environmental impact statement.

Another option is to reinject the water into the wells. This removes the necessity for treating it and can replenish aquifers. In addition to the quality of water discharged, depleted aquifers also concern environmentalists because if an aquifer loses too much water, it could collapse and cause the ground to resettle, move or become depressed in areas, even causing dams or other expensive infrastructure to collapse. Depleted aquifers also cause wells to go dry.

In addition to surface discharge and reinjection, other options exist for disposing of the water. "Reinfiltration" holds the water in ditches or other areas and allows it to "passively reenter" the aquifer. Cattle ranchers sometimes request stock ponds in semi-arid areas like Montana, Wyoming and the Southwest, and the water can be used to create wetlands and wildlife ponds to improve habitat and fishing.

Although coal bed methane shows great promise, there are a few environmental issues that need to be addressed prior to maximizing extraction of the coal bed methane gas from the mines.

## 4.7.3  HYDRO-EVACUATION MINING SYSTEM

Hydro-evacuation is an improved method of removing a mineable product, such as coal, from an underground seam without causing a significant surface disturbance at the mine site. No overburden removal, stripping, or shaft sinking is undertaken in this mining operation. It involves drilling several injection holes (vertical) from the earth's surface through the seam of mineable product, drilling a large diameter recovery hole, and an improved recoverable down hole tool, having a vertical auger and a hollow shaft extending from the earth's surface to the auger in the down hole tool. The injection holes are drilled in a pattern extending from the recovery hole towards the up-dip of the coal seam and small explosive charges are inserted through this injection holes and placed in the coal seam. These charges are sequentially ignited and water is injected in the same sequence through the injection holes which moves the fractured mineable product down-dip towards the bottom of the recovery hole. At the recovery hole the down hole tool and water nozzles create an upward movement of the slurry and raise the coal to the earth surface.

This technology has been developed by Coilwells, Inc. of Oklahoma, U.S.A, who owns patents on this new mining process. The company also owns patents on certain pieces of equipment necessary to use the patented mining process. The process allows the user to recover minerals or hydrocarbons that are contained in a bed or layer. Examples of these minerals and hydrocarbons are coal, bentonite, coal bed methane gas, and heavy oil. The new mining process allows recovery of these materials that are normally thought too steep for conventional mining methods. There are large volumes of these minerals and hydrocarbons suitable for recovery by use of this new mining method, which were once considered unmineable. Coilwells, Inc. plans to provide the rights to responsible entities such as individuals, companies and countries who want to use the new mining method. Certain pieces of equipment necessary to use the patented process successfully will also be leased. The company will also acquire the rights to steeply pitching materials and hydrocarbons by purchasing property on which they occur or by leasing the rights of recovery. Once these rights are secured, they will be sublet to those who lease the rights to use the patented process and equipment.

### 4.7.3.1   *Process description*

The installation, operation, and recovery of the coal bed methane using the hydro-evacuation process are described in this section. Typical operational details of the hydro-evacuation process are shown in Figure 4.23. On-site characterization of the sub-surface geology may require refinements in the procedures. Following the initial round of boreholes, investigators will prepare preliminary thickness and structure maps to illustrate the distribution of the coal seam across the recovery site. Producing wells will be placed at the down-dip edge of the seam to be mined with locations planned for the blast holes progressing up-dip away from the producing wells to intersect the coal seam between 100 to 300 ft below the ground surface. The recovery [production] borehole is drilled through the coal seam and roughly one foot into the floor rock. This borehole will be cased through the soil horizons. The down-hole assembly with the rotary auger, nozzles, and lower bearing assembly will be advanced by welding end-to-end 12-inch oilfield casing. The auger shaft is a hollow, tubular shaft connected to the ground surface. Water is pumped from water supply wells and delivered to storage tanks. During operations, water is pumped down the hollow auger shaft and out the down-hole nozzles.

The coal-water slurry is pumped up the annulus and out of the recovery hole by a surface-mounted dredge-pump. The slurry with the coal bed methane gas extraction movement is aided by the rotation of the auger mounted to the rotating hollow shaft. The first blast hole is drilled near the recovery well and a suitable explosive charge with electrical detonating cap is applied to the bottom. An inflatable packer is placed in the hole at the water fill-up point in the hole. The packer is set so that no explosive energy is directed up the hole, but rather is directed into the coal seam. The nearby recovery hole represents the path of least resistance. Second and third blast holes are drilled into the coal seam and detonated. Water is pumped into the blast holes which drives the fragmented coal toward the recovery well. The augers are installed and water is pumped down the recovery well through the hollow auger shaft which is reversed at the bottom of the hole and pumped up through the annulus bringing with it the coal slurry along with any coal bed methane gas extracted during the process. After the first blast holes are cleared and coal recovery declines,

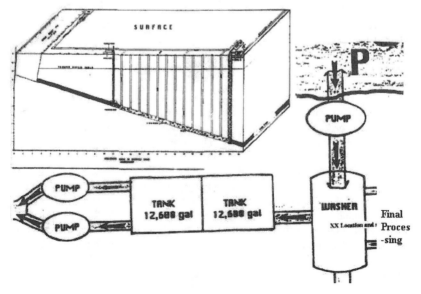

Figure 4.23.   Operational details of hydro-evacuation mining system (Source: Jackson et al. [4]).

the next row of up-dip blast/injection wells are staked, blasted, and injected. Once each blast/injection hole is cleared, it is plugged with an inflatable packer, which is left in place until the entire project is abandoned. Subsequent holes are blasted one at a time during which time all other holes are plugged except for the active recovery wells whose casing and annulus is controlled at the surface. The extraction continues until all mineable coal is recovered from the area and no further movement of coal is observed down-dip to the recovery well.

### 4.7.3.2   *Benefits/advantages of hydro-evacuation mining systems*

1. Hydro-evacuation requires an 11-man crew for base operations. This number is approximately 1/10th of the personnel needed to operate an underground mine and 1/3rd of the personnel needed to operate a surface mine for similar recoveries of coal. Hydro-evacuation process requires less unit cost for both labor and materials/equipment compared to surface mining. This reduction in cost for Hydro-evacuation is due to the minimal use of explosives to extract the coal vein. Figure 4.24 shows an exploratory mining operation using the hydro-evacuation mining system near Keota, Oklahoma.
2. A huge amount of overburden has to be removed to get to the coal seam if a surface mining method instead of a hydro-evacuation system. For example, in a surface mining operation near Wilburton, OK, they had to remove as much as 120 ft of overburden. Due to the high cost of overburden removal many operations have to be stopped and shut down unless more commercial/shallow deposits are found.
3. Most of the jobs in hydro-evacuation system require no special skills and are of lower risk than the jobs associated with other forms of mining.
4. Hydro-evacuation needs approximately 1/10 of the explosives and ½ of the capital equipment cost needed to operate a surface or underground mine for similar recoveries of coal. This is because of the obvious reduction in blasting expenses, and the reduced cost of material movement from the mine site.

Figure 4.24.   Exploratory mining operation using hydro-evacuation (Jackson [4]).

5.  Most of the equipment needed for hydro-evacuation systems are standard oil drilling gear. Effective performance of this equipment in the field has been well tested and proven.
6.  Surface disturbance in the mine area is minimal for Hydro-evacuation systems, since no surface removal or clearing of the area is needed. Only a small portion of the mine area is disturbed during a Hydro-evacuation mining operation. Thus, this system is considered to be environmentally friendly. This system can also be used to remove minerals from environmentally sensitive areas such as wetlands or prime farmlands.
7.  Surface mines have to reclaim the disturbed land. However, this liability is minimal for the hydro-evacuation mining system.
8.  Another advantage of this system is the lower cost of leasing large tracts. These tracts with unrecoverable resources by conventional means are considered non-commercial and are much less expensive.
9.  This system also entails lower cost of permitting for large tracts. Permitting can be performed on a more long-term basis. One environmental report can cover large areas. Water quality reports can be prepared on a regional basis with a basin-wide analysis. Mine plans could be prepared for contiguous tracts, which will result in lower costs for design and planning of resources and equipment.
10. The average yield could be 130 to 200 tons per hour.
11. Areas with coal reserves that are considered unrecoverable with conventional mining technologies, such as steeply dipping coal, and coal in environmentally sensitive areas, can be suitably mined using Hydro-evacuation mining system.

## REFERENCES

1.  USEPA, Technical Document, Background for NEPA Reviewers, Non-coal Mining Operations, Office of Solid Waste, December 1994.
2.  Gas Technology Institute (GTI), The Unconventional Route to the Natural Gas Future, Spring 2003, Gas Technology Institute, Des Plaines, IL, USA.
3.  Henry, Natalie, Coal bed Methane Energy Source Attracts Industry, Greenwire, March 2001.
4.  Jackson, D. and Dalton, D. Summary of Hydro-evacuation Mining Systems, presented at the George Washington University, Washington, DC, March 2003.

# CHAPTER 5

## Mineland reclamation

Subijoy Dutta[1], Raj Rajaram[2] and Bonnie Robinson[3]
[1]S&M Engineering Services, Crofton, MD, USA
[2]Complete Environmental Solutions, Oak Brook, IL, USA
[3]Geologist, Yamuna Foundation for Blue Water, Inc, Crofton, MD, USA

### 5.1  INTRODUCTION

Commitment to sustainable development necessitates integration of environmental policies and development strategies so as to satisfy current and future human needs, improve the quality of life, and protect resources. This commitment is most noteworthy in the insistence on mine land reclamation and closure plans by all regulatory agencies around the world. However, this was not the case even a generation ago when mined land reclamation was left for future generations. The mining legacies we are reclaiming in the United States and other parts of the world at present are due to the lack of policies that planned for future human needs while exploiting the mineral resource. The United States Department of Agriculture Forest Service is one of the first organizations in the world that pushed for multiple land use in exploitation of mineral resources on forest lands. It is currently taking the lead in the Sustainable Mining Roundtable (SMR) in the United States (Shields [1]). The Forest Service has committed to use the Criteria and Indicators (C&I) of sustainability generated by the Montreal Process as part of their comprehensive monitoring program in managing the U.S. forest lands. The goal of the SMR is to develop a set of national scale indicators of sustainability for mineral resources.

Any land containing mineral resources has multiple uses, and these uses have to be considered in planning for mine reclamation. For example, in the fertile lands of the Midwest United States underlain by coal, the reclamation plan returns the land to productive agricultural uses immediately after mining is completed. However, in urban settings where many quarries extraction construction aggregate, the post mining use can range from recreational lakes and golf courses to residential, commercial or industrial properties. The challenge for the mine planner is to determine the most beneficial use to which the mined land can be reclaimed to. Cooperative planning by mine operators, land developers, government land use planners, and citizens is key to successful reclamation of mined out properties.

The term "reclamation", as it applies to mine reclamation, has been used for over 60 years. It used to imply that land, disturbed by mining activities, could be rehabilitated to some productive post-mining land use. Today, however, the term "reclamation" has taken on a new meaning; "to restore" the mining disturbance to a condition as close as possible to the pre-mine conditions. The term "reclamation" is now used interchangeably with "restoration", much like the confusion between conservation and preservation.

"Restoration", especially of mining disturbances, is essentially impossible (Bengson [2]). No matter how much money is spent on the "reclamation", the complete restoration of the previous ecosystem is impossible. Man cannot recreate (no matter how much money is spent) what nature has taken eons to establish. Reclamation should be aiming at the establishment of a stable & self-sustaining ecosystem that will, in time, re-create a productive and suitable ecosystem that will replace the pre-mine ecosystem and achieve the desired post-mining land use (PMLU).

"Reclamation" should be the re-establishment of a self-sustaining ecosystem that can and will evolve into a desirable ecosystem to meet the pre-selected PMLUs. The art and science of today's "reclamation" is in reality "applied ecology", developing man-made ecosystems on sites that would not or could not develop a vegetative ecosystem on their own within a reasonable time frame. Mountain top mining, with valley fill, is an example of good reclamation for steep unusable landscapes reclaimed to flat landscapes for a variety of beneficial uses. Good, successful, reclamation costs are often considerably less than the cost to "restore" an ecosystem. Reclamation just gives nature a jump-start in starting the ecological processes evolving the ecosystem to the desired end. Reclamation can use the best plant species adapted to the site conditions, that are the most productive, that are better at stabilizing the site, and can actually enhance the future land uses to the highest PMLU possible. Once stabilized the reclamation site can then start to evolve into a climax ecosystem that is adapted to that site. It may not be a complete "restoration" of the previous ecosystem, but it will be the most successful for the particular conditions at that site. Besides, the "new" ecosystem will add to the overall bio-diversity of the region.

Mining in steep areas requires disturbing the slopes during overburden removal. During reclamation, several options are available to achieve the PMLU. With angle of repose slopes (1.33 to 1.5:1) the slope may be steeper, but the total slope length can kept to approximately 40 to 45 ft with a 30 ft elevation lift. Whereas, the same lift with a 3:1 slope results in a 90 ft long slope. Although the slope is less steep, there will be significantly greater soil loss on the longer slope. Also, the steeper, shorter slopes results in actually increasing the acreage of flat surface area available for PMLU's. If the slope is regraded more land will ultimately be disturbed, or the flat surface area atop the slope will be significantly reduced. For livestock grazing and for some species of wildlife, steep slopes may not be a detriment. In going from an angle of repose slope of 1.5:1 to a 3:1 slope, more than one-half of the flat surface area would be lost. The cost of regrading from an angle of repose to approximately 3:1 will cost significantly higher.

An effective reclamation should focus on 3-major objectives:

1. reclaim the mining disturbed land to a satisfactory level of productivity that provides for stability and public safety;
2. achieve the predetermined PMLU; and
3. establish the final reclamation as aesthetically pleasing as possible.

Reclamation goals must not only be technologically feasible, they must also be economically attainable. Where good reclamation can be done for $2,000 to $8,000/acre, restoration may cost as much as $15,000 to $75,000/acre. The PMLU values may be as little as only $200 to $500/acre. It makes very little sense to spend so much to achieve so little. With more reasonable reclamation costs, more actual reclamation may be accomplished.

Government permitting agencies use the financial assurance mechanism to ensure that the reclamation and mine closure plan submitted by the mining company is successfully completed. The financial assurance can be in the form of a surety bond, a letter of credit, or an insurance policy with the government agency as the beneficiary. United States mining regulators calculate the cost of reclamation if completed by a third party as the amount of financial assurance that is required from the mining company. In cases of acid mine drainage, the cost of ameliorating the mine drainage for as long as 50 years is calculated for determining the financial assurance requirement. This kind of guarantee provided by the mining company ensures that future generations are not saddled with the cost of mine reclamation.

## 5.2    SUSTAINABLE SURFACE MINE RECLAMATION

The Surface Mining Control and Reclamation Act (Department of Interior [3]) was the first major legislation passed in the United States to control the adverse effects of surface mining operations, and provide for the reclamation of abandoned mines. The Act specified performance standards for mines, permitting and inspection procedures for coal mines, requirements for reclamation and post-mining land use, and bonding requirements covering the period of 10 years after mine closure. The Act also imposed a small tax on surface and underground mined coal to fund reclamation of abandoned mines. The Act has been responsible for promoting sustainable mining practices in surface mines, and managing the mining legacies from past coal mining.

In the United States, coal mining is done by one of the three methods mentioned below, depending on topography. These are:

1. Area Mining;
2. Contour Mining; and
3. Mountain Top Mining.

The area mining method is used for large areas of flat or rolling hills, and immediate reclamation to post-mining use is completed by the mine operator. Contour mining is done on hill sides where high wall reclamation is done to a specified contour. Mountain top mining is done in mountainous regions of the eastern United States, and the overburden is filled in the valleys to reclaim the land. In contour mining and in mountain top mining, sediment control with terraced slopes and stream water diversion is practiced to minimize impacts to surface water bodies.

TXU Mining has reclaimed lignite mined areas in East Texas to a productive farmland by replacing tough soils with a silty loam. With research and development efforts, they developed selective overburden handling to help improve the soil. Using this method, the overburden is removed and mixed before it is replaced and reclaimed in order to make the soil more productive. In 1999, the mine received the Office of Surface Mining Director's Award for Reclamation. An estimate made by the mine shows that the productive soils can boost the county's agricultural economy by $21 million dollars over the next 20 years.

Major strides have been made in the reclamation of aggregate and limestone quarries since these are generally located near urban centers. There are several examples of mined out quarries being used as recreational lakes, parkland development, housing subdivisions, and office parks (Mineral Information Institute [4]). The Rudersdorf limestone quarry near Berlin, Germany, has developed a parkland atmosphere with several trout ponds, and an extensive network of underground tunnels running through the quarry to convey clean water to nearby waterways. The water from dewatering in the quarry (to facilitate mining) is collected and pumped via an automatically controlled pumping station to nearby Kriensee Lake, and several watercourses around Berlin.

Another reclamation success story is the Harbison-Walker Refractory Company's fireclay mine in Greenup, Kentucky, USA. Mining operations at this mine were conducted with the reclamation plan in mind. This allowed for efficiency in the handling of materials and reclamation activities. Timber was logged from the site before mining started. Topsoil and overburden were removed from the site, and properly stored to prevent wind and water erosion. Dams and silt fences were placed below the disturbed areas to control the sediment. Diversion ditches were used to redirect surface water away from the active pit areas. One of the pits was not completely backfilled so that a lake could be created. Reclamation activities such as backfilling, grading, and topsoil placement were combined with mineral extraction to increase the overall success of the operation. The land has been returned to hay land and pasture use (Mineral Information Institute [4]).

## 5.3 RECLAMATION PRACTICES FOR NON-COAL MINING

Non-coal mining operations include industrial mineral quarries and open pit metal mines. Because of the many diverse mining circumstances that could be called "non-coal," a process is described that can be applied to any operation.

In general, a non-coal surface mine is any open-pit mine, quarry, or other pit operation. Some miners insist that quarries and open pits are vastly different. It should be noted that except for size, they are very similar, particularly when it comes to their reclamation. With respect to the distinction between quarries and open-pit mines, there is only one difference of significance. This difference is that the location of quarries often create the greatest conflicts with existing or growing populations (Univ. of Missouri [5]).

It may be presumed that non-coal strip mines (Florida phosphate, Georgia kaolin, and some uranium mines) can be reclaimed concurrently as are coal strip mines. It is recognized that the land-use pressures also operate in these reclamation circumstances. A major difference between coal and non-coal reclamation is that open-pit mines are often long lived. Therefore, planning for a subsequent use has to have a time-line measured in decades instead of a few years. As a consequence, while it may be difficult to plan for an explicit post-mining use, it has to invariably deal with the consequences of changes in neighboring uses. With the present history of population growth, it can be assumed that conflicts will grow on property boundaries.

### 5.3.1   Reclamation zones

An open-pit mine (quarry) can be divided into three primary zones for reclamation purposes: the pit, all waste-disposal zones, and surface plant (mill, shop, offices, water treatment, etc.). Again, although every case has its differences, there are some general practices that can be applied everywhere.

The pit is perhaps the most delineated zone and it includes all lands upon which there has been excavation. Because of its long life potential, it could also include lands that are proposed for excavation. Mine engineers need to consider the location of possible extensions to the ore zone. Calculations, including present value of future costs, should be done on locating waste zones some distance from the current pit versus moving a waste zone that is close to the original working but that may, in time, prevent extensions to the pit.

Waste disposal zones include overburden, mill tailings, and, in the case of quarries, unsold or unmarketable product. In addition to stability and aesthetics concerns, there is a need to think about recovery of values from the waste. This is one reason why the miners generally do not to spill wastes helter-skelter into old pits. Overburden may be useful for the construction of impoundments while tailings represent a problem that has to be carefully controlled. Unfortunately, current practice often locates abandoned stockpiles near to the entrance to a quarry and they can be unsightly.

The operating plant must be laid out to be inconspicuous and attractive. However, from an engineering point of view, it has to be economical and efficient. For quarries, in particular, the operating plant may be within a planned or regulated zone (municipality or regional planning district) and needs to conform to zoning, fire-protection, and waste disposal codes.

### 5.3.2   General techniques

#### 5.3.2.1   Barriers and shields
During the planning phase, it is useful to set up sight lines from any available viewing point. Features that could be said to be disruptive should then have a permanent barrier (landscaped and vegetated) placed in the line of sight so as to shield the disruption. Obviously, this technique is harder to implement when there are large differences in relief. Barriers should be constructed with natural-looking contours; straight lines are a give away that the mound had been constructed. Trees can reduce sight lines, but thought should be given to the seasons if deciduous trees are used. Thick vegetation can reduce noise but a single line of trees, particularly if there are no leaves near the base of the trunks, will not do much good as a sound barrier.

Permanent waste piles can be set up so as to be effective barriers. However, they should be built such that revegetation can take place quickly. This implies building outer edges promptly to their design height so that they can be covered with clean material and then soil for revegetation. Terraces and mini-terraces are important for vegetation stability.

#### 5.3.2.2   Pit slopes
Final pit wall slopes should approximate natural slopes in the area. Since this is usually shallower than a working slope, care must be taken not to leave non-compliant slopes which can cause problems

when trying to close out a permitted site. Benches need to be removed so as to look natural. However, some safety terraces should be maintained to reduce the threat of falling or rolling rock.

Pennsylvania requires that quarry slopes be reduced to 35 degrees or less but that if the abandoned pit were to be used as an impoundment, providing that there was a beach terrace for swimmers to exit the pond, the underwater portion of the slope could be vertical. They require a groundwater survey to show the lowest possible water table. The beach terrace has to extend some distance, such as ten feet, below that minimum. This is to ensure that swimmers, legal or not, can walk out of the pool.

### 5.3.2.3   *Pit boundaries*
The usual mine plan starts at the center of the property and then works radially outward. It is suggested that the pit be started against one wall so that, as the pit expands, some walls will be available for final reclamation. This schedule allows for reclamation costs to be balanced against operating profits instead of having them come after production ceases.

### 5.3.2.4   *Water*
Using the pit as a lake or pond is a popular choice. However, provisions need to be made to keep the water attractive. Eutrophication and oxygen depletion can rapidly turn a water reservoir into a nuisance. Slopes into the water have to be both natural looking and scalable by an advertent or inadvertent swimmer. The impoundment should look good at conceivable high water and low water periods. If water is flowing, provisions for appropriately designed and constructed spillways must be made. Any impoundment will become a groundwater recharge zone. Provisions have to be made to ensure that toxic or hazardous substances are not carried into the groundwater. All debris and hazards that will be covered by the water must be removed. In one tragic accident, a swimmer was impaled on large steel hooks that were inserted into the highwall to carry electrical cables into the active pit. It was actually a roof bolt that had been inserted into a short horizontal bore hole and then bent upward.

All in all, the best impoundments are those where subsequent usage will ensure their survival, safety, and environmental acceptance. The worst are where no thought has been given to subsequent use and the pit has been uncared for and allowed to become a nuisance or, even worse, a death trap.

### 5.3.2.5   *Waste piles*
The need to have a naturally looking waste pile is well known and over emphasized. It is important that the piles be structurally sound to support construction. In other words, in addition to slope stability, the upper surface of the pile should have foundational stability. Casually built piles can suffer from random and uneven settling (differential settling) that is harmful to structures built on them. This is particularly true for quarries and sand and gravel pits, where they are close to urban or suburban areas, since the land is at a premium for building. The same land-use pressures have been seen in the steep mountain land of the southern Appalachian coal fields. Consequently, the same need for stable backfills exists.

### 5.3.2.6   *Other considerations*
Pits can be backfilled, particularly if there is excess waste from another operation. However, such filling needs several pre-fill planning steps. There is a need to check the following:

1. Is the pit truly exhausted or is it shut because today's prices and technology do not allow mining to be profitable?
2. What economic penalty will accrue if we cover potential incremental ore?
3. Where will the fill material come from?

A nearby operation is often the case. However, if used for refuse disposal, further permits will be needed, which are difficult to obtain. Design plans for sealing and prevention of groundwater contamination will be needed to obtain permits for using pits for refuse disposal.

### 5.3.3    *Land-use*

Land use planning is equally important in non-coal reclamation as it is for coal mine reclamation. The reclamation managers of non-coal surface mines need to pay continuing attention to the developing land uses around their pits. If the essential reclamation elements have been considered in the original plan, then a change in use for the post-mining pit is easy to effect. For example, if the original plan called for a landfill but population growth now demands a planned residential community with lake and water recreation, it should be possible. This is because the original investigations mapped the groundwater regime and the geotechnical investigations were utilized to design stable slopes. The mine plan and waste disposal zones allow for a lot of flexibility of post-mining use. Continued interaction with the local land-use management agency (city council, county commissioners, regional land-use planners, Council of Governments, etc.) is highly recommended in developing the mining and reclamation plans to yield a sustainable social and economic growth of the community.

## 5.4    RECLAMATION OF MINING LEGACIES

Mining legacies are those properties that have been abandoned after the valuable resources have been mined out. According to the U.S. Bureau of Land Management (BLM), there are about 80,000 small and medium scale mining properties that have been abandoned, and pose physical safety hazards and environmental risks. Of these, about 11,000 are hard rock mines covering an area of 261 million acres. The BLM estimates that 25% of these sites pose physical safety hazards, and 5% pose risks of watershed pollution. Since the passage of the SMCRA in the U.S. in 1977, an abandoned mine reclamation (AML) program has been instituted and many of the properties near urban areas have been reclaimed. Whenever, a viable responsible party has been identified, the party has worked with the governmental agencies to reclaim the abandoned property and remove the physical safety and environmental risks. The AML program, funded by taxes on current producers, has been used to reclaim or rehabilitate abandoned properties where no viable responsible party can be identified.

Two case studies are presented below, and represent how close interaction between mining companies (where viable) and the public and governmental agencies can produce benefits to society while reclaiming mining legacies.

### 5.4.1    *A case study of the Brooklyn Mine Reclamation Project* (Yates and Bullock [6])

In 1994, the Montana Department of Environmental Quality, Abandoned Mine Reclamation Bureau (MDEQ/AMRB) and the United States Department of Agriculture/Forest Service, Deerlodge National Forest entered into a pilot cooperative agreement to reclaim the Brooklyn Mine site. The Brooklyn Mine is an abandoned hardrock mine site located in the Flint Creek Mountains in Granite County, Montana. The primary objective of the Brooklyn Mine Reclamation Project was to protect human health and the environment from heavy metal contaminated mine wastes in accordance with guidelines set forth by the U.S. Environmental Protection Agency National Contingency Plan (NCP) in conjunction with newly developed procedures for abandoned mineland reclamation (AML). Site investigation and sampling activities led to the development of site reclamation engineering plans and specifications. The resulting reclamation construction project consisted of providing all labor, materials, earthwork, and incidentals to construct a waste repository; demolish and dispose of one metal building and miscellaneous debris; excavate, and dispose of approximately 18,300 bank cubic yards of waste rock and mill tailings in the waste repository; recontour and revegetate the excavation areas; recontour, apply soil amendments, topsoil, and revegetate three additional waste rock dumps; and reconstruct approximately 525 ft of stream bank with fish habitat features.

The Brooklyn Mine site is located in the South Boulder Mining District of Granite County, Montana. The mine site is located on the north side of the Boulder Creek drainage approximately five miles upstream from the town of Maxville. The site ranges in elevation from approximately

6,200 to over 6,500 ft above mean sea level. The terrain surrounding the mine is generally rugged, consisting of relatively steep slopes (15–20 degrees). Historical accounts indicate that one shaft and multiple adits were located at the Brooklyn Mine. Before reclamation, however, the shaft and all but one adit had collapsed or had been completely backfilled. A culvert and locking gate had been installed on the only open adit.

The Brooklyn Mine site waste sources consisted of four waste rock dumps and one tailings impoundment. Two of the four waste rock dumps were located very high up on a ridge (up to 500 ft higher in elevation than the stream bed) in steep terrain. One waste rock dump and the tailings impoundment were located on a flat bench approximately 150 ft higher in elevation than the stream bed. The remaining waste rock dump, and by far the largest, is located directly adjacent to Boulder Creek and was actively being eroded into the creek. The total volume of waste rock that was present at the site totaled 38,000 cubic yards. The volume of tailings that was present on-site totaled 4,800 cubic yards. These waste materials contained elevated levels of: arsenic, barium, cadmium, copper, mercury, lead, antimony and zinc. The purpose of this reclamation project was to limit human and environmental exposure to the contaminants of concern and reduce the mobility of these contaminants to mitigate impacts to local surface and groundwater resources.

### Reclamation procedure

The Brooklyn Mine was an abandoned hardrock mine site listed on the Montana Department of Environmental Quality, Abandoned Mine Reclamation Bureau Priority Sites List. The Brooklyn Mine site was inventoried, ranked and identified as a hazardous hardrock mine in 1993. MDEQ/AMRB selected the site for reclamation and planning activities in 1994. Figure 5.1 provides an overview of the clean-up process for the Brooklyn site. This process has been designed to comply with the requirements of the NCP; the Comprehensive Environmental Response, Compensation and Liability Act (CERCLA); and the Montana Comprehensive Environmental Clean-up and Responsibility Act (CECRA), while streamlining certain aspects of the process to better meet the regulatory and functional needs of cleaning up relatively small abandoned mine sites that are generally situated in remote locations.

The Brooklyn Mine site removal action is supported by the following documents:

A. Preliminary Site Inventory Report
B. Owner/Operator History Report
C. Community Relations Plan (CRP)
D. Reclamation Work Plan
E. Reclamation Investigation Report (RI)
    1. Field Sampling Plan (FSP)
    2. Quality Assurance Project Plan (QAPjP)
    3. Laboratory Analytical Protocol (LAP)
    4. Health and Safety Plan (HSP)
F. Expanded Engineering Evaluation/Cost Analysis (EEE/CA)
G. Environmental Assessment (EA)
H. Construction Bid Package/Engineering Design.

### Site Investigation

Site investigation field activities conducted at the Brooklyn Mine focused on collecting sufficient data to perform a human health and ecological risk assessment, as well as a detailed analysis of reclamation alternatives. Data collected to support the risk assessment included the following:

- Characterization of heavy metal concentrations, both vertically and laterally in each waste source, and isolation of the 0–6 inch depth zone for direct contact and wind erosion (air release) exposure evaluation.
- Establishment of background soil concentrations with multiple (five) background soil samples.
- Evaluation of the physical and chemical properties of the source materials that effect contaminant migration, including: pH, buffering capacity, organic carbon content, and particle size distribution.

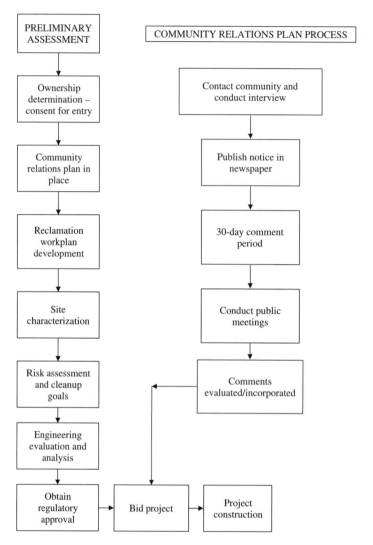

Figure 5.1.   Generalized abandoned mine cleanup procedure.

- Evaluation of groundwater.
- Characterization of impacts to surface water (Boulder Creek) with regularly spaced surface water and corresponding stream sediment samples located upstream, adjacent to, and downstream from the site.

Data collected to support the detailed analysis of reclamation alternatives (Feasibility Study) included the following:

- Accurate areas and volumes of the contaminant sources (waste rock and mill tailings);
- Contaminant concentration variations and leaching characteristics of the wastes [Toxicity Characteristic Leaching Procedure (TCLP), porosity, hydraulic conductivity, pH];
- Revegetation parameters for waste rock dumps and cover soil sources including: liming requirements; soil texture and particle size; fertilizer recommendation; organic matter content; and identification of suitable native plant species; and
- Soil characteristics of potential repository site locations.

All site waste sources were sampled by collecting solid-matrix samples for in-the-field analyses of total metals utilizing a field portable X-Ray Fluorescence (XRF) spectrometer. Approximately 10% of all samples taken were split, with half of the sample sent to a laboratory for metals analyses. This sampling methodology allows: greater field sampling flexibility, a greater number of sample analysis, reduced cost and redundant laboratory analysis to provide XRF accuracy verification.

Using the collected data, a baseline human health risk assessment was completed. The risk assessment examined the effects of taking no action at the site. The assessment involved four steps: hazard identification; exposure assessment; toxicity assessment; and risk characterization. These four tasks were accomplished by evaluating available data and selecting contaminants of concern, identifying potentially exposed populations and exposure pathways, estimating exposure point concentrations and intakes, assessing toxicity of the contaminants of concern, and characterizing overall risk by integrating the results of the toxicity and exposure assessments. The baseline human health risk assessment calculation for the Brooklyn Mine identified soil ingestion of arsenic and lead, and water ingestion of lead as the highest risk concern for the human health.

An ecological risk assessment was also completed. This assessment involved:

1. identification of contaminants, ecologic receptors, and ecologic effects of concern;
2. exposure assessment;
3. ecologic effects assessment; and
4. risk characterization.

These four tasks were accomplished by evaluating available data and selecting contaminants, species and exposure routes of concern, estimating exposure point concentrations and intakes, assessing ecologic toxicity of the contaminants of concern, and characterizing overall risk by integrating the results of the toxicity and exposure assessments. The ecological risk assessment calculation for the Brooklyn Mine identified: aquatic life effects from lead in water, and arsenic, lead and zinc in sediments; elk ingestion from lead salts; and plant phytotoxicity from arsenic, cadmium, copper, lead and zinc as presenting the highest ecological risks.

The risk assessment helped to identify the following site cleanup goals:

1. Carcinogenic goal – reduce arsenic concentration to background (20 mg/kg);
2. Noncarcinogenic goal – reduce lead concentration to 1,500 mg/kg.

**Reclamation alternatives**
The following reclamation alternatives were developed and evaluated for the Brooklyn Mine site:

| | |
|---|---|
| Alternative 1 | No Action |
| Alternative 2 | Institutional Controls |
| Alternative 3 | In-Place Containment |
| Alternative 4a | On-Site Disposal in a Constructed Repository |
| Alternative 4b | On-Site Disposal in a Constructed Modified Repository |
| Alternative 4c | On-Site Disposal/Containment Using a Modified Cover |
| Alternative 5 | Off-Site Disposal in A Permitted Hazardous Waste Disposal Facility |
| Alternative 6 | Off-Site Treatment or Reprocessing. |

The alternatives were assessed based on the following 7 NCP criteria:

• overall protection of human health and the environment;
• Applicable Rules and Appropriate Requirements;
• long-term effectiveness and permanence;
• reduction of toxicity, mobility, or volume through treatment;
• short-term effectiveness;
• implementability; and cost.

Based on a detailed and comparative analysis of alternatives, Alternative 4b: "Disposal in a Constructed Modified (Single Lined) Repository" was selected as the preferred alternative by the MDEQ/AMRB and the US Forest Service. This alternative involved construction of a single-lined, mine waste repository with a multi-layered cap. This alternative was deemed the most appropriate and cost-effective means to reduce risk to human health and the environment to an acceptable level.

**Project construction**
A stream diversion structure was necessary to isolate Boulder Creek from the waste rock excavation activities. Due to the tight time frame associated with this project, the stream diversion structure was procured and installed by Pioneer with the assistance of the Montana Conservation Corps, before the construction contractor mobilized to the site. The diversion structure was constructed from half-round 36-inch (1 m) diameter corrugated metal pipe. The pipe was raised slightly above the creek bed on jack-leg supports at ten foot centers to maintain a constant seven percent grade. A small temporary diversion was constructed at the inlet to divert the majority of the flow of Boulder Creek into the pipe. Figure 5.2 shows a pre- and post-construction picture of a part of the mine area.

The construction contract work involved removing those waste sources at the Brooklyn Mine which were the principal sources of concern (the tailings pond and waste rock dump #5) and disposing of these wastes in a constructed repository. The repository was constructed on the bench area located directly south of waste rock dump #3. The repository bottom consists of a single geosynthetic clay (GCL) bottom liner with an integral drainage layer and a low maintenance leachate collection and removal system. A secondary layer of B-Grade (off-specification) GCL was installed to protect the primary liner from potential chemical reaction. The total surface area of the repository required to contain the specified wastes is approximately 1.5 acres.

Approximately 3.0 acres of timber was removed to accommodate repository construction and removal of tailings dispersed throughout sparsely timbered areas. After the specified wastes were loaded and compacted in the repository, a multi-layered, lined cap was constructed overlying the wastes, and the cap was fertilized, seeded, and mulched. The excavated areas (tailings and waste rock dump #5 locations) were backfilled to contours matching the surrounding topography, fertilized, seeded, and mulched. The stream channel adjacent to waste rock dump #5 was reconstructed in a step-pool configuration similar to the undisturbed stream channel located upstream of the project. Excess soil originating from the repository excavation was amended with compost and used as cover soil in the excavated areas. In addition, a subsurface limestone French drain was constructed below the adit located at waste rock dump #5 to treat a minor intermittent adit discharge. The other large waste rock dumps located at the site (waste rock dumps #1, 2, and 3) were graded out to match the surrounding topography. Lime was incorporated into the upper 12-inches (0.3 m) of the dump material, as needed. The dumps were covered with soil (previously amended with compost), fertilized, seeded, and mulched in place. Excess soil from the repository excavation was also used to cover the waste rock dumps. Slopes flatter than 2.5:1 were mulched by crimping straw and drill seeded. Slopes steeper than 2.5:1 were hydroseeded with a mixture of wood fiber mulch, seed, fertilizer, and tackifer. Additionally, a biodegradable erosion control mat (straw/coconut fiber woven mat) was installed on slopes steeper than 2.5:1 following the hydroseeding procedures. Ditches were constructed to divert run-on away from each of the reclaimed areas and the repository. Temporary fences were constructed to surround each of the reclaimed source areas as well as the repository cap to allow for the establishment of vegetation without interference from livestock or wildlife. Several of the temporary roads constructed at the site were removed and reclaimed; however, some of the roads remain intact to allow access for monitoring the progress of the reclaimed areas (and maintenance when necessary) for a period of one to several years.

**Summary**
The Brooklyn Mine Reclamation Project was completed on schedule and within the engineer's cost estimate. All project objectives were accomplished. The project took 67 consecutive calendar days to complete (from August 1, to October 6, 1995).

The construction cost for this project was $767,895.00. The total engineering cost for this project was $236,380.00.

The Brooklyn Mine was a successful pilot project which tested newly developed procedures for: (1) establishing working cooperative agreements, and (2) conducting abandoned hardrock mine/mill site reclamation meeting the requirements of the NCP.

*Brooklyn Mine before reclamation. Outhouse overhangs creek shore.*

*Brooklyn Mine with reconstructed stream after reclamation.*

Figure 5.2.   A pre- and post-construction area at the Brooklyn Mine (MDEQ/AMRB [7]).

### 5.4.2   *Atlantic Richfield's southwest Montana sites* (Bullock, R. et al. [8])

Atlantic Richfield Company has been reclaiming several historic mining and smelting sites in Butte and Anaconda, located in southwestern Montana, USA. The Superfund program in the United States provides a mechanism for the potentially responsible party (PRP) to work with the Federal, State and local governments to plan for and complete the remediation of mining legacies. Based on their experience at these legacy sites, the mining company has developed a Handbook for the Redevelopment of Former Operating Sites [9]. Reclamation projects completed by the company in southwestern Montana over the past 10 years include:

- Warm Springs Ponds – a 3,000-acre wetland complex developed from a tailings-laden flood plain which includes a diverse waterfowl habitat and a State Wildlife Management Area;
- Old Works Golf Course – a championship golf course developed from a 250-acre smelting waste site;
- Rental Storage Units – a one-acre mining site developed into rental storage units by a third party developer;
- Early Child Care Center – a four-acre site developed into a child care center by a third party developer;
- Copper Mountain Park – a multi-sport community park developed as an integrated redevelopment of a mining site by a third party developer.

Elements to consider in considering redevelopment alternatives for a legacy site include:

- Location;
- Environmental condition;
- Community's and other stakeholder's interests;
- Technical issues in reclamation to allow a desired land use; and
- Economics.

Atlantic Richfield Company's experience has helped develop a process to integrate the governmental agency and community input into the negotiation, planning and implementation phases of a legacy site reclamation project. The key elements of the process, illustrated as a spoked wheel, involve the outside wheel (negotiation, planning and implementation), the spokes (information needs for completing the process), and hub (governmental agency and community input). Of all the elements, the most important is the process of continuously getting governmental agency and community input in the planning and implementation of the redevelopment project. A critical factor for the success of a redevelopment project is a strong partnership between the private companies, local communities, and the Federal and State governments.

Critical issues in the redevelopment of legacy sites include the following:

1. Environmental Issues: Environmental concerns related to former metals mining and smelting sites include disturbed landscapes, and contaminated soil, surface water and groundwater. The redevelopment should be compatible with the environmental remedy for the site. For example, if the remedy is to cap mine waste, many caps (soils, asphalt, gravel) can serve as the base component of the various redevelopment options. Redevelopments on capped mine waste range from recreational parks to industrial parks to commercial businesses.
2. Economic Issues: In Southwest Montana, the challenge of funding comes from limited government funds and a slow economy, which often deters third party developers from participating in the process. However, if site closure which address future liability comprehensively, a PRP like Atlantic Richfield Company participate in the process with significant amounts of funding. Sources of funding include:
   - Federal and State government grants to communities;
   - Local government funds for community redevelopment projects;
   - PRPs who have been identified for the site; and
   - Private, third party developers who will benefit from the redevelopment.
3. Regulatory: For the remedy and redevelopment plan to get full regulatory approval, an integrated reclamation/redevelopment plan must be developed with involvement from all critical stakeholders, including governmental agency and community representation. The redevelopment plan should be consistent with the regulatory requirements for site cleanup and should be properly funded. Regulatory "closure" regarding site liabilities must be defined, and include affirmative liability transfer from the owner/operator of the former operating site to the owners of the new redevelopment. The agreements between the parties to the reclamation plan should

be comprehensive and reflect the roles and responsibilities of the parties. Examples of such agreements include the Record of Decision for Superfund projects, prospective purchase agreements between private and government entities, and private/public contracts.

In summary, the Atlantic Richfield Company has demonstrated that by following a proven process to reclaim/redevelop legacy sites, the land can be returned to productive use for the local community.

REFERENCES

1. Shields, D.J., U.S. Sustainable Minerals Roundtable: Measuring Mineral Resource Contributions to Sustainable Development, Paper Presented at the SME Annual Meeting, Cincinnati, Ohio, USA, February 2003.
2. Bengson, S.A., Innovative Reclamation Techniques, ASARCO LLC, Phoenix, Arizona, 1998.
3. U.S. Department of Interior, Surface Mining Control and Reclamation Act, U.S. Government Printing Office, 1977.
4. Mineral Information Institute, Reclamation Case Histories, Denver, Colorado, 2001.
5. Univ. of Missouri, Rolla, "Non-Coal Reclamation", http://web.umr.edu.
6. Yates, Jack O. and Bullock, W.J., 1996. "A case study of the Brooklyn Mine Reclamation Project", Paper Presented at the 18th Annual Conference of the Association of Abandoned Mine Land Programs, Kalispell, Montana, September.
7. Montana Department of Environmental Quality/Abandoned Mined Reclamation Bureau (MDEQ/AMRB), Brooklyn Mine Reclamation Project Final Report, December 1995.
8. Bullock, R., Kerner, M., Stilwell, C., Lee, P. and Booth, D., 2003. Returning the Land to Productive Use: Reclamation and Redevelopment in the SW Montana Area, April 2003.
9. Atlantic Richfield Company and EMC2, 2002, Handbook for the Redevelopment of Former Operating Sites, February 20.

# CHAPTER 6

## Waste management

Raj Rajaram[1] and Robert E. Melchers[2]
[1]*Complete Environmental Solutions, Oak Brook, IL, USA*
[2]*Centre for Infrastructure Performance and Reliability, The University of Newcastle, Australia*

### 6.1    INTRODUCTION

The large volume of waste created during mining operations has been discussed in Chapter 3. This waste has to be properly managed during the operation, and during post-closure to minimize environmental impacts from the waste. Past mining operations have left severe environmental problems, and these have adversely impacted the environment in the vicinity of the mine site. Problems related to acid mine drainage, loss of land productivity, and impact on land values have been experienced with legacy wastes in many parts of the United States and elsewhere. The United States government is spending millions of dollars in correcting the adverse impacts of legacy wastes. Therefore, it is imperative that a mining company have and implement a waste management plan to make the operation sustainable. With the increased environmental awareness in all parts of the world, a mining permit cannot be obtained without a proper waste management plan.

This chapter will consider the following aspects of waste management:

- tailings management
- waste rock management
- maintenance waste management.

# 6.2 Tailings management

Robert E. Melchers

*Centre for Infrastructure Performance and Reliability, The University of Newcastle, Australia*

Tailings management is an integral part of a mining plan since tailings ponds occupy a large area to accommodate the tailings during several decades of mine's life, and failure of a tailings impoundment will have catastrophic impacts on downstream populations and property. The International Council on Metals and the Environment (ICME) and the United Nations Environment Program (UNEP) collaborated on a useful book on tailings management (ICME/UNEP [1]). Case studies from that book and other case histories of tailings management are described in this section.

Tailings are the residuals from the processing of ores, and contain small amounts of the mineral (not economically recoverable) and chemical reagents used in the processing. According to Professor Richard Jewell, the term tailings is derived from the fact that the process generates a concentrate at the top (or "head") and a waste called tailings at the end (or "tail") (ICME/UNEP [1]). Tailings differ from naturally occurring soils since the density and strength of tailings are initially low and increase relatively slowly with time. Soil mechanics principles are used in designing tailings dams, with due consideration given to the differences in consolidation, relevant drainage conditions, and the differences in slurried tailings and natural soils. The nature of the tailings material in a tailings dam depends on the solids contents in the slurry, and the discharge method and rate. Permeability of the tailings varies within the pond, being highest at the point of discharge and progressively lower toward the edge of the water. Consolidation of and seepage from the tailings is dependent on the permeability of the tailings. As a low-strength material in a loose, saturated state, tailings are vulnerable to liquefaction from dynamic loading processes.

Tailings dam failures in several parts of the world have provided valuable lessons in what is important in tailings dam construction, operation, and maintenance. The International Commission on Large Dams (ICOLD) was formed in 1928 to promote safety and proper design and operation of large dams built to retain water. With the size of tailings dams increasing in size to over 100 m high by the 1970s, ICOLD established a Committee on Tailings Dams and Waste Lagoons in 1976. In 1996, they published a joint report along with UNEP entitled "A Guide to Tailings Dams and Impoundments – Design, Construction, Use and Rehabilitation (ICOLD/UNEP [2]). According to Dr. Penman, Chairman of ICOLD, two major tailings dam failure illustrate the damage to life and property from tailings dam failures:

- failure of two dams in Italy in 1985 destroyed the villages of Stava and Tereso that were located about 4 km downstream of the dams – the tailings rushed at speeds of up to 60 km/h, and destroyed everything on its path
- a 31 m high tailings dam in South Africa collapsed during heavy rains in 1994, and 1.2 Mt of tailings flowed a distance of 1,960 m – the 250-house suburb of Merriespruit was destroyed, and 17 people were killed.

These failures and other tailings dam failures have occurred due to lack of application of known design methods, poor supervision during a long period of construction, and improper maintenance. Iain Bruce of Canada has compiled the statistics of dam failures since 1909 to 1999, and reports that water supply dam failures were more than tailings dam failures till 1959, and since 1969, tailings dam failures have been more than water supply dams. This could be due to the increased regulation and public attention paid to water supply dams since the 1960s. The good news is that mining companies have learned from past failures, and the number of tailings dam failures in the 1990s has reduced to 7 from 27 in the 1980s and 44 in the 1970s.

Several factors have to be considered in the design of safe tailings dams (ICOLD [3]). These include:

- hydrology and the prediction of flooding and water erosion
- seepage control with respect to minimizing excess pore pressures and preventing solid migration (piping)
- plugging of filters and drains by fines and/or precipitates
- seismicity and earthquake response of the structure
- geotechnical properties of the tailings dam foundation and dam fill material
- tailings discharge method and rate, and variations during the life of the operation.

## 6.2.1 TAILINGS DAM DESIGN

The primary focus in the design of a tailings dam is the long-term stability of the structure. The major difference between the design of a tailings dam and any other water retaining structure is that the size and capacity of the tailings dam increases as the mine production increases. Therefore, tailings dam design should be flexible and consider changes in material characteristics, climatic variations over several decades of the mining operation, and closure and reclamation of the tailings site.

The outer wall of a tailings dam is usually built with the coarse fraction of the tailings, and the other natural soils in the area. Tailings dams are built to retain solids in the tailings, and so are usually pervious. However, if the tailings are hazardous, the embankment and the bottom liner are made of low permeability clay to prevent leakage of contaminated water to the surface water or groundwater. The embankment should be designed to withstand the shear and dynamic forces resulting from seismic activity. Failures due to overtopping the embankment and erosion should also be considered. Generally, soil mechanics stability analyses are undertaken to ensure a factor of safety of between 1.1 and 1.3. Pseudo-static or dynamic analyses are undertaken to determine the factor of safety under anticipated seismic loadings at the site.

### 6.2.1.1 *Environmental considerations in tailings dam design*

Impacts from tailings dams include transport of tailings to the environment due to wind and water erosion, pollution of groundwater and surface water by hazardous substances in the tailings slurry, and downstream effects from catastrophic failure of the dam. These impacts and their considerations in design are discussed below.

Wind erosion is a problem in arid areas, and in other areas where freeze-drying or prolonged periods of drying occur. Dust from base metal tailings can cause adverse health effects on the surrounding communities. The design should include a cover over the tailings utilized for embankment building, or use of an encrustation agent that will minimize wind erosion. The operation and maintenance plan for a tailings dam should provide for wetting of the tailings to prevent wind erosion. The design of the closure plan should utilize a natural cover with the native species for successful revegetation.

Water erosion and pollution of groundwater and surface water can be minimized with a long-term water management plan. A complete understanding of the hydrology of the area and assessment of the magnitude of the rare storm events that could occur at the site is important for proper design. A complete water balance for the tailings pond will allow for the provision of adequate storage, and design of sufficient spillway capacity. Standard dam design methods can be used for design of the flood management system (ICOLD Bulletin [4]). Placement of the tailings facility at the upper end of a catchment area or in a side valley impoundment will minimize water management problems. With hazardous tailings, a design of a proper double liner system (see Ovacik Gold Mine case study) will prevent pollution of groundwater.

Designing for long-term stability and making changes to the design to accommodate changed conditions during facility operation will minimize the probability of catastrophic failure of the tailings facility. However, strict quality assurance during construction, and monitoring and verification of the stability during operation and post-closure periods is essential to prevent failure of a tailings dam. Minimum requirements of a quality assurance (QA) plan during operation and maintenance of a tailings dam include:

- measuring water content and particle size distribution of the tailings
- monitoring rainfall and evaporation
- freeboard measurement at the dam wall, and water level measurement at the spillway or the decant tower
- quantity of tailings dam deposited and volumes of water decanted
- monthly measurements of pore pressure in the dam, and movement of the dam with instruments buried in the dam
- annual monitoring of the water and wind erosion of the tailings facility.

For proper closure and reclamation of the tailings facility, the physical and chemical stability of the tailings should be studied. Chemical stability is ensured by preventing acid rock drainage, and leaching of the heavy metals in the tailings. The post-mining land use should be considered in the reclamation plan for the tailings facility. Regular inspections by trained personnel, and compliance with the approved reclamation plan are critical to the success of the long term management of tailings.

### 6.2.2   RISK MANAGEMENT

Tailings dams are sequential constructions, commencing with a starter dike. As the tailings build-up behind the dike, a new dike is created to increase storage capacity. The whole assemblage is a "tailings dam." Evidently, the tailings themselves can play a significant role in the overall stability and safety of the dam.

An important question for a tailing dam is its functional reliability or safety over its intended lifetime. This is particularly important where there are serious consequences that may arise from dam failure. There have been many cases of tailings dam failure with serious economic and human life consequences. Increasingly, this is becoming a matter of public concern. For this reason both the consequences of failure and the probability of failure within some time-frame is of interest to regulatory authorities.

Tailings dam safety ought to be of increasing interest also to operators and owners. Not only may there be serious economic and business consequences should failure occur but there may be legal implications also for responsible individuals.

The risk issues for tailings dams are, in concept, not very different from those for other potentially hazardous facilities and industries (Healy [5], Stewart and Melchers [6]). While subjective, simple assessments have been used in the past, increasingly the regulatory requirements are for more sophisticated, quantitative analyses with sound empirical and theoretical justification (Stewart and Melchers [6]). For this reason it is appropriate to review the modern approach to safety and risk assessment as applied already to other potentially hazardous industries. This will allow an indication to be given as to apply it for tailings dam risk assessment.

Underlying all risk assessments is the reasonable assumption that no system can be completely safe forever. The important question then becomes what level of risk is tolerable (and by whom). To deal with this requires in the first instance means to estimate the risks. It is with this issue that the present discussion is concerned. Some comments are made about the setting of tolerable risk levels.

#### 6.2.2.1   *Risk assessment processes*

The risk assessment process may be summarized as in Figure 6.1. Most of the steps will be obvious. Both consequence and consequence probability (failure probability) estimates are involved. These are each subject to uncertainty. They are combined to estimate "risk".

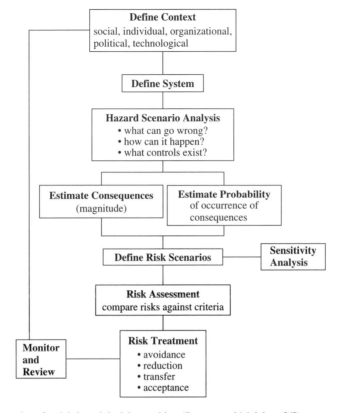

Figure 6.1.    Flow chart for risk-based decision-making (Stewart and Melchers [6]).

A critical step in the process is Hazard Scenario analysis. It attempts to elucidate all possible ways in which the system might "fail". Various techniques have been developed in practice, in different industries, and hence with different names (Stewart [6]).

Some central tools in the risk process are fault trees and event trees (or consequence diagrams). A fault tree provides an overview of the connectivity of the causes of failure. It indicates the events contributing to system failure. An event tree represents the chain of consequences arising from a failure of a logically earlier part of the system. Simple examples are given in this section.

Qualitative analysis using Figure 6.1 does not attempt to assign numbers to magnitude of consequences or to probabilities. Only subjective assessments of consequences and of their occurrence probability are employed. These may be obtained in a variety of ways, including interaction between risk analysts and operators. The results can be tabulated in a "risk-consequence" matrix. This is easy to use and understand for non-technical audiences. However, usually it is not sufficient for regulatory or for economic decision-making purposes. For these, quantitative analyses are required.

### 6.2.2.2  *Quantitative analysis*

There are two types of quantitative analysis commonly employed (Stewart and Melchers [6]). QRA (Quantified Risk Analysis) has all consequences and probability estimates as numerical "point" estimates. This allows the process in Figure 6.1 to be employed and for a risk estimate to be obtained. Unfortunately, typically much of the input information is uncertain with the result that analysts select conservative best estimates at each stage in the analysis. The result is that the assessment outcome can be highly conservative although the degree of conservativeness is not known. For tailings dams, this may be very important if major decisions hang on the risk assessment outcome. Despite this draw-back, QRA is still widely employed in the chemical and petrochemical industries, particularly for "first-pass" risk estimation.

The "state-of-the art" risk assessment procedure is PRA (Probabilistic Risk Analysis). In this the elements in the analysis are treated as random variables. Occurrence rates also may be given as random variables. Although relatively complex, it has the advantage that more explicitness is required in the analysis and that outcome uncertainty is recognized.

### 6.2.2.3   *Tailings dam application*

For illustration, consider a section through a typical tailings dam (Figure 6.2). Some potential failure modes are indicated.

The potential failure modes for a tailings dam will include slip circle failure, piping, overtopping, down-stream water table issues, ground-shaking due to earthquake activity, poor maintenance or operational problems, etc. (Vick [7]). This information can be represented in a fault tree (Figure 6.3). Typically it consists of AND and OR gates. For independent events the probabilities of occurrence are through an OR gate and are added (i.e. either event will contribute to the upper

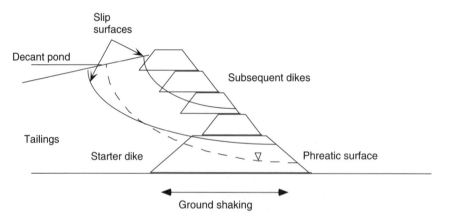

Figure 6.2.    Schematic tailings dam showing some potential failure modes and influences.

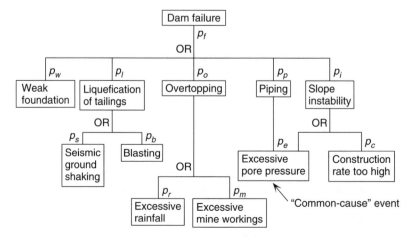

Figure 6.3.    Typical fault tree.

level event). For AND gates the probabilities are multiplied (i.e. both/all events must occur for the upper level event to occur). More complex rules apply for dependent events and for common-cause situations (Melchers [8]).

For each failure mode the associated probability of that mode occurring is required for the risk assessment. This is the most difficult part of the analysis. An indication of how this might be estimated is given in this section.

It is necessary also to consider the consequences of each form of failure. This may involve criteria related to stream and ground water pollution, downstream impacts, or potential effects on public health, etc. Usually, it is the performance of the whole system (of which the dam may be only a part) that will be of interest.

The chain of events leading to a failure outcome can be represented by an "event tree". A simple example for a tailings dam is given in Figure 6.4. It shows as an example the development of the outcome event probability for one of the chain of events, commencing with the probability of occurrence of the initiating event (slope instability). While "yes", "no" branches are common, more complex branching may occur, such as shown on Figure 6.4. Evidently, in this simple case the event probabilities are the product of the previous probability and the branching probability. More complex rules exist for dependence between events and for common-cause situations (Melchers [8]).

### 6.2.2.4   *Estimating risk probabilities*

Figures 6.1, 6.3 and 6.4 show that failure probabilities are important components of system risk estimation. However, for systems subject to stochastic processes (such as the loads tailings dams must resist) the simple tools in standard QRA and PRA need to be refined to deal with such processes. A short overview of this relatively complex subject follows.

The system shown in Figure 6.2 may fail under applied load in any (or more) of the slip surfaces shown. The probability of this occurring will depend on the applied load, the strength properties of the soil in the different layers, water table levels, etc. Any one slip circle will involve different soil layers. Typically, a slip circle must pass through several layers to cause failure. Hence, failure

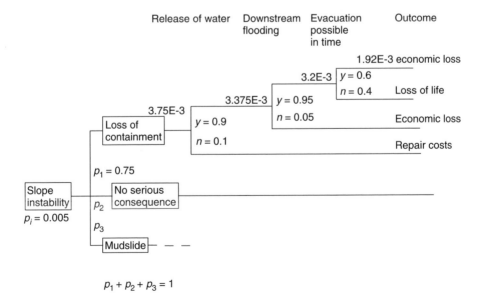

$$p_1 + p_2 + p_3 = 1$$

Figure 6.4.   Typical event tree for a failure mode of a tailings dam showing example event probability calculation.

in any one slip circle can be considered as failure of a system with (say $n$) parallel components. Slip circle failure $F_{ss}$ then requires failure $F_i$ ($i = 1, ..., n$) through each soil layer

$$F_{ss} = F_1 \cap F_2 \cap F_3 \cap etc. \tag{1}$$

Moreover, one or more slip circles can occur. Failure of the dam (or a significant part of it) now is $F_s$, brought about by failure through one or more slip circles $F_{ssj}$

$$F_s = F_{ss1} \cup F_{ss2} \cup F_{ss3} \cup etc. \tag{2}$$

where the component terms $F_{ssj}$ refer to any slip circle and may be defined by (1). Here (1) and (2) define the "failure domain" for the system.

In principle, what is of interest in dam risk assessment is the time $t_1$ to the first occurrence of a failure event (the "first exceedence event"). This time should be long for safe dams. It could be estimated readily if the exact trace of (fluctuating) loading on the dam is known. But since the loading is a stochastic process, the "first exceedence time" will be a random variable. It follows that the probability that the system fails in a given time period $[0, t_L]$ (e.g. the "design life") is the probability that the dam will fail when it is first loaded (say immediately after construction), denoted $p_f(0, t_L)$, and the probability that it will fail subsequently, given that it has not failed earlier. The necessary computations for this type of approach are often too complex and a simplified analysis generally is used. It has been applied very successfully in many practical examples.

Assume that the resistance $r$ of the dam to failure mechanisms remains constant for reasonably long periods of time. (A step-wise approximation could allow for deterioration.) Also, let the load processes be assumed "stationary", that is, their statistical properties do not change with time. As a result, probability of failure per unit time interval remains constant. Moreover, provided the dam survived its construction process (not always guaranteed, as history shows), any initial failure probability may be ignored.

For dams, there is usually only one extreme load process of interest at any one time (e.g. the water level, ground shaking, etc.). The extreme (worst) value of this load $Q$ during the dam's lifetime is of interest. It is best represented by an "extreme value" distribution $f_Q()$ (for the maximum). Evidently, it is applied only once in the dam's lifetime.

When the maximum load is applied, failure can be represented by $r < Q$ which may be given as

$$Z = r - Q < 0 \tag{3}$$

where $Z$ is the "safety margin". The probability of failure is then the probability that the safety margin is less than zero, or, more formally

$$p_f = \text{Prob}(r < Q) = \text{Prob}(Z < 0) = \int_r^\infty f_Q(x)dx \tag{4}$$

Allowing also for the fact that the strength of the dam (in any of its failure modes) is uncertain and can be represented by the associated cumulative distribution function for $R$ given by

$$F_R(r) = \text{Prob}(R < r) = \int_{-\infty}^r f_R(x)dx \quad F_R() \tag{5}$$

the failure probability becomes

$$p_f = \text{Prob}(R < Q) = \int_{-\infty}^\infty F_R(x) f_Q(x)dx \tag{6}$$

Expression (6) may be interpreted loosely as follows. Under the integral, the first term, given by (5), denotes the probability of failure given that the actual load has the value $Q = x$. The second term is the "probability" that the load takes the value $Q = x$. This is then integrated over all possible values of $x$, a dummy variable.

This simple form must be extended somewhat to cater for realistic systems since usually there are a number of random variables involved in the resistance and also, the performance of the system is more complex than considered so far. This can be done as follows, noting that (6) can be written as

$$p_f = \text{Prob}(R < Q) = \text{Prob}(Z < 0) = \text{Prob}[G(\mathbf{X}) < 0] \tag{7}$$

where now $G()$ is known as the performance function, being a generalization of $Z$. The vector $\mathbf{X} = (R, Q)$ denotes the (random) loads and resistances. This approach also allows extension of (6) to situations where there are several performance functions to be met, for example, each of the slip circles in Figure 6.1. For this case, for example, (7) becomes

$$p_f = \int \cdots \int_{\cup_i G_i(\mathbf{X}) < 0} f_{\mathbf{X}}(\mathbf{x}) d\mathbf{x} \tag{8}$$

where $f_x()$ is the joint density function of the random variables $\mathbf{X}$. One obvious possibility to solve (8) is through Monte Carlo simulation. However, in its elementary form this is highly inefficient. An alternative is use the so-called First Order Second Moment (FOSM) method. It is widely used, even though it is not always very accurate. Details are available in the literature (Melchers [8]) and an example of its application follows.

### 6.2.2.5 *Example of probability of slope instability*

For a slip circle it is conventional to write the factor of safety for the $k$-th slip circle as (Tan et al. [9]):

$$FoS_k = \frac{\sum_{i=1}^{n}(c_i + \sigma_i \tan\phi_i)\Delta L_i}{\sum_{i=1}^{n}\tau_i\Delta L_i} \tag{9}$$

where, as usual, $c_i$ is the cohesion of the $i$-th segment and $\phi_i$, $\sigma_i$ and $\tau_i$ the corresponding angle of internal friction, normal stress and shear stress respectively. $\Delta L_i$ is the arc length of the $i$-th segment and there are $n$ such segments contributing. To conform to the convention for performance functions, (9) is expressed in terms of the performance function $G()$ (see (7) above) as

$$G_k = \sum_{i=1}^{n}(c_i + \sigma_i \tan\phi_i)\Delta L_i - \sum_{i=1}^{n}\tau_i\Delta L_i = 0 \tag{10}$$

From observations of the variability of field data, it has been found that the coefficient of friction $\mu_i = \tan\phi_i$ is more closely described by a normal distribution. This will be adopted here as FOSM theory can be used for this example. Further, finite element analyses (FEA) show that the stresses $\sigma_i$ and $\tau_i$ at any point in the dam are approximately linearly related to the bulk density $\gamma_i$ through constants $k_{\sigma i}$ and $k_{\tau i}$ respectively, obtained for each segment from the FEA. These may be substituted in (10) to provide a simpler performance function.

Statistical properties for the variables $c$, $\phi$, $\mu$ and $\gamma$ for both total and effective stresses could be obtained from the literature and from in-situ tests prior to commencement of the design. In other cases, it may be possible to use literature estimates, to be confirmed after construction commences. It is common to neglect correlation between the random variables and to ignore spatial

Table 6.1.   Assumed statistical parameters.

| Material | Variable | Mean | Coefficient of variation |
|---|---|---|---|
| Clay | $c$ | 53 kPa | 0.40 |
| | $\mu$ | 0.26 | 0.385 |
| | $\gamma$ | 18 kN/m$^3$ | 0.059 |
| Filter | $\mu$ | 0.63 | 0.06 |
| | $\gamma$ | 20.4 kN/m$^3$ | 0.025 |
| Transition | $\mu$ | 0.63 | 0.06 |
| | $\gamma$ | 20.4 kN/m$^3$ | 0.025 |
| Rockfill | $\mu$ | 0.63 | 0.06 |
| | $\gamma$ | 20.4 kN/m$^3$ | 0.025 |

Table 6.2.   Results of slip circle probability analyses.

| Slip circle no. $k = 1, ..., 6$ | Probability of failure |
|---|---|
| 1 | $5.2 \times 10^{-7}$ |
| 2 | $7.0 \times 10^{-8}$ |
| 3 | $7.55 \times 10^{-8}$ |
| 4 | $1.3 \times 10^{-5}$ |
| 5 | $2.68 \times 10^{-5}$ |
| 6 | $7.7 \times 10^{-9}$ |
| Overall system | |
| Upper bound | $2.76 \times 10^{-5}$ |
| Lower bound | $2.78 \times 10^{-5}$ |

variability. To include these matters adds very considerably to the computational effort required, usually for relatively little gain in accuracy compared with the high uncertainty attached to the estimates for mean and variance (or for the coefficient of variation) for each random variable. Table 6.1 shows values of parameters adopted for the analysis of a typical dam (Conrad [10]).

Using these values in the performance function (10) requires that for each slip circle, proper account is taken of the geometry to have the appropriate contributions to each segment. Results of the FOSM analysis for each of 6 slip circles are given in Table 6.2.

Evidently, in this case the overall failure probability is closely controlled by slip circle 5 since it makes the largest contribution. Typically, sensitivity analyses would be performed to check on how the probability estimate varies with changes to the parameters.

### 6.2.2.6   *Monitoring and safety updating*

Monitoring of the state of a dam can provide useful information about its continuing safety. Evidently, planning of the inspection processes and the integration of the results with the safety assessment is important. Usually, the interpretation of the monitoring observations requires appropriate expertise.

As new understanding of the system becomes available through monitoring, the parameter values for the random variables used in an earlier risk analysis may need to be modified. There are theoretical procedures in probability theory such as Bayes theorem available for doing this (Melchers [8]). In many cases, however, it is simpler to use the newer estimates in a revised safety analysis. This approach is referred to as a "Living Risk Assessment".

Inspection and monitoring can improve the estimate of the safety of the dam. This is sketched in Figure 6.5 in terms of reliability. The estimate of the reliability goes up as a result of favorable observations at the point "inspection time point".

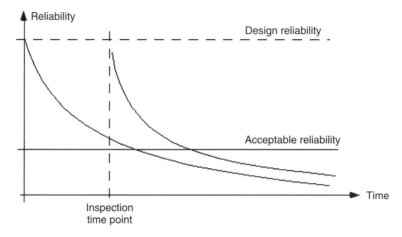

Figure 6.5.   Schematic variation of reliability showing effect of an observation at "inspection time point".

The above suggests the possibility of managing dam safety to optimize some desirable criterion. Examples might include minimizing the total expected costs, minimizing environmental impact or reducing to a minimum the likelihood of regulatory breaches and fines. The decision for the choice of criterion will depend on the regulatory environment and/or company objectives. The possibility also exists to optimize the times between monitoring or the intensity of monitoring and to optimize repair and maintenance policies.

### 6.2.2.7   *Acceptance criteria*

"Satisfactory" dam performance demands that one or more criteria are available against which to measure it. The setting or derivation of performance criteria for hazardous facilities is a problem with a fascinating and complex history, which has probed some deep philosophical questions (Conrad [10], Royal Society Study Group [11]). Psychological aspects such as risk perception and risk aversion, social and cultural preferences as well as political processes and risk communication will play a part also.

There are essentially two categories of decision criteria – economic and regulatory. These would be expected to apply also to dam risk assessment.

### 6.2.2.8   *Economic or pseudo-economic criteria*

Economic evaluation has the advantage of forcing all parties to evaluate their objectives in monetary terms. (Pseudo-economic criteria, such as utility analysis, require a similar approach but in terms of a different unit of measurement.) In principle, the idea is to maximize the expected net present value of the system as a function of all information about the system and the decisions to be made. In principle, most organizations will perform some form of assessment along these lines but the more general procedure requires inclusion of all potential consequences and associated risks. Details are available in the literature (Stewart and Melchers [6], Royal Society Study Group [11]).

### 6.2.2.9   *Regulatory criteria*

Typically regulatory criteria are developed by public authorities acting, broadly, on behalf of the public and mandated by government. Typically, regulatory authorities develop general safety goals and set specific safety standards, monitor system performance, and prosecute if specified safety standards are violated.

Table 6.3.   Typical safety targets for land use.

| Land use | Individual fatality risk ($\times 10^{-6}$ per year) |
|---|---|
| Hospitals | 0.5 |
| Residential, hotels, motels, holiday resorts | 1 |
| Commercial (inc. retail centers, offices and entertainment centers) | 5 |
| Sporting complexes and active open space | 10 |
| Industrial | 50 |

There are a number of possible "measurement units" available, including expected number of deaths, injuries or cost equivalents, short and long term expected health implications and various environmental criteria. None are completely satisfactory measures. Typical sets of "safety targets" for different industries are shown in Table 6.3.

A somewhat different but related approach is that of the "safety case". It consists of a document describing how the regulatory safety goals have been met. Such a document is reviewed or audited by the regulatory authority to ensure that (i) the study deals in sufficient depth with the facility under discussion (completeness requirement), (ii) appropriate event probabilities and consequences have been considered, and (iii) compliance to the relevant regulations has been achieved and documented. An important advantage is that the onus of proof is put back on to operator. It is used extensively in the off-shore, chemical and petro-chemical industries, particularly in Europe.

### 6.2.2.10   *Mixed criteria*

A mixed economic and regulatory framework for decisions is the so-called ALARP (As Low As Reasonably Practical) or the ALARA (As Low As Reasonably Attainable) approach. Although terms as "low", "reasonably", "possible" and "attainable" are highly subjective and difficult to define, ALARP has been adopted widely, for example, by the U.S. Nuclear Regulatory Commission and others and in the off-shore industry.

### 6.2.2.11   *Summary*

Safety estimates depend heavily on good quality understanding of the system being considered. They also depend on the information in the "tails" of probability distributions. Although simplified models can be used in both cases, it is obvious that the quality of the safety assessment will depend on the quality of the input. For quantified, probabilistic risk assessment as outlined above, this may make severe demands on analysts.

Unfortunately, there are no real alternatives. Simpler methods of safety or risk assessment, which do not rely on such detailed understanding, modeling and information about stochastic processes and probabilistic variables essentially "hide" the difficulties and they do this by ignoring the uncertainties. This can easily lead to a false sense of security.

The incompleteness of data and of probabilistic models need not, however, be an insurmountable obstacle in applications. Experience gained "in-service" through monitoring and observation can be used to refine understanding of the system and of the characteristics of the processes involved. As indicated above, there are recognized procedures for so doing.

The second approach is to recognize that "subjective" estimates and probabilities can be used in engineering safety assessment. Engineers often have no choice but to be quite pragmatic about what can be measured and what must be assumed and to use whatever is available and justifiable in safety assessment procedures. There is theoretical justification for this, even if not accepted by all theoreticians.

6.2.3   CASE STUDIES OF TAILINGS DAMS

The following are case studies of tailings dams.

6.2.3.1   *Ovacik gold mine, Turkey*

The Ovacik gold mine is located 100 kilometers (km) north of Izmir, and about 15 km inland from the Aegean Sea in western Turkey. Annual ore production of 300,000 tons from both open pit and underground mines is processed to recover about 100,000 ounces of gold. The tailings contain cyanide and heavy metals, and the challenge in the design of the tailings facility is to prevent these hazardous substances from adversely impacting the environment. The mine is located in an area of seismicity with a peak ground acceleration of 0.4 g (where g is the gravitational acceleration constant), as recorded by the Turkish Earthquake Research Institute. The probability of an earthquake with this acceleration is 1 in 500 years, and of an earthquake with an acceleration of 0.6 g is about 1 in 3,000 years (ICME/UNEP [1]).

The International Nickel Company (INCO) cyanide destruction process is used to reduce the cyanide in the tailings to less than 1 part per million (ppm). The tailings dam is designed to withstand an earthquake of 0.6 g acceleration. The unique feature of the tailings dam is the composite lining system to prevent the infiltration of contaminants into the groundwater. The composite lining system consists of 500 mm of compacted clay overlain by 1.5 mm of high-density polypropylene, 200 mm of compacted clay and 200 mm of gravel. The embankments are made of compacted rockfill, and provide storage for 2 million cubic meters of tailings. Features of the tailings dam facility include a decant water system to recirculate the water, a leachate collection and recovery system to prevent water build up and liner damage, and a surface water diversion system to prevent wall erosion.

The construction of the tailings facility was managed by qualified engineers, and supervised by the Turkish Dam Safety Inspection agency. The rockfill was constructed by placing rock in 1-m lifts and compacting the lifts by eight passes of a 15-ton compactor. The liner construction was completed under international standards, with special emphasis on testing the seams in the polypropylene liner. Holes found in the seams were repaired and vacuum tested. The tailings facility was independently audited and was found to comply with both national and international environmental standards.

6.2.3.2   *Strathcona mine, Ontario, Canada*

The Strathcona mine of Falconbridge Limited is located about 50 km north of Sudbury, Ontario, and processes 10,000 tons of nickel-copper ore per day. Since 1968, the mill has produced a total of about 30 Mt of tailings. The tailings are deposited hydraulically into a 100 ha basin located several kilometers from the mill (ICME/UNEP [1]). The tailings are split into a slimes fraction containing less than 1% sulfur, and a coarse fraction which is used as backfill in the underground mines. The tailings are also split into two streams, one containing iron and sulfur and the other with low sulfur. The streams are segregated into cells within the tailings basin. The tailings are maintained in a physically and chemically stable condition by following management procedures listed below:

- a preventive maintenance plan
- a spill response plan
- an emergency preparedness plan
- subaqueous disposal of high sulfur tailings to prevent oxidation and acid generation
- construction of engineered containment structures and tailings transportation systems
- a tailings depositional plan
- an operations manual for all systems
- documented water management plan
- identified skill requirements and adequate training for staff

- an environmental monitoring and reporting plan
- meticulous record keeping on operational conditions, construction detail, maintenance completed, and any unusual occurrences
- a documented roles and responsibility structure
- tailings closure considerations in all aspects of operation and maintenance
- long-term financial requirements for proper closure and post-closure monitoring.

The inspection and monitoring program followed by the Strathcona tailings facility includes the following:

- daily tailings line inspections, water level and quality monitoring
- annual visual inspection of the perimeter containment structures
- corrective action for areas needing improvement
- an external audit by a geotechnical engineering firm, as needed
- scheduled environmental audits by qualified personnel
- extensive reviews of older containment structures, as needed.

### 6.2.3.3   *Emergency response at Omai tailings dam failure, Guyana*

The Omai gold mine in Guyana is 160 km southwest of Georgetown, the country's capital. The mine, started in January 1993, is currently producing at the rate of 345,000 ounces of gold per year. The tailings facility is located north of the plant site in a small valley, and is constructed as an earth fill embankment. On August 20, 1995, the earth fill dam cracked and tailings containing cyanide wastes were discharged into the Omai River. The mine crew immediately excavated a diversion ditch to carry a major portion of the flow, approximately 1.2 million cubic meters, to the main mine pit.

This action minimized the impact to the Omai River, and the large Essequibo River downstream of the Omai. However, about 2.5 million cubic meters of tailings reached the Omai River over the 4.5 days it took the mine crew to build a coffer dam to contain the flow. Company personnel notified people living along the Essequibo River of the incident, and made sure that they were not adversely impacted by the tailings flowing into the River. The environmental damage documented by the authorities was a fish kill in the Omai River, and a plume of suspended clay particles in the river for several days following the spill. An extensive investigation by the government and international experts was conducted to understand the reasons for the tailings failures, and suggest measures to prevent such incidents.

Lessons learned from the emergency response to this tailings failure include the following:

- a local team of competent personnel should take immediate action instead of waiting for permission from management personnel
- the measures described in the emergency response plan should be implemented without wasting any time
- an effective communication system should be used to convey the information to management personnel, the local community, government, media, and others impacted by the spill
- the management should confront the crisis in the best way possible, and restore calm to the affected population
- the remediation of the problem should involve all competent personnel and any experts to determine the most effective procedures to fix the problem and restore any damage caused by the spill
- the management should handle the public relations involved with honesty and build trust with the local community and government.

The excellent response to the accidental spill and extensive investigation of the problem resulted in the company getting permission to resume operations within six months of the accident.

# 6.3  Waste rock and soil

Raj Rajaram

*Complete Environmental Solutions, Oak Brook, IL, USA*

## 6.3.1  INTRODUCTION

Mining produces a lot of waste rock and soil from the removal of overburden (in open pit mining) and rock that does not meet the required ore grade for processing. In heap leaching operations, the spent ore pad has to be properly managed to prevent surface and groundwater contamination. Mining legacies have been created by past mining operations that did not consider the environmental impacts of waste rock and soil, and disposed of this material in low areas surrounding the mine site. This has resulted in acid mine drainage and heavy sediment loads into surface water bodies and adverse impacts to groundwater underlying the waste piles.

Efforts in the United States and Europe have been undertaken to address these mining legacies, and methods have been developed to minimize the environmental impacts of these waste piles. By incorporating these methods into current mining operations, legacy waste will not be left for future generations to manage. This section describes some of the methods available to manage waste rock and soil piles.

## 6.3.2  PLANNING TO CONTROL ACID MINE DRAINAGE

A detailed mapping of the overburden strata above a coal deposit can help identify soils and rock that produce acid mine drainage (AMD), and soils and rock that produce alkaline mine drainage. Acid base accounting methods have been used by the coal industry for several decades. The U.S. Bureau of Mines developed a pre-mine prediction methodology to determine the acid generation potential from a planned mining operation, and suggested methods to minimize the generation of AMD (USBM [12]). Use of this prediction methodology, and devising overburden handling methods to neutralize AMD potential can significantly decrease the adverse impact from overburden waste piles.

Acid base accounting can also be performed for sulfide metallic ore bodies, and waste rock disposal methods planned using this information. Waste disposal site selection to minimize infiltration of AMD into the groundwater, and designing methods to collect and treat the AMD resulting from the waste piles can prevent adverse impacts from AMD. In the past decade, paste backfilling has been used to reduce the AMD potential from waste rock left underground and from surface waste rock piles. This technology creates a dense paste with the finer portion of the tailings, water and cement, and pumps the paste to the location where it is needed to seal the voids in the waste rock. It has been effective in minimizing AMD, and in underground applications, stabilizing the waste rock and supporting underground openings.

## 6.3.3  TREATMENT OF AMD FROM WASTE ROCK AND SOIL

The adverse impacts from AMD has been demonstrated over the last 30 years through measurement of water quality downstream of mining operations. From the unsightly red staining in creeks and rivers to the adverse impacts on fish and other biota, the environmental and economic impacts of AMD have been studied in many watersheds in the United States. This has led to a significant body of research on low cost treatment options for managing AMD. Different methods of lime application for active and passive treatment of AMD, and treatment technologies to remove the manganese and other minerals in AMD have been developed in the United States. A prime driver in the development of these technologies is the financial assurance that the mine operator has to provide the regulatory agency when obtaining a mine permit. Many agencies are requiring that the mine operator provide sufficient financial assurance to ensure that the AMD resulting from a mine will not have adverse impacts for 50 years after mine closure (Office of Surface Mining [13]). This has resulted in proactive measures to prevent AMD, and active treatment of AMD during mine operations.

Monitoring requirements are also spelled out in the permit for the post closure period. The financial assurance provided by the mine operator ensures that if the mine operator abandons his/her responsibility, the regulatory agency can select a contractor to perform the monitoring and take any corrective measures necessary to prevent adverse impacts from AMD. Such requirements are making the mining industry take responsibility for their actions, and making mining a sustainable enterprise.

## REFERENCES

1. International Council on Metals and the Environment/United Nations Environment Programme (ICME/UNEP), Tailings Management, ISBN 1-895720-29-X, Nov. 1998.
2. International Commission on Large Dams/United Nations Environment Programme (ICOLD/UNEP), A Guide to Tailings Dams and Impoundments: Design, Construction, Use and Rehabilitation, Bulletin 106, New York, New York, 1996.
3. International Commission on Large Dams (ICOLD), Tailings Dams and Seismicity, Bulletin 98, 1995.
4. ICOLD, Selection of Flood Design, Bulletin 82, 1992.
5. Henley EJ and Kumamoto H.: *Reliability Engineering and Risk Assessment.* Prentice Hall, New Jersey, 1981.
6. Stewart MG and Melchers RE.: *Probabilistic Risk Assessment of Engineering Systems.* Chapman & Hall, London, 1997.
7. Vick SG.: *Planning, Design and Analysis of Tailings Dams.* John Wiley & Sons, New York, 1983.
8. Melchers RE.: *Structural Reliability Analysis and Prediction, Second Edition.* Chichester, John Wiley & Sons, 1999.
9. Tan CP, Donald IB and Melchers RE (Eds): *Probabilistic Slip Circle Analysis of Earth and Rockfill Dams* [in] *Probabilistic Methods In Geotechnical Engineering.* KS Li and SCR Lo, Rotterdam, Balkema, 1993, pp. 281–290.
10. Conrad J (Ed.): *Society, Technology and Risk Assessment.* London, Academic Press, 1980.
11. Royal Society Study Group: *Risk: Analysis, Perception and Management.* London: Royal Society, 1992.
12. U.S. Bureau of Mines, Pre-Mine Prediction of Acid Drainage Potential, U.S Government Printing Office, Washington, DC, 1986.
13. Office of Surface Mining, Methodology for Estimating Treatment Costs of Acid Mine Drainage, Pittsburgh, Pennsylvania, 2001.

# 6.4 Maintenance wastes

Turlough Guerin
*Shell Oceania, Rosehill, New South Wales, Australia*

## 6.4.1 INTRODUCTION

Maintenance wastes, if not managed properly, represent significant environmental issues for mining operations. Petroleum hydrocarbon liquid wastes present a specific maintenance waste issue. This section provides an overview of the issues arising from maintenance activities that can impact upon the sustainability of a mining operation. Treatment technologies and practices for managing oily waste water, used across the broader mining industry in the Asia-Pacific region, are described. The importance of waste stream characterization and a framework through which a mine can evaluate the effectiveness of their existing oily waste water management systems is provided. Source reduction, through proper housekeeping, equipment and engineering modifications, and segregation and/or consolidation of hydrocarbon waste streams, minimizes treatment costs, improves safety and reduces environmental impact.

Globally, the mining industry is undergoing changes in how it approaches the management of both process and non-process wastes. Hydrocarbon contaminated wastes, a major non-process waste, generated from maintenance practices, pose a very important challenge in this regard.

Mining maintenance involves the generation of wastes, many of which contain petroleum hydrocarbons. It is therefore important that these be managed properly so that potential adverse environmental effects are avoided. Maintenance activities at mining operations can generate waste streams that contain spilt fuel, used oil, grease, silt, degreasers and detergents. These wastes include solid materials such as oil filters, oil-soaked rags and liquid wastes containing waste oils, oily waste water and degreasing agents. There are other non-waste related impacts from maintenance activities and these include energy use, impacts on soil and water quality. Environmental implications from maintenance in activities in the mining industry are highlighted in Table 6.4.

A maintenance strategy for equipment in the mining industry therefore needs to be a key component of a sustainable mining operation.

There are now a number of trends evident across the mining industry in Australia, which indicates that mining maintenance, fluids management, and contractor management areas are now being seen as important contributors to sustainable mining. These are as follows:

- Mining companies are looking for solutions for managing their hydrocarbons and other fluids from the point of supply through to disposal and re-use options.
- Mining companies are seeking out suppliers to their industry that can provide greater stewardship of the products they supply including petroleum hydrocarbons and the resultant wastes that may be generated from use of their products.
- Mining companies are consolidating the number of contractors on their sites including suppliers and waste management contractors.

These issues are central to petroleum hydrocarbon management and maintenance activities at mining operations across the globe.

## 6.4.2 MANAGEMENT OF PETROLEUM WASTES

This section provides an overview of how to conduct a maintenance waste audit, provides the results of a survey of oily waste water management at mining operations across the minerals industry in the Asia Pacific region and the types of systems and approaches used for managing oily waste water streams. A review of the literature and equipment vendors is also provided to assess options for managing the hydrocarbon waste streams identified.

### 6.4.2.1   *Auditing of maintenance waste streams*

Auditing of maintenance waste is the process of identifying what wastes are present and understanding how these are generated. An audit allows a proper definition of the waste streams and helps develop a management plan.

Table 6.4.   Impacts on sustainability from maintenance activities at mining operations.

| Aspect of maintenance program[1] | Description of impact or potential impact[2] |
| --- | --- |
| Liquid petroleum hydrocarbons[3] | Groundwater contamination from spillages and leaks[4] <br> Soil and surface water contamination from spillages and leaks[4] <br> Air pollution (from uncontained volatile components if present) <br> Disposal without re-use or energy recovery is unsustainable |
| Solvents | Air pollution – uncontained vapors could adversely affect human health <br> Disposal without re-use or energy recovery is unsustainable <br> Groundwater contamination from spillages and leaks[4] |
| Waste grease | Surface soil contamination from spillages and leaks[4] <br> Soil and groundwater contamination from petroleum hydrocarbons and metals impregnated in grease[4] <br> Disposal without re-use or energy recovery is unsustainable |
| Detergents | Toxicity to numerous aquatic organisms; Emulsification of petroleum hydrocarbons (and reduced waste oil recovery) |
| Sediment (suspended solids) | Surface water contamination from spillages and leaks; Blockages cause flow-on effects in oily wastewater management systems |
| Contaminated rags, paper, protective equipment and gaskets[5] | Excessive use and wastage is unsustainable |
| Used oil filters, burst hydraulic hoses, empty fuel, oil and grease containers (e.g. drums) | Often landfilled which is unsustainable. Petroleum hydrocarbons within these materials have the potential to impact soil and groundwater (i.e. act as contamination sources) |
| Energy usage (powered equipment) | High levels of use can be unsustainable if not linked into an energy generation system on site |
| Water usage (for washing vehicles prior to maintenance) | High levels of use are unsustainable (may reduce quantity and quality of water supplies for other uses e.g. available for local communities near mine) |

[1] Including contaminants, co-contaminants or waste types.
[2] Note that these are qualified as "potential" impacts because these wastes will not always have a negative impact on the environment, i.e. if they are managed and disposed of properly.
[3] These include fuels and engine, hydraulic, gearbox, differential, and steering oils, off-specification fuels and oils, cutting fluids, and oily wastewater.
[4] Soil and groundwater contamination can add significantly to the environmental liability of a mining operation and site investigations, risk assessments and remediation programs may be required to address these at closure or as part of the divestment of the operation.
[5] Also include oil contaminated plastics, leather and cloth (e.g. gloves).

The main petroleum hydrocarbon waste streams at mining operations are typically oily waste water, waste oil and diesel, waste grease and contaminated sludge. Other hydrocarbon waste streams include oily rags, oily gloves, breathers, oil filters and spill response absorbents. Spills, leaks and lubricant change-outs are the primary sources of waste oil, whilst overflows are the primary source of waste diesel. Examples of petroleum hydrocarbon waste streams from maintenance activities are listed in Table 6.5.

Table 6.5. Sources of wastes from maintenance operations in the minerals industry[1].

| Stage of mining process | Practice or specific site location | Type and source of waste |
| --- | --- | --- |
| Exploration | Drill maintenance areas | Spillages and leakages of oils, grease, and degreasers during maintenance to drilling rigs |
| | Drill mast maintenance areas | Grease and oil sand blasted from mast frame before overhaul maintenance and re-painting is carried out |
| | Drilling operations | Drilling mud with ores containing hydrocarbons |
| Mine | Shovels, excavators, scrapers, backhoes, wheel loaders and bucket loaders | Waste oil from oil changes to mine equipment, spillages from breakdown maintenance, blown hydraulic hoses, spillages from refueling, maintaining oil and grease levels on field equipment; empty drums and used protective clothing |
| Maintenance | Wash down areas | Wash down of mobile equipment, effluent containing oils, diesel, grease, detergents and soil |
| | Heavy vehicle equipment servicing | Oil and filter changes on mobile equipment, waste grease containers, blown hydraulic hoses, used protective clothing and lead acid batteries; waste tires; worn brake pads; solvent for engine parts cleaner; plastic drums; waste coolant, brake and transmission fluid |
| | Light vehicle servicing | Oil and filter changes on mobile equipment, waste grease containers and lead acid batteries; tire bay wastes; general waste around car ramps; worn brake pads; solvent brake cleaner; solvent for engine parts cleaner; plastic drums; waste coolant, brake fluid and transmission fluid |
| | Servicing pits | Spillage during vehicle servicing, regular greasing and cleaning out of sludge pits; used protective clothing |
| | Workshop floors | Spillage onto workshop floor during maintenance and repairs, and leakage and spillage from oil storage area and from wash down practices |
| | Oily wastewater separators | Incorrectly designed or poorly maintained equipment |
| | Oil filter draining | Spillages around collection vessel |
| | Waste oil storage | Spillages during storage and transfers |

*(Continued)*

Table 6.5. (*Continued*)

| Stage of mining process | Practice or specific site location | Type and source of waste |
|---|---|---|
| | Workshop drain cleaning | Sludge (from build-up) |
| | Compressor sheds | Oil changes, leakages, compressor clean down, water/oil drainage from filters and air receiver, clean down of concrete floor |
| | Drum storage areas | Leaks/spills from drums, wash down of concrete floors and drum cleaning |
| | Fuel supply depots and infrastructure | Leaks of diesel and gasoline (on- or off-site), sometimes from underground supply pipework; refueling leaks (overflows and broken seals); surface water run-off. |
| | Oil supply bays | Spills during filling of storage tanks, filling of vehicles and mobile tankers |
| | Equipment refueling | Spillage (overfilling) during refueling of equipment and servicing trucks, leaking pumps and blown hoses |
| Upstream (or primary) processing | Processing plants | Oil changes on scrubbers, screens and conveyor belts; grease |
| | Crusher areas | Dust suppression foam; grease and oils |
| Ore shipment/transport | Stackers, reclaimers, conveyors, train load out areas | Grease and leaked oil, particularly hydraulic fluid |
| Downstream processing | Milling, smelting, refining, preparation for sale | Various metals and minerals; Various hydrocarbons spills, soil and groundwater contamination in particular from lubricants, cutting fluids, and hydraulic fluids. |

[1] This is a comprehensive listing of potential sources and types of wastes observed at mining operations during site visits by the author.

Characterization of individual waste streams is an important first step in the selection of an appropriate treatment technology. This can include assessing the quality of waste oil, such as % water and/or solids in waste oil. The design calculations for oil-water separation systems require that certain characteristics of the hydrocarbon waste stream be known. The more thoroughly the waste water is characterized, the more predictable the performance of the separation process or technology. This characterization should be conducted as part of an initial assessment or review of the oily waste management system and practices at an operation.

The majority of oily waste water streams from maintenance activity contain both free and emulsified oils, and cleansers, which usually contain silt and grease. Cleaning of spills, leaks and overflows with floor washdown can generate a large proportion of an operations' oily waste water (in some instances up to 60%). The majority of washdown water passes through triple plate separators or sumps before being treated for discharge to the surrounding environment. Only a proportion of the waste oil or diesel is typically recovered from oily waste water streams.

Information on oily waste water characteristics can be obtained from site personnel including field notes, waste water chemical assessments, supply inventories, and discussion with operational personnel.

Figure 6.6 illustrates a process that can be followed to conduct a waste audit at a mining operation. This is a general guide and can be used for characterizing a range of different waste streams.

The analysis of oily waste water from the heavy vehicle washdown facility, during a typical year of operation shows that high concentrations of methylene blue active substances (MBAS), an

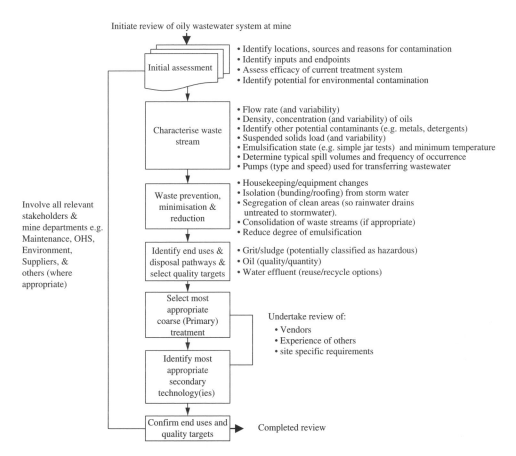

Figure 6.6.   Auditing oily wastewater management.

indicator of the presence of cleansers, and oil and grease (O&G) are typically present. These are expected from such a maintenance waste stream because detergents are commonly used and waste hydrocarbons are often dislodged during the vehicle cleaning process. There are often large discrepancies between the O&G and total petroleum hydrocarbon (TPH) values. Traces of zinc, copper and cadmium are often detected at several times the limits of detections for these compounds in oily wastes in washdown water. Increases in metal concentrations in oil normally occur during machine operation. The major metals expected in used machine oil include aluminum, copper, chromium, titanium, molybdenum, nickel, and antimony.

Cleansers (including degreasers, detergents and other surfactants) in waste water are derived from the cleaning and regular maintenance of machinery and vehicles. In addition to steam cleaning operations using only water, strong degreasing agents have traditionally been used in the industry. The older generation compounds were often organic and included chlorinated solvents. They were often petroleum-based and consequently were excellent in cleaning oily surfaces. Older generation cleansers often referred to as hard detergents are biodegraded in a range of environments. However, some degradation products, particularly from these hard detergents, can be toxic in the environment and should be limited as far as practical (Scott and Jones [1]).

Oily waste water management practices used in the mining industry across the Asia-Pacific region were reviewed from 22 mining operations. The results from the review of Asia-Pacific mining operations are as follows:

- Gravity separators are the most common techniques for separating oil from water. Nine of these were in operation across the 22 sites surveyed.

- Six of the operations surveyed used coalescing plate separators. These coalescing plate separators performed at varying levels of effectiveness.
- Retention ponds and skimmers were also commonly used with 6 of the 22 operations surveyed employing these.
- The ineffectiveness of the systems used was largely due to poor maintenance on the oil-water separation technology itself.
- It was evident that the oil-water separation technologies were generally not maintained for optimum performance.
- There was range of water quality targets for the effluent streams for each operation, reflecting differing regulatory and operational requirement.
- Common feedback from operations was that they were seeking external services for improving and maintaining their oil-water separation and oily waste water treatment systems.

The detailed findings from the survey are provided in Table 6.6.

Table 6.6. Results of survey of maintenance oily wastewater treatment technologies and practices used across Asia-Pacific mining operations and their effectiveness[1].

| Main technology applications and oily wastewater issues at mining operations[2] | Minimization (M), recovery/recycling (R), treatment (T) | Effective?[3] | Climate |
|---|---|---|---|
| *Traps on stormwater drains (1 site)* | | | |
| Used to eliminate oil sheen and retain free oil from spills | M | Yes | Temperate |
| *Centrifuge (1 site)* | | | |
| Oily sludge from triple interceptors from around the plant and power station | R | ND | Tropical |
| *Oil skimmers (7 sites)* | | | |
| An oil trap uses a skimmer prior to retention dam | R | No | Cold, wet |
| Proposed for sedimentation ponds (selected system is based on adsorbents and a collection drum) | R | ND[4] | Arid |
| For one particular sediment/oil trap, a. skimmer is proposed to recover diesel | R | ND | Arid |
| Floating weir-type skimmer used in dry weather, post grit sump but prior to interceptor | R | ND | – |
| Wash waters containing free oils, diesel and silt | R | Yes | Arid |
| Washdown water from truck wash (free oil, diesel, greases, and silt) (at 2 sites) | R | No | Arid |
| *Gravity separator systems (9 sites)* | R | | |
| Heavy equipment washdown and small vehicle refueling spills (small amounts of free & emulsified oils, fuels, and sediment) | | Yes | Tropical |
| Equipment and heavy vehicle washdown areas | R | No | Arid |
| Maintenance workshops producing wastewater (solids and free oil) | R | Yes | Cold, wet |
| Workshop and truck washdown generating detergent, free and emulsified oil | R | No | Tropical |
| All oily wastewater from mine | R | Yes | Tropical |
| Washdown water and maintenance workshop spillages (soil, oil, detergent) | R | No | Arid |

*(Continued)*

Table 6.6.   (*Continued*)

| Main technology applications and oily wastewater issues at mining operations[2] | Minimization (M), recovery/recycling (R), treatment (T) | Effective?[3] | Climate |
|---|---|---|---|
| Refueling bay for mobile vehicles | R | No | Arid |
| Washdown water from truck wash (free oil, greases, and silt) | R | Yes | Arid |
| Washdown water from heavy and light duty equipment (oily sludge, free oil, silt, and diesel) | R | Yes | Arid |
| *Coalescing plate interceptor (6 sites)* | | | |
| Hardstand runoff (potential spills). Treated effluent from primary gravity separator (used to improve the quality of oil recovered from maintenance streams). Has first flush treatment | R | ND | Cold, wet winters |
| Used on wastewater with 200–1,000 mg/L oil | R | ND | Tropical |
| Used for heavy mobile equipment washdown water (500–800 mg/L O&G, ~2,000 L/min) | R | No | Cold, wet winters |
| For washdown water after grit removal system (degreasing solvents, diesel fuel, lubricating oil and hydraulic fluids). Also for free oil from spillage and emulsified oil from degreaser usage; high concentration of grit and gravel | R | ND | Arid |
| For spillages of diesel and gasoline, which become mixed with water | R | Yes | Not provided |
| *Coal fines (1 site)* | | | |
| Residual oil is recovered & treated when recycled process water is mixed with retention dam water | R, T | ND | Temperate |
| *Drain/creek and retention dam/sedimentation pond (6 sites)* | | | |
| Primary treated effluent from sediment/oil traps and interceptors | T | Yes | Arid |
| Stormwater, untreated washdown water, "treated" effluent from oil/sediment traps | T | Yes | Cold, wet winters |
| Treated effluent from oil traps and stormwater. Bund overflow from maintenance oil storage area in high rainfall periods | T | Yes | Tropical |
| Workshop water (often >1,000 mg/L O&G) and heavy vehicle washdown (high volumes) | T | Yes | Temperate |
| Workshop oily wastewater | T | Yes | Temperate |
| Workshop and truck washdown (oil, detergents, suspended solids and solids) | T | ND | Temperate |
| *Bioremediation (3 sites)* | | | |
| Land treatment of oily wastewater and sludge | T | Yes | Arid |

[1] These sites were identified by surveying 22 operations in Australia and the Asia Pacific region. Preventative approaches to oily wastewater management were not within the scope of the site visits in which these observations were made (focus was on oily waste water treatment).

[2] Each row represents a mine site unless otherwise stated.

[3] Qualitative as assessed and provided by site personnel at each site and from observations of the author during site visits and assessment of any available data.

[4] ND = not determined or an assessment unable to be made.

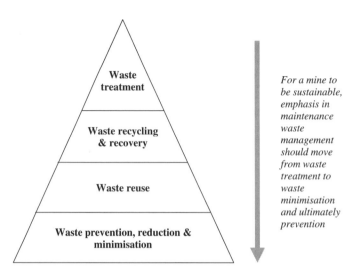

*For a mine to be sustainable, emphasis in maintenance waste management should move from waste treatment to waste minimisation and ultimately prevention*

Figure 6.7.    The waste hierarchy applies to maintenance wastes.

### 6.4.2.2    *Waste prevention, reduction and minimization*

A review of waste prevention, reduction and minimization approaches to managing maintenance wastes was completed in 2002. This included a review of technologies, improved methods for collecting, consolidating, and storing petroleum hydrocarbon wastes. The waste hierarchy principles (Figure 6.7) can be applied to maintenance waste management.

#### 6.4.2.2.1    *Prevention*

Increasing the working life of lubricants and equipment fluids present the most effective way of preventing the generation of wastes. Mining operations should seek out advice from their lubricant and fluid suppliers for strategies and approaches that can be practically applied to extend fluid life. Strategies can include optimizing the effectiveness of filtering systems and conducting a cleanliness audit to identify sources of lubricant and fluid contamination. Standards and procedures should be adopted that aim to keep fluids clean and free from any contamination.

Standard practices should be developed, along with a training and awareness program for operators, aimed at preventing the generation of maintenance wastes. These should highlight the consequences of not following appropriate practices. Housekeeping should also consider opportunities for the different waste streams to be combined, segregated or treated closer to the source (to prevent dilution effects). Housekeeping should also be added to the agenda for any on-site meetings that are routinely held. Various preventative approaches to maintenance waste management are provided in Table 6.7.

Where there is potential for significant spills to occur (e.g. planned maintenance), oil adsorbents such as matting and/or drip trays should be considered. Organic absorbent material for example can be cut to any shape, and does not generate fine particles which may contaminate "clean" systems. The absorbents can then be stored (segregated from other forms of solid waste) and incinerated. Some absorbent materials can be wrung out and reused if handled properly. A spill recovery system is also necessary where free oil is handled, and where maintenance is carried out.

Vacuum systems that allow oil spills to be recovered rather than emulsified, are very effective, and can often incorporate intensive cleaning systems to minimize the volume of contaminated water generated during spill cleanup. Sawdust or other adsorbents may also be used to adsorb the spill, which will reduce the amount of water used during washdown.

Solid wastes should be kept separate from liquid wastes. Any non-liquid and solidified greases, oily rags and other solid waste (e.g. saturated adsorbent material or sawdust used to mop up spills) should be responsibly disposed of through a licensed waste contractor with current certification.

Table 6.7. Sustainable approaches to the supply and handling of petroleum hydrocarbons at mining operations[1].

| Level in waste hierarchy | Examples of technologies, innovations and practices |
|---|---|
| Prevention | Strategies to improve maintenance planning and scheduling including predictive maintenance, condition monitoring, enhance oil life in plant and reduced wear of components |
| | Segregation of wastes as close to the point of origin as possible e.g. keeping contaminated waste oils separate and contaminated grease from uncontaminated grease |
| | Quick release couplings on storage vessels for facilitating loading and unloading from transport vehicles |
| | Bar coding technology for placement on waste drums and containers |
| | Vacuum pump to evacuate engines, gearboxes, differentials and fixed plant through access (drain) points |
| | Spill containment pallets that hold 2 × 205 L or 4 × 205 L drums for temporary storage and collection of waste oil |
| | Handling equipment for empty and full 205 L drums for improved ergonomics and efficiency |
| | Tanks can be made that have integral secondary containment, vacuum generator pump system, vacuum gauges (for transfers to collection vessel), bunding to capture leaks or spillages, and colour coding to minimize mixing oil and/or waste types |
| Reduction and minimisation | Enhance oil recovery from equipment & plant |
| | Efficient (manual) refilling of engines with oil to minimize wastage to minimize spillages |
| | Automatic dispensing of grease to specified lubrication points on plant to minimize losses and increase efficiency of use |
| | Use of bulk containers for purchase of chemicals and petroleum hydrocarbons |
| | Mobile oil draining systems (tank on wheels) |
| | 205 L drum funnels for improving handling of waste oils to minimize splashing including lockable funnels for improved security for ensuring segregation |
| | Oil transferred to trucks using pumps, e.g. vacuum pumps on trucks to avoid manual handling |
| Re-use, recycling and recovery | On-site oil filtering and oil laundering |
| | Re-use of waste oil in explosives production e.g. Ammonium nitrate fuel oil (ANFO) |
| | Re-use of waste gear oil in lower value applications such as chain oil |
| | Blending used oil into fuel e.g. using waste oil as a diesel extender for use in burners or in diesel engines and industrial and space heaters |
| | Cement, lime kilns and Brickworks as fuel to fire ovens |
| | Foundries for melting metals in a furnace |
| | Power stations as furnace light up fuel and in abattoirs and laundries as boiler fuel |
| | Reprocessing (off-site) of used oil to produce base oil (particularly for hydraulic and gear oils) |
| | Processes for recycling high-density polyethylene motor-oil bottles |
| Treatment and disposal | Separation and recovery of oil from oily wastewater (as described in detail in this chapter) |
| | Treatment of oily wastewater using biotreators |
| | Use of composting for treatment of biodegradable solid and semi-solid wastes containing petroleum hydrocarbons |
| | Use of biodegradable cellulose fibre for absorbing spills |
| | Incineration with recovery of energy |

[1] These are technologies now available commercially and commonly used in Australia and other developed countries.

Site personnel should be trained in the use of spill response kits, and these should be located at convenient points close to potential spill locations. Also, reducing the solids loading in oily waste water can also be achieved by using an industrial vacuum system prior to a final washdown. Site personnel often do not see the value in this since the washdown will achieve the same results. However, this can greatly reduce the maintenance that is required on the primary and secondary systems.

The selection and use of cleansers at mining operations can adversely affect the operation of oily waste water management systems at these operations and therefore can lead to the unnecessary generation of waste. For example, at one Australian mining operation, the site was purchasing up to 10 different degreasers. These were being used at rates as high as 1,000 L of product per day. Such procurement activity is not uncommon at mining operations. Steam cleaning can reduce the presence of cleanser that, in turn, has the potential to emulsify oils in the waste water.

Consolidation of the range of cleansers used at mining operations can assist in minimizing the amount of oil emulsification occurring in the oily waste water stream. Quick break detergents, i.e. cleansers that do not form strong emulsions when mixed with oil and water, can be used to replace the older generation hard detergents, which form very stable oil-in-water emulsions. If it is not known whether or not there is a problem with cleansers, oil droplet size analysis can be conducted on representative samples of oily waste water from the effluent stream from the site. This will provide more definitive information on the nature of the oily waste water, indicating if cleansers are contributing to the formation of strong oil in water emulsions, and can assist in determining which treatment technologies would be most appropriate for oily waste water stream.

Source reduction has the potential to greatly minimize environmental impact by decreasing the volumes and changing the characteristics of the waste waters requiring treatment. Waste reduction can be achieved through proper housekeeping, equipment modifications, and segregation or consolidation of contaminated streams, and is critical to minimize treatment costs and the potential for environmental impact. With appropriately trained and motivated people, and appropriate equipment and systems, the production of waste can be minimized.

Source reduction at mining operations can include:

- introducing equipment for ensuring oil is captured or directed from the engine or plant to an appropriate transfer and/or storage vessel such as adaptors and connectors compatible with plant oil/fluid drain outlets
- addressing people issues such as changing work practices to minimize contamination.

Management and supervisors need to be made aware of source reduction potential and then to provide specific training to personnel doing the work. Sustainable solutions require effective problem solving and solution development and this can be achieved by including petroleum hydrocarbon product suppliers and waste management contractors in regular site meetings. Mining operations personnel are unlikely to have all the necessary knowledge for maximum source reduction and usually benefit from an outside perspective. Examples of waste reduction approaches to maintenance wastes are provided in Table 6.7.

### 6.4.2.3   *Reuse of maintenance wastes*

There are selected maintenance wastes that can be re-used with minimal modifications to the waste product. These wastes include used oil, rags, drums, containers, and other metal wastes. Re-refining waste industrial oil for re-use as lubricants is best done by isolating the individual product types after use so that the re-refining process can address the specific nature of the contamination and degradation of each type of lubricant. In terms of maximizing value from the waste, keeping waste oils segregated is desirable and mixing waste oils is undesirable. Examples of re-use include on-site oil filtering (to remove grit contamination only) and re-use of waste gear oil in lower value applications such as chain oil. The simplest method for filtering is via a filter attached to a machine sump for continuous removal of solid contaminants. The removal of these contaminants improves lubricant condition as well as reducing oxidation and emulsification, both of which may be catalyzed by small solid particles. Examples of re-use of maintenance wastes are listed in Table 6.7.

### 6.4.2.4   *Recycling, reprocessing and recovery*

#### 6.4.2.4.1   *Recycling*
The quality of the used oil dictates how it is re-used. For example, used engine oils have simpler, lower value end uses because of the increased contamination and degradation. On the other hand, hydraulic oil for example, is not extensively degraded through use so it is more valuable for recycling.

Modern engine oils are compounded with a variety of additives. These include: viscosity index improvers, pour depressants, detergent dispersants, and anti-wear additives. In service, they are exposed to flame and heat, contamination from fuel and moisture and from combustion by-products, oil degradation products and airborne dirt. On the other hand, hydraulic oils, turbine oils and other lubricants drained from central systems are exposed to far less heat and contaminants during their service lives than are engine oils. For this reason the latter are much better candidates for cost-effective re-refining. Machine lubricating oils of various types are much more likely to be found in concentrated and segregated forms because they are used in industry and much cleaner than engine oils.

Transformer oil, quench oil, forging compounds, sulfurized cutting oils, highly chlorinated drawing oils and synthetic lubricants made from glycols and phosphate esters can all be reprocessed. However, the re-use of special compounds for their original use depends on the exclusion of other lubricants as major contaminants of the used oils.

#### 6.4.2.4.2   *Reprocessing*
It is easier and better to regenerate a lubricant at the earliest possible time in the life of the lubricant. Oxidation is more heavily influenced by metallic particles. Emulsification and the formation of rag layers are enhanced by the presence of organic fibers. In addition, moisture can accumulate in oil reservoirs and oil emulsions can form aided by the effects of bacterial action. Portable oil recycling equipment can be brought to the site that can remove water and dirt from the oil in recirculating systems and restore these lubricants. Such a unit would normally include a heat source, combined with vacuum to lower boiling points, and pressure filtration.

The approach with turbine, transformer and transmission oils should be to extend life, by cleaning and preventing contamination. However, if they must be disposed of, then alternative uses can be considered before recycling. For example, with the addition of a tackiness additive they can be used as rope, chain or open gear oils.

#### 6.4.2.4.3   *Recovery*
There are numerous recovery, or lower value re-use, options for waste oil and there are numerous options for energy recovery from waste hydrocarbons. When it is impossible to avoid admixtures of different waste streams, waste oil can be refined to a lower value product such as use as a fuel. For example, a machine tool uses two oils: a medium viscosity central system lubricant and a lower viscosity metalworking lubricant. These two oils mix during operations because of leakage but they cannot be re-compounded after re-refining in either service because the similar distillation curves of the two products resist their separation.

In terms of the economics of re-refining waste oil to fuel, it is important to realize that contaminant materials often provide as much fuel value as the lubricants with which they are mixed. As such, these contaminants are not contaminants at all but acceptable components. Petrol, mineral spirits, diesel fuel, animal fats and many additives are made from hydrocarbons. Waste industrial oil can be processed into marketable liquid fuels, generally after mixing with waste crankcase oil to reduce the percentage of sulfur and chlorine. Markets for reprocessed oil include asphalt plants, industrial boilers (factories), utility boilers (electric power plants for schools and homes), commercial boilers (generating heat for schools, offices), steel mills, cement/lime kilns, marine boilers (tankers or bunker fuel), pulp and paper mills.

Engine oils can come from all types of engines, petrol, diesel and gas. If kept separate from other oils, they can be used as a diesel substitute after dewatering and filtered before mixing. The oil can be blended with diesel up to 5% but the ideal should be 1% or less. It has been discussed and considered as a re-use option by engine manufacturer, but there are questions concerning exhaust emissions including excessive smoking. There are also questions about the need for extensive filtration needed to avoid excessive wear on fuelling components. This was a "hot" topic during the 1990's, but with increasingly stringent emission standards emerging globally, the idea has not gained momentum.

Most oils, after dewatering and filtering, can be mixed together for use as a diesel substitute in explosives manufacture. Current practices allow a mix of up to 75% waste oil and 25% diesel to be used in ammonium nitrate fuel oil (ANFO) instead of 100% diesel. A further use for these oils is in explosive emulsions.

Examples of waste recycling, reprocessing and recovery applied to mining maintenance wastes are given in Table 6.7.

### 6.4.2.5    *Treatment technologies*

Mining maintenance activities generate solid, semi-solid and liquid waste streams. The most common maintenance wastes are liquids from used products (waste oil) and diluted product waste streams (oily waste water). These maintenance wastes usually present the greatest challenge to a mining operation because of the large volumes generated. This section describes the management of liquid, semi-solid and solid wastes providing particular emphasis on liquid wastes.

#### 6.4.2.5.1    *Oily wastewater management*

Waste water treatment facilities are best operated with a relatively constant flow rate of water, rather than having to deal with flows that always vary. Collection structures should be evaluated to determine if the water entering treatment systems could be minimized. Waste streams, which are generated at irregular intervals, could be combined with other streams after primary treatment.

The most critical parameter in the selection of an appropriate oil-water separation technology is knowledge of an oily waste water's droplet size distribution (Gopalratnam et al. [2]). Free oil is generally considered to have droplets less than 250 µm in diameter. Oil droplet size is primarily reduced by chemical or mechanical emulsification by (1) subjecting the waste water to very strong turbulence (by high-shear pumps or mixing), or (2) by the action of chemicals. Chemicals that can cause emulsification of oils include radiator coolants/glycols (if present at concentrations of greater than 1%), degreasers (solvent-based cleaners), detergents (detergents with high pH generally are more aggressive in their cleansing action), and any number of organic and inorganic chemicals that may enter an oily waste water system such as from tank cleaning and rinsing.

The presence of cleansers does not necessarily result in the generation of stable emulsions, particularly if the cleansers are of a "quick-break" type. These types of cleansers aid the removal of oil from water. Furthermore, some hydrocarbon products such as certain hydraulic oils have additives that have emulsifying properties. The stability of the emulsion (ability to maintain a constant size distribution) is dependent on surface charge, pH, viscosity, temperature, the concentration of emulsifying agents (Kulowiec [3]) and particulates. Particulates destabilize emulsions (API [4]). Diesel can also cause an emulsion to form. It is important to note that there is a wide range of cleansers currently being used across the mining industry. Some of these are so called quick-brake detergents that phase separate shortly after initial emulsification. Others do not readily phase separate after doing their work of cleaning and it is these conventional strong detergents that are likely to be most problematic in oily waste water systems.

Emulsified oils are often resistant to separation because the droplets either do not rise to the surface (because chemicals interfere with the coalescence of the smaller oil droplets) or rise so slowly that they cannot be removed effectively with most oil/water separators. In general, the technology needed to remove smaller droplets is more expensive and maintenance-intensive than that needed for larger droplets. It should be noted, however, that for a given technology, it is possible to determine the probability of an oil droplet of a given size being removed. Smaller droplets may coalesce after a period of quiescence. Therefore, in practice, there is a tendency to rely on an understanding of the types and amounts of oil and other chemicals, the type of operation/source, the methods used to transport the waste streams, observation, and the experience of technology suppliers. Some examples of generic waste water "types" and their droplet size ranges are as follows:

• oily water from fuelling depots – about 50–150 μm
• workshop washdown water – about 20–60 μm
• some vehicle washers and workshops – less than 25 μm.

There are, however, operational methods for determining the emulsified and dissolved oil fraction (API [4]). This may assist in giving an indication as to the potential effectiveness of gravity-based separation (coarse treatment), as well as the potential for the presence of smaller droplet sizes. This test may be simplified to provide a quick field test for comparing waste waters. Such a test simply involves measuring the relative height or volume (if a measuring cylinder is used) of the oily layer that separates out over a given time period. It may also be possible to adapt the test to provide time-series results. This may assist in predicting the retention time required in certain secondary treatment systems (retention dams).

Based on the US EPA guidelines, oil concentrations allowed in waste water discharges to sewers are generally in the range of 50–100 mg/L, while oil allowed in direct discharges (to surface water) must normally be in the range of 5–15 mg/L with the added stipulation that none be floating or visible (i.e. no oil sheen visible) (Bennett [5]). Achieving an effluent oil quality of 5–15 mg/L should not be assumed as a given for most oil-water separation technologies, especially for more complex waste streams. Effluent qualities quoted in the product literature are generally for a new, optimized process and should only be used as a guide for narrowing down the technologies for use.

### 6.4.2.5.2 *Primary treatment technologies for oily wastewater*
Coarse or primary treatment close to the source of contamination is required to protect the downstream oil-water treatment system from being overloaded. Gravity separation is used for coarse treatment and as protection against major spills or leaks. Gravity based oil-water separators have been widely used in the past, as a protective device for containment of spills and leaks and as a pretreatment stage to remove free phase oil. Most other separation technologies do not deal well with "slugs" of oil. These systems also function as a sedimentation basin, capturing settleable solids and trash and may be as simple as a concrete pit. Plate separators use the principle of gravity separation due to differences in densities between oils and water. They comprise of either a single plate, three plates or may contain a nest of parallel plates (typically ~10–20 mm wide). Once the oil contacts the plates it migrates upward where it can be recovered. These are only effective when operators are using products that are susceptible to rapid phase separation or where a belt skimmer is employed (lube oils). Performance can be enhanced by selecting appropriate plate designs or by the use of a coalescer. Where an oily waste water stream is thoroughly emulsified, or solvents are present, plate separators are unlikely to be effective. Separator size is selected on the basis of waste water flow rates. Manufacturers of plate separators encourage minimal use of hard detergents and to only use small amounts of quick-breaking detergents.

Reduction of oil, chemicals (e.g. cleansers and coolants) and trash (e.g. rags, gloves, and paper towels) at source should be an integral part of the process of implementing any oil-water separation system. Reduction at source provides a means of minimizing costs and potential maintenance issues, as well as maximizing separation efficiency at this coarse treatment stage. Positioning the separator close to the source of contamination minimizes:

• the potential for environmental contamination of soils (transport pathways)
• the potential for oil emulsification due to pumping (where this occurs), and hence the increased difficulty of subsequent separation of oil from water
• the volume to be treated by the separator, by limiting the water input, which affects the size of the separator required.

Many mining operations use triple plate interceptors. However, there has been variable success with these systems for oil-water separation at mines in the Asia Pacific region (Table 6.6). The main barriers to the effective use of gravity separators are inappropriate usage, the need for frequent maintenance, and environmental conditions. These are each discussed in the following subsections.

Gravity separators should not be used as the only mechanism for separating oily contaminants from an oily waste water stream. They should be considered a coarse, or pretreatment, stage of oil-water treatment. Conventional gravity separators should not be expected to achieve effluent oil levels less than 100 mg/L (API [4]). Effluent oil concentrations from gravity separators in refineries can range from 30–150 mg/L, although deviations on either side of this range are common because of the extreme variety of waste waters to which this type of oil removal treatment is applied (Ford [6]).

Removal efficiency is a function of the minimum oil droplet size that can be removed, the density differences between the oil and water, temperature, and the efficiency of the oil removal devices (skimmers). Large droplets and density differences, low water viscosity and high temperatures favor gravity separation. As a result, conventional separators are generally referred to as the most economical and efficient means to remove large quantities of free oil (Ford [6]). They do not, however, effectively separate emulsified oils (of small droplet size) from the water phase. Therefore particular care must be made to ensure mine operators do not use detergents in waste water streams that will potentially end up in such separators.

The capacity of a primary oily water separator must be matched to the oily waste water influent flow rate. Where possible, water runoff from clean areas should be segregated from contaminated waste waters by appropriate bunding and plumbing. Catchments should be limited where rainfall is intense, in an attempt to match capacity to flow. Control of sediment build up is also critical.

Inadequate maintenance practices are the main cause of separator failure at mine sites (Table 6.6). Methods for removing oil and solids from an oil-water separator and the proper functioning of these systems (to avoid excess build up) are crucial to their effective operation. Regular inspections and maintenance should be scheduled. Oil and solids disposal options should be addressed at the design stage, including how and when these should be removed.

Free floating oils can be recovered by a number of techniques, including manual removal by bucketing out (in emergency situations), vacuum or pumping out, use of absorbent pads or socks, and skimmers. Absorbent pads or socks can be wrung out and reused if handled properly. Appropriate disposal of the absorbent materials needs to be considered. Several types of oil-skimming devices are in use including the rotatable slotted-pipe skimmer, the rotary-drum skimmer, the floating skimmer (Anonymous, 2001d), and rope skimmers. Large floating skimmers, called weir skimmers, are also commercially available for recovering free oil from large surface water areas. The efficiency of any skimmer is limited by the minimum thickness of oil that it can achieve on the water surface, the rate of oil removal, and the minimum amount of water that it withdraws with the oil. Waste oil skimmings from separators can usually be stored under quiescent conditions to allow further separation. The separated oil can then be re-used. For open gravity separators or washdown pits in refueling areas, significant build-up of fuels such as diesel, gasoline and kerosene should be prevented by the use of skimmers, to minimize the potential for occupational health and safety (OH&S) issues to arise.

Once the oil is recovered, incinerators (particularly those with an energy recovery option) offer an environmentally attractive alternative to landfill disposal, but potential air emissions must be monitored and minimized. Some absorbents are biodegradable offering the opportunity to compost these wastes.

Oil-water separators, whether intentional or not, also function as sedimentation basins. Provision must therefore be made to remove settleable solids. Typically silt and sand traps are required upstream to the separator to minimize problems associated with high solid loads. Solids can be removed manually (e.g. using a small bucket excavator) or a sludge pump, depending on the design of the primary or coarse treatment system. The solids produced by oil-water separators are potentially hazardous and appropriate disposal methods should be defined during the design phase. One option is to place the contaminated sediment on an on-site land treatment or sludge composting facility.

A description of primary oily waste water treatment technologies are given in Table 6.8.

### 6.4.2.5.3   *Secondary treatment technologies for oily wastewater*

A secondary treatment process for a given waste stream may require one or more of the technologies described in this section. Selection of the most appropriate technology is primarily dependent on characteristics of the waste stream, specifically oil droplet size distribution, influent suspended solids concentration, influent oil concentration, and treatment targets.

Secondary treatment systems include:

1. Coalescing plate separators/fibrous bed coalescers
2. Hydrocyclones
3. Chemical Treatment
4. Flotation
5. Membranes such as cross-flow filtration and ultrafiltration
6. Filter Beds
7. Retention Ponds
8. Bioremediation.

A description of oily waste water treatment technologies are given in Table 6.8.

Implementing secondary treatment systems is dependent on influent flow rate (and variability), environmental conditions (location, access, and weather), maintenance issues (amount of sludge for further treatment/disposal), and capital cost (including installation) and operating costs (utilities, labor and chemical requirements).

It should be noted that the expected effluent qualities provided by suppliers should only be used as a guide in secondary oily waste water treatment technology selection. This data provided in Table 6.8 is based on ideal operating conditions and does not necessarily hold for very complex waste streams. Site-specific testing is recommended if stringent treatment targets ($<10–50$ mg/L oil) are required.

### 6.4.2.5.4   *Coalescing or parallel plate separators*

Coalescing or parallel plate separators operate using the same mechanism as conventional gravity separators. However, the separator's effective horizontal surface area (and thus efficiency) is increased by the installation of parallel plates in the separation chamber, without requiring an increase in the size of the separator basin. This results in (1) decrease in the space requirements for the unit, (2) a potential for an increase in the flow through the unit, (3) a reduction in non-uniform, turbulent flow characteristics, thus providing more favorable conditions for separation of free oil, and (4) an increase in the removal of smaller oil droplets of free oil (resulting in a lower effluent concentration of oil). Coalescing plate separators can also be manufactured with a polyurethane foam coalescer, in a net-like structure, that is claimed to be more effective that the parallel plate design.

Parallel plates are available in modules that can usually be retrofitted into a conventional separator without major structural modifications. However, mechanical sludge removal equipment may also be required, to avoid having to raise the plate pack from the separator at regular intervals.

Parallel plate separators are efficient systems for removing oil droplets down to $30–60\,\mu\text{m}$ (compared with $150\,\mu\text{m}$ for conventional gravity separators). Care should be taken in selecting a system that will continue to operate reliably over a continued timeframe. Parallel plate units may experience clogging problems if the plate inclination is too shallow or the plate-to-plate spacing is too narrow. Flushing with water or air, or mechanical cleaning, may be required to clear blockages if the design or installation is inappropriate, if bulk solids are not removed prior to the separator, or if the unit is not maintained. The growth of bacteria, especially in high temperature environments, can also cause problems with coalescing plate systems, as can the generation of $H_2S$ in anaerobic conditions.

Table 6.8.  Summary of technologies for oil-water separation from review of literature and vendor supplied information[1].

| Technology | Acceptable influent (mg/L oil) | Effluent quality (oil in water, mg/L) | Stable emulsion handled? | Free oil removed? | Minimum oil droplet size (μm) | Can tolerate SS[2]? | Maintenance requirement | Capital costs estimates | Operating costs estimates |
|---|---|---|---|---|---|---|---|---|---|
| *Primary treatment* | | | | | | | | | |
| Sedimentation tanks | >20,000 | >150 | No | Yes | >150 | Yes | Low | $10–40k | $1–2 K |
| Gravity separators | 10,000–20,000 | 30–150 | No | Yes | 150 | Yes | Low | $5–10 K | $1–2 K |
| *Secondary treatment* | | | | | | | | | |
| Land treatment | >10,000–20,000 | NA[7] | Yes | No | NA[7] | Yes | Low-Medium | $300K–500 K | $10–30 K |
| Composting (of sludge) | >10,000[4] | NA | Yes | Yes | NA[7] | Yes | High | $60–150 K | $20–50 K |
| Drain/creek or retention dam | <10,000 | 10–100 | No | Yes | <50 | Yes | Low | $10–100 K | <$5 K |
| Parallel (coalescing) plate separators | <10,000 | 10–20 | No | Yes | 30–60 | No | Moderate | $3–9 K (1–2 kL/hr); $25 K (4 kL/hr); $200K (15 kL/hr) | $2–5 K |
| Coalescing plate separator | >5,000 | <5 | No | Yes | 20 | Yes | Moderate | $8–25 K | $1–2 K |
| Hydrocyclones | <5,000 | 10–50 | Yes | Yes | 10–20 | Yes | Low | $7–10 K (2 kL/hr)[3] | $1–1.5 K |
| Centrifuge separators | 1,000–6,000 | 70 | Yes | Yes | 10 | Yes | High | >$50K | Not obtained |
| Chemical treatment | <5,000 | 10–50 | Yes | No | <100 | – | Medium | $5–10K | ~$5/kL (Depending on dosing rate) |

| | | | | | | | | |
|---|---|---|---|---|---|---|---|---|
| Induced gas flotation | <500–900 | 1–25 | Yes | Yes | 10 | Yes | High | >$150 K | Not obtained |
| Cross-flow filtration | <100 | – | Yes | No | – | Yes | High | $65 K | Included in capital |
| Dissolved Air Flotation (DAF) | <300 | 1–20 | Yes | Yes | 5 | Yes | High | $200–300 K (195–320 kL/hr) | Included in capital |
| Electro-flotation | – | – | Yes | No | – | Yes | High | $40–60 K (20–30 kL/hr) | Included in capital |
| Filter Beds | <100 | <5 | – | – | 5–250 | Yes | Medium | <$10 K | Included in capital |
| Membranes | <100 | <5 | Yes | No | – | No | High | $30 K (2 kL/hr) | Included in capital |
| Carbon adsorption | <50 | <5 | No | No | – | No | Medium-High | $5–10 K | ~$2,500/m³ GAC[6] |
| Reed beds | 10–500[5] | <5–10 | Yes | No | – | Yes | Medium | –[6] | –[6] |

[1] Information, values and ranges of values were sourced from various vendors and literature during 2002. All costs are estimates only and are in $AUD x = $USD 0.55x.

[2] SS = suspended solids.

[3] For higher flow rates use clusters of cyclones in parallel.

[4] Oily wastewater could be applied to composting windrows as a source of water.

[5] Reed beds are being used for treating oily wastewater effluents in other industries including refineries.

[6] It depends on size of unit.

[7] NA = not applicable.

### 6.4.2.5.5    *Hydrocyclones*

Hydrocyclones increase the force for phase separation and can handle solids in the influent. These are claimed to be capable of separating droplets as small as 10–20 μm. However, it is also possible to adjust the cyclones to remove smaller droplet sizes, but at an increased cost. The concept of the hydrocyclone system (Figure 6.8) is increasingly being used for treatment of light and heavy vehicle washdown water and workshop waste waters.

Suppliers consider that the hydrocyclone units are comparable in cost to any coalescing plate system after 2–3 years, due to significantly lower maintenance and operating costs. The advantages of a hydrocyclone system include:

- They are typically less expensive to install than plate separators.
- There are usually few (if any) issues with hazardous goods zoning (i.e. from flammability).
- Floating suction skimmer ensures material is fed to the cyclone from the top of the collection pit.
- They minimize oil build up in the collection pit (most coalescing plate separators suction off the bottom of the pit).
- They are totally enclosed systems.
- The cyclone action increases the droplet size distribution of the oil, making subsequent gravity separation more effective.

A critical limitation of this technology is that the presence of fuel in the influent will significantly reduce effectiveness.

### 6.4.2.5.6    *Chemical treatment*

For very stable emulsions, chemical treatment may be the only method for ensuring that oil can be removed from the waste stream. Chemical separation can result in coalescence (formation of larger droplets) and/or precipitation (binding the oil droplets to a chemical flocculent). The former increases the volume of free oil removed, while the latter increases the volume of solids generated

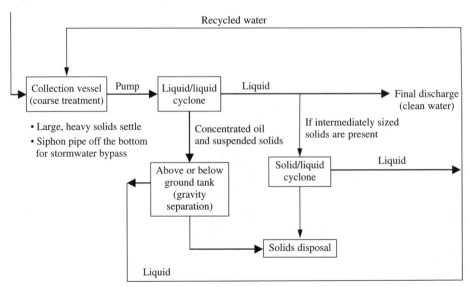

Figure 6.8.    Flow diagram of a hydrocyclone treatment system.

by the treatment process. Different chemicals are used, depending on whether the surface charge in the emulsion is anionic, neutral or cationic (depending on the dispersing agent involved). A number of proprietary emulsion breakers are available through specialty chemical suppliers. Their effectiveness in specific situations may be determined through bench-scale treatability tests, though such tests can be very limited because of the likelihood of changes in the oily waste water characteristics when taken from a site. Any discharge limitations on the concentrations of these chemicals and any affect on downstream biological processes should be assessed. The duration and degree of mixing of the emulsified waste water with an emulsion breaker are also critical to the cost-effectiveness of such chemical addition (API [4]). Chemical treatment does not need to be incorporated into an existing oil/water separation system, but it is often used with dissolved air flotation (DAF). In flotation, chemicals are added to alter the surface properties (e.g. for oil emulsions) and to enhance flotation of suspended solids (Varadarajan and Viraraghavan [7]).

Systems based on chemical addition or pH control tend to be more operator intensive and more sensitive to fluctuations in flow rate, contaminant concentration and temperature (Bennett [5]). Incorporation of an equalization tank can minimize the impact and allow operation at a constant reagent dosage. Given this potential sensitivity and/or increased complexity of operation, there is a trend to favor the use of mechanical separation systems.

### 6.4.2.5.7 *Flotation*

Flotation systems (Dissolved Air Flotation and Induced Gas Flotation) increase the difference in specific gravity between the oil and water by blowing fine air bubbles through the waste water. As the bubbles rise (at a rate governed by Stoke's law), they attach themselves to the oil droplets. The oil/air droplets have a lower specific gravity than the oil alone and rise quickly to the surface (2 to 3 times the rate achieved by gravity separation (Kulowiec [3]). The rising bubbles also transfer momentum to the oil droplets. Both of these mechanisms, which enhance the removal of oil droplets, hinder the removal of solids.

The DAF process enables the separation of fine particulate matter, oil, emulsified oil and grease from inflowing waste water. At specific times, chemical coagulants are automatically dosed into the suction line, which produce coagulation of the small particulate matter. The small bubbles, which are generated as a result of depressurization of the influent, combine with the particle agglomerates, become buoyant and float with the particulate matter. The DAF process therefore requires that a portion of the influent stream be pressurized up to 10 atm to generate the bubbles. The floating mass is continually removed by a rake or skimmer and transferred to a sludge pit. When liquid aluminum sulfate is added, the acidity of the stream increases. When this coagulant is employed, caustic soda needs to be added to offset the acidity. The DAF process can handle large volumes of oily waste water (up to 1 kL per minute, depending on the capacity of the unit). The oil in the sludge pit can then be reclaimed, treated, or recycled (Guerin et al. [8]). There are essentially five different types of flotation systems, their classification being based on the method of bubble formation (Bennett [5]). DAF and induced/dispersed air (or gas) flotation are the major commercial applications. The advantages of DAF systems are that they can achieve low effluent concentrations of oil and can remove very small droplet sizes (1–5 μm). However, the systems tend to be sensitive to changes in flow rate and influent concentrations of suspended solids and oil (generally not economical for waste streams with oil concentrations of >300 mg/L). A vapor recovery hood (or "carpet") may be also required to capture emissions (volatile organic carbons). These systems tend to be significantly more capital intensive than other oil separation systems and have high operating (chemical, labor, power and maintenance) costs.

### 6.4.2.5.8 *Membranes*

Membranes are generally regarded as polishing technologies for emulsified oils. These can result in very low effluent concentrations of oil (<5 mg/L). The presence of free oil and large amounts of suspended solids can interfere with the operation of a membrane system. Membrane systems are also very expensive.

#### 6.4.2.5.9   *Filter beds*

Filter beds have been widely used in the past as secondary treatment systems. Peat, for example, has been used in a wide range of filter bed applications (Malterer et al. [9]). Peats with low fiber contents and high lignin pyrolysis material and ash content have been shown to be the most effective peats to remove free and dissolved phase petroleum hydrocarbons from water (Cohen et al. [10]). Peat has the ability to both adsorb dissolved and free phase petroleum hydrocarbons from water, and provide a catalytic surface on which microbial activity can occur, and it has recently been reported to be effective for petroleum hydrocarbon contaminated groundwater in field applications.

#### 6.4.2.5.10   *Retention ponds*

Retention ponds are also widely used across the minerals industry for oily waste water separation. Many emulsions, given sufficient time and quiescence, will form separate oil and water layers. This is the rationale behind the use of retention dams. Retention dams provide a relatively simple concept and low cost option for secondary treatment of oily waste water. Their use is best in arid climates and where there are large areas of land for evaporation. Such systems are however, inappropriate where there are odorous or volatile compounds in the waste stream. These ponds can attract wildlife. While a positive attribute, this also poses potential for an environmental risk and the consequential need for an environmental risks assessment.

#### 6.4.2.5.11   *Bioremediation*

Land treatment, or commonly known and referred to as land farming, is a widely used bioremediation process for the treatment of organic wastes. From a minerals industry perspective, land treatment is particularly well suited to the treatment of oily waste water (where the free oil has been removed), in low rainfall high evaporation (desert) climates, and where there are large areas of land available. Such environments are often present at mining operations and hence these processes have been widely adopted in the mining industry. Land treatment processes involve the application of a waste stream (typically a liquid, solid or sludge) to a shallow treatment plot. The bed is tilled regularly to aerate and mix the soil. Nutrients are applied according to their background concentrations, and the biological demand imposed by the contaminants (Huesemann [11]). While this technology is generally supported by environmental regulatory authorities in Australia for use in arid climates, under specific licensing requirements for a particular site, its deployment should only be used where alternative secondary treatment technologies cannot be used.

The limitations of land treatment processes, as applied to treatment of oily waste water, need to be stressed to provide an overall perspective of this technology in relation to environmental management at mining operations. The limitations of land treatment technologies are as follows:

- Successful applications are typically in desert-like environments where evaporation rates are high and rainfall is low and soil is well drained.
- Loss of volatile contaminants applied in the wastes – this is only an issue where light hydrocarbon fractions (e.g. from gasoline, solvents, kerosene) are present in the waste stream. However, it is often difficult to prevent lighter hydrocarbon wastes entering the land treatment influent stream and some contamination via this route is likely to occur.
- Potential for contaminants to leach from the treatment plots to underlying groundwater – this includes mobile hydrocarbons, detergents and metals, depending on the underlying soil type and hydrogeology.
- The establishment of such processes has also been viewed as forming further contaminated areas. The contaminated soil may contain metals in addition to organic contaminants. Once the mining operation has ceased, there will be contaminated soil remaining that will require responsible disposal and therefore will have an associated closure cost.
- Labor intensive – land treatment facilities require ongoing supervision, maintenance and materials handling works. Contractors need to be managed and these activities incur substantial costs.
- Potential for saline conditions to develop in the land treatment soil bed – this could occur where high concentrations of salts enter the influent and are left unchecked. This has not been found to be an issue (to date) from observations of land treatment facilities visited and studied by the author.

Detailed operational aspects of land treatment as applied to oily waste water and soil contaminated with organic wastes including petroleum hydrocarbons is beyond the scope of the current paper, however, these issues are covered in detail in other scientific and technical papers (Elbagouri et al. [12], Guerin [13], Guerin et al., 1994a [14]).

## 6.4.3   OTHER WASTES

In addition to petroleum (liquid) wastes, the following maintenance wastes are also commonly generated at mining operations:

- used oil filters
- used hydraulic hoses
- used greases
- ethylene glycol

Oil filters are usually processed by separating the oil and the metal. In some systems, the oil is recovered from the filters using a centrifuge. In others, the filters are drained and crushed. The oil is recycled and metal is returned to steel mills. Other options include the recovery or reuse of the filter media. In these cases, filters are shredded. The metal can be sent to a scrap metal dealer, the filtering media to a cement kiln, and the oil to a re-refiner. Alternatively, the gaskets and media can be burned for energy recovery.

Rubber hoses are considered as waste material similar to urban solid waste. Prior to disposal, it is important to ensure that any excess oil or fluid is drained from the hose. These should be disposed of in a permitted waste management facility in accordance with the relevant state, and/or local regulations. State environmental regulators should be consulted prior their disposal. If the hoses are contaminated with toxic substances, special disposal considerations may be required.

Oil filters, hydraulic hoses, greases and related wastes can be treated using incineration. There is a range of incineration technologies currently available. The critical issues with these are the emissions that they generate and how these emissions are controlled in the process. The major benefit of incineration technology is its ability to destroy the hydrocarbon component of hydrocarbon contaminated wastes and therefore completely remove the risks associated with the hydrocarbons. Residual material from the incinerator can then be disposed of as clean fill or recycled where this is metal. The negative aspect of this technology, in addition to the risks associated with air emissions, is the energy requirements and capital costs.

Landfilling is the most common method currently used for disposing of hydrocarbon contaminated wastes. There are two main types of landfills: conventional landfills for domestic wastes and secure landfills for hazardous wastes. The concept of landfilling is unsustainable and all other alternatives should be investigated before using this approach.

Ethylene glycol is distinguished from waste oil in that it is not a petroleum hydrocarbon. Glycols are best described as di-alcohols, that is, a hydrocarbon molecule with two OH groups attached. Automotive antifreeze/coolants are often made from ethylene glycol or propylene glycol because these substances depress the freezing point of water and also elevate the boiling point. Ethylene glycol antifreeze is water soluble, classified as combustible and is toxic by ingestion. All glycols have high boiling temperatures. Used coolants may contain heavy metals (copper, lead, zinc) or other contaminants from gas or oil that increase the level of hazard posed by pure glycol and could be a characteristic hazardous waste. Ethylene glycol can be filtered, neutralized, and distilled to recover the pure glycol, which can be reused with no compromise in performance. The reprocessing equipment is simple to build and operate, so small systems are available to allow on-site reprocessing at garages and service centers. For antifreeze, a fresh charge of corrosion inhibitors is typically required. It is critical when disposing of these wastes that the "pure" product is diluted to at least 10% of the original concentration because of the potential contamination caused by these fluids.

# REFERENCES

1.  Scott, M.J. and Jones, M.N., The Biodegradation of Surfactants in the Environment, Biochim Biophys Acta, 1508, 2000, 235–251.
2.  Gopalratnam, V.C., Bennett, G.F. and Peters, R.W. (1988). The simultaneous removal of oil and heavy metals from industrial waste water by joint precipitation and air flotation, Environmental Progress, 7, 84–92.
3.  Kulowiec, J.J., Techniques for Removing Oil and Grease from Industrial Wastewater, Pollution Engineering, Feb. 1979, pp. 49–52.
4.  American Petroleum Institute (API). Design and operation of oil-water separators, American Petroleum Institute, Washington, D.C., 1995.
5.  Bennett, G.F. (1988). The removal of oil from waste water by air flotation: A review, CRC Critical Reviews in Environmental Control, 18, 189–253.
6.  Ford, D.L. (1978). Technology for removal of hydrocarbons from surface and groundwater sources, in "Oil in freshwater: Chemistry, biology, countermeasure technology" (Vandermeulen, J. H. and Hruey, S. E., Eds.), Pergamon Press, New York, NY.
7.  Varadarajan, R. and Viraraghavan, T., Dissolved air flotation in water and wastewater treatment, Indian Journal of Environmental Protection, 13, 1993, pp. 103–108.
8.  Guerin, T.F., Rhodes, S.H., Leiner, C., Hammerschmid, K., Roden, S., McAllister, P.J., Peck, P.C. and Kelley, B.C. (1994a). Management and treatment of wastes from maintenance operations in the mining industry, in "Maintenance in the mining and metallurgical industries" (Hargreaves, A. and Montegner, J., Eds.), The Australasian Institute of Mining and Metallurgy and The University of Wollongong, Wollongong.
9.  Malterer, T., McCarthy, B. and Adams, R., Use of peat in waste treatment, Mining Engineering, Jan. 1996, pp. 53–56.
10. Cohen, A.D., Rollings, M.S., Zunic, W. M. and Durig, J.D. (1991). Effects of chemical and physical differences in peats on their ability to extract hydrocarbons from water, Water Resources Research, 25, 1047–1060.
11. Huesemann, M.H. (1994). Guidelines for land-treating petroleum hydrocarbon-contaminated soils, Journal of Soil Contamination, 3, 299–318.
12. Elbagouri, I.H., Elnawawy, A.S., Abdal, M., Al Daher, R. and Khalafawi, M.S. (1994). Mobility of soil and other sludge constituents during oily sludge treatment by landfarming, Resources Conservation and Recycling, 11, 93–100.
13. Guerin, T.F. (1996). Selecting the appropriate technology for treating a contaminated site or waste stream, Australian Journal of Mining, 11, 82–86.
14. Guerin, T.F., Rhodes, S.H., Kelley, B.C. and Peck, P.C. (1994b). The application of bioremediation to the clean-up of contaminated sites: The potential and limitations, in "Contaminated Soil", Royal Australian Chemical Institute (RACI) Industrial Chemist Commercial Services Group in conjunction with the University of Newcastle, Newcastle.

# CHAPTER 7

## Best practices for sustainable mining

Subijoy Dutta[1] and Raj Rajaram[2]

[1]S&M Engineering Services, Crofton, MD, USA
[2]Complete Environmental Solutions, Oak Brook, IL, USA

### INTRODUCTION

Sustainable mining has been practiced by some mining companies since the 1980s. The primary driver for this change of attitude by mining companies was the environmental legislation that has been enacted in the United States, Europe, and several other countries around the world. Concurrently with the legislation related to environmental impacts, the United Nations (UN) has launched several international initiatives to integrate environmental considerations in to development planning. Founded in 1991, the International Council on Metals and the Environment (ICME) has worked with the UN and large mining companies to promote the development and implementation of sound environmental and health policies and practices in the production, use, recycling and disposal of non-ferrous and precious metals. Member companies of ICME have helped ICME and the UN Environment Program Industry and Environment (UNEP IE) disseminate the best mining practices for sustainable mining through several publications. This chapter will describe some of the best mining practices for small and large scale mines, both surface and underground, and provide some case histories. More case histories are provided in Chapter 8.

Sustainable mining requires interdisciplinary skills and team work. A team of mining engineers, environmental specialists, economists, and community relations specialists has to work together to develop sustainable mining plans, and obtain the acceptance of the public and the governmental agencies for the mining plan. For example, the Nicolet Mining Company in Northern Wisconsin, United States, employed mining engineers, environmental consultants, and community relations specialists to draw up their mining plan for an underground copper, nickel and zinc mine, and produce an environmental impact report for review by the governmental agencies. While the report was being reviewed, they took people living near the proposed mine to their operations in Canada where successful mining and reclamation had taken place. This educational effort by the company helped many reluctant neighbors of the mine to understand the environmental issues and how they have been successfully handled by the mining company in the past, and thus support the mining operation. The mine has not yet received the permits since the governmental agency is weighing all the environmental risks with the proposed benefits from the mining operation.

Experience gained by mining companies over the past century has led to safer and sustainable mining operations. In addition, research by governmental agencies and the mining companies have produced technologies to better manage the toxic substances used in mineral processing, and take proactive steps to managing environmental impacts during and after mining. Best mining practices that have emerged from industry and government, are described in the following sections. A general guidance on best management practices (BMPs) for a few common mining operations is also provided in detail under Section 7.1.

# 7.1   Best management practices (BMPs) in mining operations

## 7.1.1   INTRODUCTION

Similar to any other operation, any surface or underground mining operation involves various phases and activities. How these activities can be conducted in a manner to cause minimal or no damage to human health and environment are considered as the best management practices for a mining operation. Use of such practices is also a key factor in the overall sustainability of the mining operation. These practices must be based upon sound engineering principles which are a combination of general and specific to mining operations. Proper management BMPs of mining wastes involve careful management of all site clearing, digging, blasting, collecting, processing, and storing operations. To perform these activities in an environmentally safe and friendly manner, a close parallel to a proper hazardous waste management program could be drawn. The primary hierarchy of proper environmental practice for waste management generally involve (Dutta, 2002 [1]):

   i.  Waste identification and characterization
  ii.  Recordkeeping and manifestations
 iii.  Safe storage and disposal practices
 iv.  Inspection and assessments of hazardous waste sites
  v.  Remedy selection and design
 vi.  Remedy implementation and site closure
vii.  Post closure and site restoration/reuse.

Similarly, in a mining operation, waste identification and characterization should be the first step. Periodic assessment of hazardous and mining wastes should also be conducted. The waste disposal/management techniques should be properly selected and designed so that no future contamination problems arise out of the mining site. Proper implementation of the selected management techniques is one of the most important steps in the best management of mining operations. Lastly, a successful and productive reclamation and reuse of the land should provide the ultimate touchstone of a sustainable mining operation.

It is an insurmountable task to provide the complete guidance on the best management practices for all different technologies and processes used in mining operations. In this Chapter we will try to provide a general guidance on best management practices (BMPs) for a few common mining operations. The following general processes are listed as a few examples of common processes involved in mining and mineral processing operations.

- Excavating, Dredging, and Conveying
- Solvent Extraction
- Debris Washing
- Separation
- Wet and Dry Screening
- Gravity Concentration
- Froth Flotation.

## 7.1.2   AIR EMISSION AND OTHER ISSUES INVOLVING MINING OPERATIONS

Dust emission from the mining operation is one of the major issues raised by the neighboring communities at various mining sites. However, many different control options (BMPs) are available for mitigating and suppressing fugitive dust emission as specified later in this Chapter. Proper use of such BMPs will also help in building community alliance and sustainability in the mining operation.

During implementation of any mining technology the following steps are generally undertaken: (a) Site preparation and staging, (b) Pre-mining activities, (c) Mining activities, and (d) Post-mining activities. General cross-media transfer potential for contaminants (mostly during the site preparation, pre-mining, and post-mining activities) are identified below.

- There is a probability of inaccurate site characterization with any environmental evaluation of a mining operation. The material encountered at the site may not be like the ore studied in pilot-scale tests. Additional contaminants may be encountered, and the percentage of the fine-grained fraction may be significantly different from that expected. These factors may lead to a long-term storage or generation of high residual volume, and thus increase the potential for cross-media transfer (Dutta, 2002 [1]).
- During several different activities associated with proper implementation of waste management, including staging and site preparation (e.g. clearing, grubbing); drilling, well installation and trenching operations; mobilization and demobilization of equipment; excavation; transport of materials across the site; and some mining activities, there is high potential for fugitive dust emissions due to movement of equipment at the site.
- During premining operations such as excavation, storage, sizing, crushing, dewatering, neutralization, blending, and feeding, there is the potential for dust emissions.
- Migration of contaminants to uncontaminated areas may occur during mobilization or demobilization.
- Leaching of contaminants to surface water can occur from uncovered stockpiles and excavated pits.
- Improper handling and disposal of residues (e.g., sediment/sludge residuals or post-washing wastewater) may allow contaminants to migrate into and pollute uncontaminated areas.
- Post-mining discharges of wastewater, if improperly managed, can cause migration of contaminants.

## 7.1.3    GENERAL BEST MANAGEMENT PRACTICES FOR MINING OPERATIONS

Various general control practices to prevent potential cross-media transfer of contaminants during mining operations have been identified below. Also, proper design is recommended prior to implementation of the mining activities and processes to avoid cross-media transfer problems during different steps.

### 7.1.3.1    *Site preparation and staging*

Prior to movement of equipment on site the following activities are most commonly undertaken (USEPA, 1997 [2]):

- Site inspections; surveying; boundary staking; drilling and trenching; sampling; demarcation of hot spots for cleanup; and construction of access roads, utility connections, and fencing.

The following BMPs are generally recommended:

- Avoid entering any contaminated waste area. In unavoidable circumstances, build a temporary decontamination area, which could be later used during cleanup activities. Any above-ground and underground source of contaminants should be identified and located prior to starting any excavation of contaminated media.
- Special attention and care should be taken during site preparation activities so that any potentially contaminated media are not disturbed. In case of unavoidable circumstances, the contaminated media should be subjected to a very minimal disturbance/alteration during these activities.
- Any soils and soil/coal-gas sampling, field air permeability testing, and demarcation of hot spots should generally be followed by plugging/covering of any holes or depressions created during these activities to prevent intrusion of water. It would also be appropriate to install relevant signs at the same time so that repeated entry to the site is not required.

- Contaminated drilling mud from any drilling operations should be collected in a lined/contained system. This will prevent the contaminants from mixing with the normal surface water runoff from the area and the surrounding natural watercourse.
- Contaminated waste generated during site preparation or further site characterization activities should be managed protectively.
- Site investigation and operational plans should take into account the presence of permeable zones and account for potential pre-existing underground sewers and electrical conduits.
- Surface drainage and any subsurface utility systems should be identified.
- Local watershed management goals and priorities should be incorporated into the surface water management plan for the cleanup activities.

### 7.1.3.2   *Pre-excavation activities*

Prior to beginning the actual mining operations the following activities are most commonly undertaken.

- Excavation, transportation, storage, sizing, crushing, dewatering, neutralization, blending, installation of feeding systems for ore to be processed etc.
- During the above activities, measures should be taken to control fugitive dust emissions and to prevent releases of contaminated media to the natural environment.

To prevent cross-media transfer of contaminants, the following BMPs are generally recommended for the above activities:

- Any aboveground and underground sources of contaminants, such as storage tanks, should generally be removed.
- Any offsite runoff should be prevented from entering and mixing with on-site contaminated media by building earthen berms or adopting similar other measures.
- Provisions should generally be made to capture on-site surface water runoff by diverting it to a controlled depression-area or lined pit.
- Sizing, crushing, and blending activities should be conducted under an environment where the off gases, volatiles, dusts, etc. are all captured inside a hood or cover, or controlled using other options. The dust and VOC emissions associated with these activities that exceed acceptable regulatory limits should be controlled by capturing these emissions and then treating the captured vapor/air to the extent practicable.
- When mixing or dewatering, the contaminated aqueous stream should be collected in a lined/contained system. This will prevent the contaminants from mixing with the normal surface water runoff from the area and the surrounding natural watercourse.
- Protective management/disposal of contaminated debris is recommended to prevent cross-media transfer.
- The technology design should be checked to ensure that the corrosion factor has been taken into account in the design for all appropriate pipes, valves, fittings, tanks, and feed systems.
- Entry to the active site should be limited to avoid unnecessary exposure and related transfer of contaminants.
- The temporary decontamination area should be used as recommended to keep the site-related contaminants within the active cleanup area.
- Fugitive dust emissions should be controlled during excavation by spraying water to keep the ground moist.
- Consideration of climatological extremes/high winds. When conducting any of the mining activities. Real-time weather data could be used to monitor weather conditions and accordingly control mining operations. The weather monitoring stations are reported to have nominal cost and are found to be highly useful in controlling weather-related cross-media transfers. To determine possible extreme conditions, local weather data for the past 10 years could be reviewed from publications (NOAA, 1995) of the National Climatic Data Center, 151 Patton Avenue, Asheville, NC 28801-5001.

- During excavation, blending, and cleanup of contaminated soils, VOC emissions should be monitored and appropriate emission control measures undertaken.
- Operational plans should include adequate inspection procedures that look specifically for corrosion and wear.
- It is also critical to check that the air pollution control devices are designed for the corrosive nature of the hot gases that are expected to enter these devices, if used in certain operations.
- As an effective erosion control practice, scheduling of construction activities should be arranged to limit the time of exposure of disturbed segments of the site. This entails directing work to one area of a site, then completing and stabilizing that area before moving on to other areas of the site.

### 7.1.3.3  *Mining activities*

During active mining operations the highest priorities are generally given to health and safety issues during operation. All of the requirements of Mine safety and health administration (MSHA) have to be met first by the mine operators in U.S. In addition to the MSHA requirements there are other State and Federal requirements which are required to be met by the mining companies in U.S. Some of the BMPs suggested above fall within those requirements. However, to nurture a sustainable growth and development with minimal environmental impact to the community the following two BMPs are generally recommended during active mining operations in any part of the world.

- Minimize fugitive dust emission by spraying water or applying other dust suppressing material, such as foam.
- Use noise control methodologies during the active mining operations by providing temporary noise barriers/fences utilizing the brushes removed during site clearing and by limiting operations of machinery generating high noise levels during the daylight hours. In addition to the safety requirements of blasting, similar practices to minimize disturbance to the community are recommended.

### 7.1.3.4  *Post-mining activities*

During the post-operational period the following activities are most commonly undertaken:

- Collection or destruction of organics/wastes
- Collection of particulates
- Removal of any hazardous materials
- Treatment or disposal of aqueous wastes
- Disposal of dusts collected as a result of emission control during materials handling, stabilization, or any other activities/post-mining.

During the conduct of the above activities, measures should be taken to prevent release of contaminated media to the natural environment. The following BMPs are generally recommended for the above activities:

- Treated wastes should be checked for leachability prior to disposal in a landfill or other similar systems. Possibilities of long-term degradation and migration of contaminants to groundwater should be carefully evaluated and checked prior to disposal of stabilized/treated material.
- Contaminated debris, soils, and liquid wastes resulting from excavation and installation of wells should be properly handled, either treated on-site or trucked away for off-site disposal. Berms should be built around the active excavation, storage and treatment areas, if necessary, to prevent migration of contaminated runoff away from the area.
- If solid materials such as granulated carbon filters are used to collect emissions, they should be removed carefully from the emissions system to avoid rupturing them and dissipating the contaminated carbon materials. They should be placed into tightly covered containers until they can be recycled or properly disposed of.

- Containers that hold residual liquids should be stored where they cannot be disturbed or ruptured by large equipment. This may require construction of a residuals management unit separate from the treatment and storage areas.
- All dust or other particulates that are collected during emissions control activities should be tested for contamination levels and handled and disposed of properly.
- Air stripping or other treatment of extracted (contaminated) water/liquids should meet all applicable surface water discharge standards for post-treated water.
- When residual mining wastes are obtained in the form of pure waste/liquids (e.g., condensate from steam regeneration of carbon beds), the recycling/reuse option for such residual waste should be considered.

REFERENCES

1. Dutta, S., *Environmental Treatment Technologies for Hazardous and Medical Wastes – Remedial Scope and Efficacy*, Tata McGraw-Hill Publishing Company, ISBN 007-0435863, New Delhi, March 2002.
2. USEPA, *Best Management Practices for Soil Treatment Technologies*, EPA-530-R-97-007, Washington, DC 1997.

# 7.2 Tailings pond design

Kelvin K. Wu[1], L. Owens[2] and John W. Fredland[3]
[1] *Mine Safety and Health Administration, Pittsburgh, PA, USA*
[2] *U.S. Department of Labor, Pittsburgh, PA, USA*
[3] *Mt. Hope, WV, USA*

## 7.2.1 INTRODUCTION

The mission of the Mine Safety and Health Administration (MSHA) is to eliminate mining accidents and promote improved safety and health conditions in the U.S. mines. An important part of this mission is the safety of the many impoundments constructed by the mining industry. MSHA has jurisdiction over approximately 860 impoundments at coal mines and 695 impoundments at metal and non-metal operations. Many of the dams are over 300 feet high and are some of the highest dams in the U.S. Dams operated by metal and nonmetal mining companies are required by Section 56.20010 of Title 30, Code of Federal Regulations, to be of substantial construction and inspected at regular intervals, if they can present a hazard. This paper will concentrate on MSHA's program with respect to impoundment safety at coal mining operations.

MSHA's standards pertaining to the construction of impoundments at coal mines are found in Section 77.216 of Title 30 of the Code of Federal Regulations (US Dept. of Labor [1]). The current impoundment regulations were promulgated largely as a result of the failure of a coal waste dam along Buffalo Creek, in West Virginia, on February 26, 1972. That failure resulted in the deaths of 125 people and left over 4,000 homeless. As a result of the regulatory changes made in the aftermath of the Buffalo Creek failure, coal companies are required to obtain MSHA approval of design plans before an impoundment of significant size or hazard potential is constructed. Additionally, these dams are required to be examined at intervals not exceeding seven days, or at other approved intervals, by a qualified coal company inspector. Annually, a professional engineer must certify that each impoundment was operated, maintained, and constructed according to its approved plan.

In general, coal mining companies must submit a complete engineering design plan for a proposed impoundment, with supporting analyses and documentation. Plans are required for constructing a new impoundment, and for increasing the size of, or otherwise modifying, an existing site.

## 7.2.2 MSHA'S IMPOUNDMENT PLAN REVIEW PROCESS

MSHA has eleven districts that cover all of the coal mining operations in the U.S. When an impoundment plan is submitted to one of MSHA's district offices, it will either be reviewed by engineers on the District Manager's staff, or, more typically, be forwarded for review to the Mine Waste and Geotechnical Engineering Division, in the Pittsburgh Safety and Health Technology Center. The goal of MSHA's plan review process is to ensure that the plan is complete, that all potential failure modes have been adequately investigated and designed for, and that the design and specifications are consistent with accepted and prudent engineering practice.

If plan deficiencies are found during the review, a letter is sent to the coal company explaining what issues need to be clarified or resolved before the plan can be approved. MSHA review engineers may meet with representatives of the coal company to assist in resolving outstanding issues.

### 7.2.3   PLAN REVIEW TECHNICAL ISSUES

An impoundment must be designed to have adequate factors of safety against the different ways that failure can occur. That is, the dam, basin, and appurtenant structures must be capable of withstanding the different forces they may be subjected to over their life. For impoundments with high-hazard potential, that is, where failure would likely cause loss of life or serious property damage, the impoundment must be designed to withstand rare events, such as large storms and earthquakes. The key points of MSHA's review of impoundment plans are summarized here. Some comments are included on issues that have frequently been areas of contention with mine owners during reviews.

#### 7.2.3.1   *Design storms*

Impoundments must be capable of handling large storms without failure or significant damage. MSHA requires that dams with high-hazard potential be designed to withstand the probable maximum flood (PMF). The PMF represents the maximum runoff condition resulting from the most severe combination of hydrologic and meteorological conditions considered reasonably possible for the impoundment's drainage basin (USBR [2]). It is based on the runoff from the probable maximum precipitation (PMP), with antecedent moisture condition III (AMC III). AMC III corresponds to a watershed that is saturated from antecedent rains and represents the highest runoff potential. Impoundments with moderate-hazard potential must be designed to handle one-half of the PMF, while low-hazard potential sites must be designed to handle at least the 100-year frequency rainfall of 6-hour duration.

Plans submitted to MSHA must show that the impoundment can withstand the design storm while maintaining adequate freeboard, and without significant erosion of discharge facilities. A minimum freeboard of at least 3 ft is required, unless a wave-runup analysis is performed to justify that a lesser freeboard is adequate. Besides allowing for wave runup, the freeboard accounts for uncertainties, such as in the hydrologic analysis, and in the settlement of the crest of the dam.

Specifics of the design storm used for an impoundment depend on the type of spillway facilities to be used. Where an open-channel spillway is used, the main concern is that the combination of the inlet elevation, and the size and slope of the spillway, is adequate to safely pass the peak outflow, while providing adequate freeboard. In these cases, the designer typically needs to show that the facility will handle the peak discharge from a 6-hour duration storm. In some cases, instead of an open-channel spillway, the storm inflow is discharged only through a decant pipe. The approach in these cases is to provide sufficient capacity in the impoundment to store the storm runoff, and then release the stored water through the decant system. If the runoff is to be stored and passed, then the design storm must be extended to at least 72-hours. To allow for consecutive storms, outlet facilities are required to remove 90% of the design storm inflow within 10 days. Facilities that do not meet this drawdown rate can be designed to store the runoff from two design storms, if the second storm is evacuated within a reasonable period of time, typically 30 days. For slurry impoundments, where the crest elevation changes frequently, the plan should include charts that show the maximum allowable normal pool level, and spillway and decant inlet elevations, for given crest elevations.

Pumping systems cannot be considered in routing a storm through an impoundment, and the use of pumps to meet reservoir drawdown criteria after a storm is discouraged. If pumps are proposed because of special circumstances, then a number of measures need to be taken to ensure that sufficient pumping capacity will be available when needed. These measures include: substantiation of pumping capacity; weekly activation checks; provisions for an independent fuel or power source that will be available under adverse weather conditions; and contingency backup plans in the event the primary pump fails.

*Common review issue*: To use a design storm of less than the PMF, the plan must substantiate, through a conservative dam-break analysis, that failure of the impoundment is unlikely to cause loss of life.

### 7.2.3.2  *Flood routing*

Plans must include a hydrologic routing analysis to determine the minimum freeboard during the design storm and the discharge rates through outlet facilities. Any accepted method can be used to generate the inflow hydrograph and route the design storm through the impoundment.

*Common review issues*: Submitted plans must include explanations and substantiation of all parameters used in the routing, along with topographic maps that show the entire watershed area. Hard copies of both input and output files must be submitted for computer analyses. It is helpful to the review process if input files are also provided to MSHA on an electronic medium, such as a disk.

### 7.2.3.3  *Hydraulic design of spillways and decants*

Whatever type of outlet works are used in an impoundment, the plan must show that the facilities can handle the flow velocities and any other structural loads imposed on them. Spillways must be analyzed to show that the channel freeboard will be adequate and the lining material will have sufficient strength and anchorage to resist the flow velocities. This typically requires water surface profile and tractive force analyses. On solid linings, issues of uplift pressure, and measures to relieve them, must be addressed. Because of uncertainty in how riprap may perform, its use as a spillway liner under high flow velocities is discouraged. Plans proposing the use of riprap need to include analyses, using conservative methods, to support the proposed stone sizes. Riprap specifications need to address stone gradation, layer thickness, and bedding and filter requirements, and stone durability.

Several issues need to be addressed in plans with respect to the design of decant pipes. These include their hydraulic capacity, durability, corrosion protection, joint details, water-tightness, seepage-control measures, trash rack, anti-vortex device, thrust blocks, bedding, backfill, and structural design. Decants that will flow under pressure, and pipes where there may be concern for loss of material into joints, need to be pressure tested, to at least the maximum expected design pressure, to ensure that the joints are watertight.

Concerning seepage control measures along conduits, the preferred method is to incorporate a diaphragm, or zone of drainage material, designed to collect seepage along the pipe and discharge it through material that meets filter and permeability criteria. Typically, recommendations from the Natural Resource Conservation Service, U.S. Department of Agriculture (SCS [3]), are used for locating and sizing the diaphragms.

*Common review issues*: An issue that is unique to mining impoundments sometimes arises when a decant pipe discharges into a groin ditch. When this approach is taken in a plan, then the portion of the groin ditch downstream of the decant pipe outlet becomes part of the outlet spillway and must be designed to the same standards. This means that this portion of the groin ditch must be designed so that no significant damage would be expected to occur with the combination of design storm discharge from the decant system, and design storm runoff from the contributing watershed above the groin ditch.

Plans sometimes do not adequately take into account the added freeboard needed at bends in spillway channels. Adequate substantiation of the ability of spillway linings to resist uplift and tractive forces is not always provided. Where grouted riprap is proposed, the depth of grout, and the method to be used to ensure proper grout penetration, needs to be specified.

### 7.2.3.4  *Site exploration*

Impoundments must be designed to be stable under all of the conditions they will be subjected to, including maximum water pressures and earthquake shaking. To ensure that all potential modes of failure are identified and properly analyzed, plans must include a thorough subsurface site investigation. Complete boring logs, information on geologic conditions, groundwater and permeability conditions, and any other pertinent information, such as geophysical studies, need to be included in the plan. The number, location, and spacing of borings depend on site conditions and should be consistent with accepted geotechnical engineering practice. As with most geotechnical projects, the amount of site evaluation work depends on the complexity of the conditions, the degree of uncertainty, and the level of potential hazard or impacts. The site evaluation program should be conducted by experienced geotechnical engineers or geologists, who evaluate all the available information and make adjustments to the exploration plan as the work progresses.

*Common review issues*: Problems include failure to locate old mine openings or auger holes until construction is underway, and failure to verify the extent of mining near impoundments. These issues are discussed in more detail in a later section.

### 7.2.3.5   *Material testing*

To properly analyze for stability and deformation, the relevant engineering properties of the materials must be determined. Depending on the conditions, this may include information on density, moisture content, shear strength, permeability, compressibility, and shear wave velocity. Any zone with low blow counts needs to be identified, tested, and accounted for in the design. For engineering property tests, sufficient information must be included in the plan to document that the tests were appropriate and were conducted properly.

*Common review issues*: For shear strength tests on soils and refuse, appropriate sample sizes and strain rates must be used. Since coarse refuse generated in the preparation plant may be used over an extended period of time, and coal from various sources may be processed, plans need to include provisions to check the engineering properties on a periodic basis to ensure that they remain consistent with design assumptions.

A conservative approach should be taken to selecting shear strength parameters, especially where there are any concerns about the testing program, such as the sample size and the representativeness of the samples. For example, a good conservative practice is to assume that the cohesion is zero for granular materials.

### 7.2.3.6   *Static slope stability*

The stability of the embankment must be analyzed for all critical conditions, and normally for each stage of construction. For static stability, worst case conditions with respect to seepage and pore-water pressures need to be analyzed. If applicable, analyses for rapid drawdown conditions need to be performed.

*Common review issues*: The location of the phreatic surface used in the stability analysis is not always substantiated in the plan. Wedge type failures are not always considered where there is a layer of material of lower strength. When storage is relied on to handle the design storm, the stability needs to be analyzed based on the highest phreatic level that can develop during the design storm, rather than the phreatic surface based on the normal water level.

### 7.2.3.7   *Seismic slope stability*

In any case where there is a layer of loose, saturated material within the structure or foundation, the potential for liquefaction under seismic loading must be investigated. This would apply for all slurry impoundments built using upstream construction (Figure 7.1), unless it can be shown that, even if the fines liquefied, adequate stability is maintained. Impoundments with a high-hazard potential need to be analyzed for the ground motions that would result from the maximum credible earthquake that could potentially impact the site. Nearby blasting could also subject the site to seismic loading and needs to be analyzed.

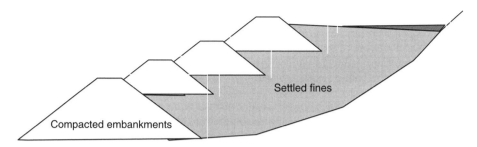

Figure 7.1. Simplified cross-section illustrating "upstream construction" (not drawn to scale and internal drainage provisions not shown).

When subjected to shaking, loose material tends to be contractive, or move into a denser packing. If the material is saturated, then the volume cannot change until drainage can occur, and the stresses are transferred to the water in the pore spaces. The increase in pore water pressure reduces the effective stress, and can lead to liquefaction, or a significant loss of strength. Material needs to be tested to determine its strength when subjected to shaking, and seismic stability needs to be investigated to show whether the embankment will have an adequate margin of safety under such conditions.

One way to perform a seismic stability analysis is to determine how the different materials in the embankment and foundation will respond to the shaking. For this approach, the appropriate ground motion parameters for the site, indicating the intensity, duration and frequency of the shaking, need to be determined. This is true even though coal slurry impoundments are not necessarily located in areas of the country that are thought of as being prone to earthquakes. For preliminary analyses, the ground motion information used on other projects in the area, such as seismic studies on Corps of Engineers dams, could be consulted.

Analyzing seismic stability is much more complex than a static stability analysis. The dynamic response properties of the embankment and foundation materials need to be determined to assess how the seismic waves would be attenuated or amplified within the structure. These include shear modulus, damping ratio, and their relationships to strain. Even if stability is found to be adequate, deformation, that is, how much the crest of the dam may settle due to strain softening must also be determined.

An alternative approach to seismic analysis is to determine the undrained shear strength that the material will exhibit as a result of an earthquake, and design the embankment to be stable at this reduced strength value. The testing needed to determine this strength is much more complex than normal shear-strength testing. The undrained shear strength is highly sensitive to the void ratio of the material, and without painstaking care to track the void ratio throughout the sampling and testing process, and correct the strength for the in-situ void ratio, the results are uncertain. To avoid the complexity of the seismic analyses, another approach that can be considered is to assign the fines a shear strength value of zero, and then design the cross-section to prevent loss of the dam crest under this condition.

*Common review issues*: Seismic stability is an extremely complex problem. It is one of the most difficult issues dealt with by MSHA plan review personnel. Problems include: obtaining undisturbed, representative samples of material potentially subject to liquefaction; defining and following proper testing protocol; defining the ground motion parameters applicable to the site; determining what margin of safety is reasonable; and dealing with the questions of the applicability of available technical information to fine coal waste. Since there are no documented cases where a coal slurry impoundment has been subjected to significant earthquake shaking, a conservative approach must be taken. Yet plans are often submitted with minimal factors of safety, and without mitigating design features that would enhance seismic stability, such as provisions for internal drainage of the fines. Just as seismic stability analyses increase the complexity, time and cost of the design process, they also add considerable complexity and time to MSHA's review process.

### 7.2.3.8   *Upstream construction – excess pore water pressure*

A special concern that must be addressed on upstream construction sites is the development of excess pore-water pressures during the construction of embankment stages founded on settled fines. Several upstream slope failures have occurred for this reason. Plans need to address this issue because of concerns for both the safety of the dam, and the safety of the equipment operators working on the pushout material.

When an embankment is raised by upstream construction, material is pushed out onto settled fines. Because the fines were hydraulically placed, they are loose and saturated. Under these conditions, and depending on the permeability and drainage conditions, the weight of the pushout material is initially transferred to the water in the pore spaces of the fines. Until drainage can occur to relieve this condition, the excess pore-water pressure reduces the effective shear strength of the fines and can cause a bearing capacity or slope stability failure.

*Common review issues*: Plans need to address the issue of excess pore-water pressures during upstream construction, and, when necessary, limit the rate at which material is placed. Where the development of excess pore pressure is a concern, plans should include the installation of quick-response piezometers, so that the pressures can be monitored to ensure that they stay within pre-defined acceptable limits.

For any plan that includes upstream construction, it is recommended that the design engineer meet with the construction personnel to ensure that they understand the significance of excess pore-water pressures. As has happened in some of the failures that have occurred, some equipment operators may mistakenly think that when it comes to upstream pushouts, "the thicker, the better," while not realizing the detrimental effect of placing too much fill, too soon.

### 7.2.3.9   *Seepage analyses*

The stability of an earthen dam is dependent on how seepage through the embankment is handled. Impoundment plans should include internal drainage measures and must include substantiation for the anticipated maximum phreatic level. This needs to be based on a seepage analysis that considers the permeability of the materials, the horizontal to vertical permeability ratios, and the position and size of the underdrains. To account for stratification, compacted fill should be modeled with the horizontal permeability being at least nine times greater than the vertical permeability, unless otherwise documented. The other main concern with seepage is that the water is collected and discharged in a manner that prevents internal erosion from occurring. Plans need to show that the filter criterion is met whenever flow occurs from one material into another.

Seepage analyses performed using finite element methods must show the basis for all properties, and the program input and output. Since significantly different results can sometimes be indicated by relatively small changes in permeability values, the sensitivity of the results to permeability changes should be examined, and a conservative approach needs to be taken.

*Common review problems*: The presence of surface mine "spoil" in the foundation of an impoundment site introduces considerable uncertainty with respect to seepage since spoil is not necessarily compacted in a systematic manner and there is the likelihood of large voids in the material. These conditions can allow internal erosion to occur. Impoundment plans need to include measures to remove such material, or compensating design features must be provided.

### 7.2.3.10   *Internal drains and filters*

Plans need to show that the size and position of internal drains is adequate to control the phreatic level, filter criteria are met where flow occurs into the drains, the filter medium is not subject to clogging over the long term, and suitable materials are used with respect to durability. Drains used to control the phreatic surface should be designed with a capacity to carry at least 10 times the anticipated seepage flow. When a geotextile is proposed for a filter around a drain, long-term flow tests should be conducted, with representative samples of soil or refuse from the site, to demonstrate whether the filter will be subject to clogging. Because of concerns for clogging, designers should specify geotextiles having the largest opening size and maximum flow capacity, while maintaining the soil retention requirements. Piezometers and weirs should be installed and monitored to verify that the fabric does not clog over time. The issues of filter clogging, soil retention, permeability, constructability, and monitoring, all need to be addressed in the plan.

*Common review problems*: Proposed drains do not always have sufficient horizontal size, or may not be properly positioned within the cross-section, to ensure that the phreatic level will be drawn down into them. Adequate substantiation may not be provided to address the issue of a geotextile becoming clogged over the life of the filter. Durable stone, which will not be subject to degrading from acidic seepage or weathering, is not always specified for drain material. For granular filter zones, filter criteria should be checked for the worst case, that is, at the extremes of the proposed gradations.

Sometimes geotextiles are proposed to be placed around drains which consist of large pieces of rock. In such cases, plans need to address the strength of the geotextile with respect to puncturing, and the measures to be taken during construction to ensure that the geotextile is placed on a uniform surface which will prevent the fabric from being damaged.

### 7.2.3.11  Cutoff trenches

Seepage through a permeable zone under an embankment can cause instability from excessive seepage pressures or internal erosion. Cutoffs of compacted, fine-grained soil are typically designed to prevent this condition. Cutoff trenches generally extend to the top of rock, or to the level of low permeability material in the foundation. Plans should include specifications to clean the bottom and sides of cutoff trenches in rock so that a good bond is obtained with the compacted material. Plans should call for the trench to be examined by an engineer, familiar with dam safety and the plan requirements, before and during material placement. Geologic features, such as open joints in the trench, which could cause seepage or piping problems, should be examined and properly treated, such as by grouting or filling with dental concrete.

### 7.2.3.12  Compaction of fill

Proper compaction increases the strength and reduces the permeability of embankment material, and is an important element in the construction of a safe impoundment. Plans need to include compaction specifications that place acceptable limits on minimum dry density, range of placement water content, and lift thickness. Best practice is normally to compact to at least 95% of the maximum Proctor dry density, with moisture content from $-2$ to $+3\%$ of optimum. Lift thickness for coarse coal refuse should not exceed 12 inches, and for fine-grained materials should not exceed 8 inches.

*Common review issues*: Proctor moisture-density relationships can present a problem when dealing with coarse coal refuse. The results from Proctor testing need to be compensated for, if the material contains a high percentage of plus ¾-inch material. In such cases, the results of a Proctor test in a 6-inch diameter mold may misrepresent the maximum Proctor density. Either a rock-correction factor should be used, or tests should be conducted in a larger diameter mold. Laboratory strength testing of embankment materials should model anticipated compaction specifications.

Compaction specifications should call for hard or smooth compacted surfaces to be scarified before additional fill is placed to ensure that adequate bonding is achieved between layers and to prevent the creation of horizontal seepage planes. Specifications should also require that when new fill is to be placed on an existing slope, the material be tied in by horizontal benching.

### 7.2.3.13   *Structural design of decant pipes*

An important design consideration for decant pipes is their ability to support the structural loads over their design life. The pipes used for decants in slurry impoundments are commonly subjected to depths of cover that far exceed what is commonly found in highway embankments or other culvert or sewer applications. Plans need to show that the pipes have adequate factors of safety with respect to wall crushing, buckling, and excessive deflection. The potential for corrosion to weaken the pipe material must also be analyzed. And in situations where a pipe crosses different foundation materials, the potential for damage to the pipe from differential settlement needs to be addressed.

When there is uncertainty about the performance of a pipe under higher fills, plans must include either provisions to install a replacement pipe at a higher elevation in the fill, and grout the original pipe, and/or measures to monitor the performance of the pipe as the fill level rises.

*Common review issue*: Available design methods for the amount of fill that can be placed over flexible pipes typically are aimed at lower fill applications, and there are questions about their applicability to the higher fills found in impoundments.

### 7.2.3.14   *Slurry cells*

Some mining companies have elected to use a system of cells in lieu of a major impoundment for disposal of slurry. Most of the same design considerations apply to a slurry cell disposal facility as an impoundment. These include: a breach analysis to determine the hazard classification and the appropriate design storm; a geotechnical investigation to determine embankment and foundation characteristics; appropriate materials testing; static and seismic slope stability analyses; underdrains to control the phreatic level; monitoring devices to confirm design assumptions; construction specifications; and construction monitoring. The number of cells that can be active (not capped with backfill) at any given time is usually limited in size and volume to the amount of storage that will result in an low-hazard class facility, otherwise, diversion ditches and cell spillways have to be designed based on a rainfall greater than the 100-year frequency, 6-hour duration storm.

It has been common practice to design a "structural shell" as the downstream containment structure for the cells (Figure 7.2). The structural shell is designed much in the same manner as a dam with the required width, slopes, benches, internal drainage system, and embankment-material strengths, to achieve the required safety factors for slope stability. Slurry cells seem to work most efficiently when the depth of fines in the cells is kept relatively shallow, preferably to 5 ft or less. At this depth, the fines can usually drain and dry out enough that they can be efficiently covered with coarse refuse.

For the purpose of determining the storage volume, cells are considered collectively by MSHA unless it can be clearly demonstrated that they cannot fail in series. If the collective slurry cell volume is greater than twenty acre-feet, the site is classified as an impoundment and the MSHA

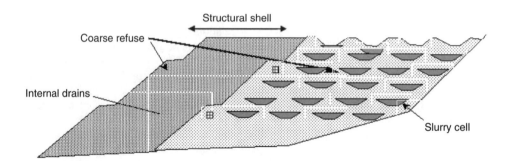

Figure 7.2.   Simplified cross-section through structural shell and slurry cells (not drawn to scale).

impoundment regulations apply, otherwise the plan is approved under 77.215(h) of Title 30, Code of Federal Regulations, and treated as a refuse pile with an approved plan.

Slurry cell refuse disposal sites can have advantages over a large impoundment. These include a decrease in potential liability due to less slurry and water being available for release downstream in case of a failure. Cells can usually be designed and constructed to mitigate breakthrough potential into underground mine works, and, for preparation plants which normally produce combined refuse, cells can provide for emergency disposal of thickener sludge.

Some disadvantages of cells are: frequent construction of diversion ditches, new cells, and cell spillways is necessary as the site increases in height; a relatively large ratio of coarse to fine refuse is required; close planning and supervision of the site is needed to ensure that the construction, filling, and backfilling of cells is accomplished in the proper sequence to make the system work as intended; they are not usually suited to large operations due to slurry volume considerations with regard to hazard classification, diversion ditch and spillway design requirements; and market changes, mining method changes, or other factors which change the strength characteristics of the refuse, or the ratio of fine to coarse refuse, may require re-design of the facility.

*Common review issue*: In attempting to keep sites small enough to avoid MSHA and state agency impoundment regulations, designers and coal operators sometimes end up with sites that don't provide sufficient disposal capacity to meet operational requirements. This usually results in operational and compliance problems and eventual re-design of the site.

### 7.2.3.15    *Construction specifications*

The completeness of construction specifications is frequently a major source of review problems in impoundment plans. Basically, the specifications need to address every aspect of the construction and clearly define the requirements. This includes foundation preparation; cutoff trench; fill placement; rock-correction factor for Proctor tests; underdrains, filters, treatment of mine openings; decant installation and backfilling; pipe joints; pipe corrosion protection; seepage diaphragms; pipe pressure testing; spillways; erosion protection; and concrete work.

The plan needs to specify how quality control will be maintained during construction to ensure that the plan requirements are met. Plans should require that the designer, or his representative, inspect critical phases of construction. This is especially important because designs are typically based on the extrapolation of a limited amount of information, e.g. borings, test samples, and laboratory results. The mining company needs to ensure that any conditions encountered that are not consistent with the original design assumptions and analyses are taken into account, and the plan is modified appropriately.

Coal refuse disposal sites are often constructed over decades, and the material used in the construction usually comes from various mining areas, and may, over the life of the facility, be subjected to different preparation processes. For these reasons, plans need to include provisions to verify the engineering properties of the refuse at regular intervals, and to check such properties anytime there is reason to suspect that a change has occurred in the characteristics of a construction material. This includes properties such as shear strength and Proctor moisture-density relations. For example, plans typically require that field density tests be performed, at random locations, for each 2,000 cubic yards of embankment fill. A common plan requirement is to conduct supplemental Proctor tests for each 20 field density tests.

*Common review issues*: Construction specifications are a frequent source of questions during plan reviews. The most frequent problems are that the specifications are not complete, are not clearly written, or are not consistent with MSHA's design guidelines. Often this applies to critical phases of construction, such as the installation of drains, or pipe backfill. Sometimes the specifications are not consistent with an item indicated on a drawing. For facilities constructed over several years, plans should include provisions for periodically verifying the shear strength and other pertinent engineering properties of the fill material.

### 7.2.3.16   *Instrumentation and monitoring*

Impoundments need to be monitored to ensure that the structure is performing as designed, and to provide warning of potential problems. Plans should include detailed information on any aspects of the site that should receive special attention during normal examinations.

Because the level of saturation within the embankment is so critical to stability, plans should include a sufficient number of piezometers. Plans should also include provisions to monitor the discharge from underdrains by the installation of weirs, so that accurate and consistent readings are made. Combined with the piezometer readings, this information will indicate how the underdrains are performing. Where compressibility is an issue, or there is any threat of subsidence affecting the structure, movement should be monitored by survey monuments, or, where conditions warrant, by other measures, such as inclinometers or settlement gauges.

For any impoundment, it is good practice to install a rain gauge in the immediate area. One benefit is that the performance of the discharge works can be better evaluated. On sites where the discharge from mine workings is monitored, a rain gauge is needed so that changes in discharge can be evaluated to determine whether they correlate to rainfall, or represent increases in seepage from the impoundment.

Cautionary and "red flag" levels should be established in the plan for all monitored values. For example, acceptable levels for piezometers readings, based on seepage and slope stability analyses, should be established by the designer and included in the plan.

*Common review issues*: For the monitoring information to be useful, it must be plotted and analyzed by an engineer familiar with the principles of dam safety and with the plan requirements, and this evaluation must take place in a timely manner. The plan should specify how the data will be collected and evaluated, appropriate to the site conditions. The instrumentation proposed in a plan is often insufficient to obtain the needed monitoring data.

### 7.2.3.17   *Mine workings near an impoundment*

It is not unusual for the impoundments constructed by the coal industry to be located in the vicinity of underground mine workings. In many areas of Appalachia, it may be difficult to find a valley that hasn't been undermined, especially near the coal preparation facilities that the impoundments serve. In designing an impoundment for these conditions, the plan must include the measures taken, or to be taken, to evaluate and remedy the potential impact of the mining on both the safety of the dam, and the structural integrity of the basin area. When there is active mining in the vicinity, the impact of the impoundment on the safety of the mining operation must also be evaluated.

The deformation and tensile strains created by mine subsidence can cause cracks in the embankment or foundation, and differential settlement can damage decant pipes and underdrains. Tensile strains ranging from 0.15 to 0.30% are sufficient to cause cracking of some earthen materials. A crack in a dam, an open joint in the foundation, a sinkhole opening, or an opening along a decant pipe, can cause a concentration of seepage and a loss of material from internal erosion. Over 30 percent of all dam failures occur due to seepage or piping problems. Plans with mine workings located in the vicinity of the impoundment need to address these issues.

The presence of mine workings requires that a more extensive site investigation be conducted to verify the extent of the mining, determine the engineering characteristics of the overburden and pillars, and evaluate the strata disturbance that has occurred. Analyses then need to be conducted to determine the potential impacts on the dam and reservoir. Sites may not be suitable for the construction of an impoundment without extensive remedial measures, such as backfilling of portions of the mine workings.

The area affected by mine subsidence can be larger than the area of the mine workings. The larger area is defined by a draw angle, which can vary depending on the nature of the overburden and the local topography. The most prudent and recommended design approach is to locate impoundments far enough away from mine workings that they will not be affected by subsidence, or, to backfill or grout mine openings that are close enough to influence the impoundment site. Information Circular 8741, published by the U.S. Bureau of Mines (Babcock and Hooker [4]), recommends "safety zones" for the areas near dams and reservoirs where no mining should take place. All information used in determining how close to the impoundment mining can safely occur, or the location of the "safety zone," needs to be fully documented in the impoundment plan.

With room and pillar mining (first mining only), plans need to address sinkhole potential, pillar stability, and the margin of safety against pillars punching into the floor or roof, especially with shale floor strata that may soften due to moisture. If auger mining has occurred, the potential impacts from deformation and seepage need to be evaluated. Auger holes in the abutment areas should be handled by backfilling the openings and then covering the area with drains and filters. Full extraction mining (either longwalls or second-mined pillars) will affect the surface in virtually all cases, with the surface strains generally increasing as the mining depth decreases. Multiple seam mining complicates the analyses by creating stress concentrations and increasing the uncertainty about how the overburden load is distributed to lower seams. Plans need to provide the technical basis for the draw angle and strains considered in the design.

In cases where there are mine workings near an impoundment, the level of design uncertainty is increased. Under these circumstances, compensating and redundant design features should be incorporated into the plan. The redundancy is appropriate so that the disruption or failure of one feature does not jeopardize the safety of the dam. Such measures might include the following:

- conducting a more thorough foundation investigation to verify the location of the mine workings and locate zones of increased permeability;
- backfilling or grouting mine openings in critical support areas to minimize the amount of movement that can occur;
- taking special precautions during foundation preparation to ensure that any openings or open cracks in rock foundations are filled and properly sealed off;
- specifying a wide dam crest and cross-section to provide increased mass and greater resistance to piping failure;
- maintaining an ample amount of freeboard to compensate for potential subsidence;
- specifying larger cross-sections for filters and underdrains so that these features can continue their function with the maximum likely subsidence;
- locating any decant pipes over unmined or backfilled areas;
- compacting materials at water contents slightly wet of optimum to increase their ability to deform without cracking;
- incorporating design features, such as impermeable zones, to minimize the amount of seepage through the dam and its foundation;
- incorporating design features, such as chimney drains, to collect seepage and discharge it in a controlled manner;
- using wide zones of material with "self-healing" characteristics, to act as crack stoppers;
- specifying a comprehensive monitoring program for the dam to provide early indications of potential problems;
- training dam inspectors about the potential affects of the underground mining so they know what signs to look for during their inspections.

Once a dam is constructed, any mining that is proposed near it must be carefully planned. Due to the uncertainty of long-term support, the development of entries near or under dams should be avoided. Only under favorable conditions, and where entry development is required for ventilation or haulage safety, should limited mining be considered near an existing dam. Since full extraction mining, i.e. longwall and second mining of pillars, affect the surface in virtually all cases, such mining is normally not acceptable either under or within a zone of influence of an existing dam.

### 7.2.3.18   *Note on mine map accuracy*

In evaluating the potential impact of underground mining, mine mapping is obviously a key element. The accuracy and completeness of the information found on mine maps can vary widely

Table 7.1.    Example of mine map inaccuracy.

| Expected horizontal distance to workings, based on mine map (feet) | Actual horizontal distance to works based on drilling (feet) | Discrepancy (based on horizontal drilling – workings actually closer or farther from highwall compared to map) |
|---|---|---|
| 40 | 37.5 | 2.5 feet closer |
| 122 | 92 | 30 feet closer |
| 137 | 89 | 48 feet closer |
| 27 | 67 | 40 feet farther |
| 67 | 41 | 26 feet closer |

(Table 7.1). There may be mining that was not mapped at all, mining that is accurately reflected on mine maps, or anything in between.

Many mining accidents have occurred where an active underground mine broke into old mine workings that either had not been shown on a map, or were shown as being hundreds of feet from their actual location. Since 1995, there have been over 100 inundation accidents at coal mines. These incidents may have occurred because coal was mined and the full workings not surveyed, or the area could have been surveyed, but the survey may have been in error, or the mine mapping was improperly transposed from the original coordinate system to a revised coordinate system.

When mining occurs near an outcrop, conditions may make problems with mine map accuracy more likely. It is not uncommon for roof conditions to deteriorate as the outcrop is approached. This occurs as the mining encounters lower cover, more weathered roof strata, more frequent jointing, and possibly "hillseams" (Sames and Moebs [5]). Furthermore, the last cut made toward an outcrop would typically not be provided with roof support. Since surveyors would not have access to the unsupported entry, its depth would be estimated rather than directly surveyed. If a roof fall occurs, it may not be accurately surveyed. This is more likely to occur near a coal outcrop.

As an example of inaccuracies in mine maps, drilling was performed, in 2001, at an impoundment site to check the accuracy of the mine map. A 46-inch-thick seam had been surface mined in the impoundment area in the 1970s, leaving a highwall. Room-and-pillar mining, with second mining of the pillars, was then performed from 1985 to 1995. Twenty-six horizontal holes were drilled from the highwall location to verify the mine map. Five of the drill holes encountered mine voids as shown in Table 7.1.

Discrepancies like these between the "mapped" and "actual" amount of barrier width could have serious implications in the safety of an impoundment. This example points out the need to verify mine maps – even more recent maps – when performing an impoundment site evaluation.

### 7.2.3.19    *Lessons learned from slurry breakthroughs*

Breakthroughs of slurry into underground mine workings have occurred in recent years, the most notable one being at the Big Branch Impoundment, near Inez, Kentucky, in October of 2000. That failure resulted in over 300 million gallons of slurry discharging into and through an underground mine, and ultimately polluting miles of creeks and rivers.

Impoundment designers must consider that a combination of unfavorable conditions can occur that include: shallow cover over the mining; more fractured and weathered rock near the outcrop, or near the bottom of the valley; the possibility of deeper soil in portions of the valley; the presence of open joints or "hillseams;" unsupported roof in the final cuts made in the underground mine; deterioration of the mine's roof, ribs and floor over time; softening of the floor from water; and, especially near the back end of the reservoir, the presence of a pool of water and the finest, most flowable slurry.

In evaluating breakthrough potential, mining companies should do the following:

• Research all available information on the geologic and mining conditions, including mine maps, topographic maps, aerial photos, geologic reports, other drilling in the area, etc.

- In the preliminary design stage, have a geotechnical engineer, or an engineering geologist, walk and examine the entire impoundment site, for signs of mine openings, auger mining, old mine facilities, subsidence, hillside movement, hillseams or other geologic features, springs or water seepage, any unusual or suspicious conditions, and any activity (roads, ditches, etc.) that may have reduced the thickness of the overburden/barrier.
- Consult with miners, especially retired miners, who worked in the area, about the mining conditions, the practices used when mining near the outcrop; and whether the available mine maps reflect their knowledge of the mining, including auger mining, that took place.
- Confer with mine surveyors who may have worked in the area, long-term residents, and mineral rights holders about their knowledge of mining in the basin and surrounding area.
- For cases where accuracy is critical, physically verify the location of the coal outcrop line instead of relying on large-scale geologic maps or projections.
- Perform sufficient exploration (drilling, test pits, geophysical) to determine the geologic conditions, especially the amount and nature of the overburden (rock quality, joints, hillseams, weathering, etc.).
- Perform sufficient testing to determine the pertinent engineering properties of the materials.
- Collect enough information to evaluate the consequences, if a breakthrough were to occur. Examples of questions to be considered include: could a breakthrough occur directly into an active mine? Is there an active mine in a seam below the level of the seam where the breakthrough could occur? Could a breakthrough occur into an abandoned mine but then create pressure on a barrier or bulkhead adjacent to an active mine? In the event of a breakthrough, in which direction does the coal dip and where would the inflow end up? Could the inflow discharge out of a mine opening, or break through an outcrop barrier or bulkhead? How much hydrostatic pressure have any bulkheads been designed to withstand?
- Account for variations in geologic conditions and their impact on safety, such as differences in the soil/rock thickness at the heads of hollows and in old landslide areas, versus elsewhere in the basin.
- Verify the accuracy of mine maps at a sufficient number of places to provide confidence that the barrier widths indicated on maps are accurate. This may require extra drilling or the use of geophysical methods, supplemented by drilling.
- Have the conditions evaluated by a professional engineer familiar with the concepts of subsidence, pillar strength, dam safety, etc.
- Identify, and evaluate the margin of safety for, each of the ways that a breakthrough could occur.
- Ensure that the worst case conditions are evaluated. Several areas where the amount of cover, or the thickness of the outcrop barrier is more critical, may need to be investigated.
- Incorporate measures to provide strength, control seepage and reduce hydrostatic pressures, as appropriate, in the areas of potential breakthroughs.
- Provide for monitoring of the conditions needed to show whether the measures taken work as intended and stay within the design parameters. Such monitoring could include piezometric levels, discharge or seepage rates, water levels in the mine, ground movement, rainfall, etc.

### 7.2.3.20 *Comments on breakthrough evaluations*

Once the basin conditions have been defined by the site evaluation program, the conditions should be evaluated by a professional engineer for the margins of safety against each way that a breakthrough could occur. The following comments apply to this evaluation:

- The professional engineer performing the evaluation needs to be familiar with the concepts of geotechnical engineering including dam safety, subsidence, seepage, piping, rock mechanics, pillar strength, etc.
- Failure modes considered should include: roof fall; sinkhole development; punching or shearing of a plug of material in horizontal and diagonal orientations; floor heave or softening; pillar instability; internal erosion; progressive unraveling; dispersion of shale strata; and any other

scenario that the conditions dictate. With internal erosion, material can be lost into subsidence cracks, sinkholes, auger holes, and punch-outs.

• The impact of the weight of accumulated slurry and water in the impoundment should be included in the analyses.
• Previous breakthroughs have occurred in areas where the underground disposal of slurry had occurred, so the potential impact of underground slurry disposal, or accumulations of mine drainage, in deteriorating roof, rib and floor conditions, should be taken into account.

### 7.2.3.21    *Comments on some basin remedial design measures*

Where the barrier or the overburden is found to be inadequate, engineered measures must be incorporated to compensate for the existing conditions. These measures could include some combination of the following: providing support by backfilling portions of the mine; improving the in-situ materials by grouting; constructing an engineered barrier; isolating the reservoir from the area of influence of the mining; constructing secondary defense measures, such as bulkheads, to contain a breakthrough in the mine; and other engineered measures that would control seepage and reduce pressures in the areas of potential breakthroughs. The following comments are offered on basin design measures:

• Backfilling mine workings can be a costly and difficult process. However, where the overburden cannot be relied upon to prevent sinkhole development, or subsidence cracks, there may be no other alternative for a particular site. When backfilling is proposed, subsidence and/or pillar analyses need to be conducted to substantiate that the backfill strength and area will be adequate. The angle-of-draw needs to be considered so that support is provided to all critical areas. The design needs to specify the strength of the backfill material; the area to be backfilled; and the backfill methods to check that the design intentions for strength and areal extent are met.
• One design approach, when the problem is high hydrostatic pressures against the outcrop barrier, is to construct a fill-barrier around the inside of the reservoir. The hydrostatic pressure problem is handled by installing drains in the fill-barrier which collect the seepage, reduce the pressure on the outcrop barrier, and discharge the seepage to a safe point. Drains in such barriers should be designed and constructed with the same type of specifications as used for the internal drains of embankment dams. That is, analyses should be provided to show that the drain is adequately sized and sloped to handle the discharge, and to demonstrate that filter criteria are met wherever flow occurs from one material into another. This approach should always include a monitoring plan, with provisions for recording the outflow from the drain, sufficient piezometers to verify the hydrostatic pressures, measurement of water levels in the mine, and monitoring of water discharge from the mine, to verify that the drain and barrier are working as intended.
• Where the combination of cover and overburden characteristics indicate that the potential for the development of subsidence cracks or a sinkhole exists, the design needs to provide a positive method to prevent them from developing, or the impoundment needs to be located far enough away that it would not be adversely affected. The potential for sinkholes and cracking can be minimized by backfilling the mine workings to provide support for the overburden. The approach of placing fill material on the surface above a potential sinkhole location, with the idea that the fill will collapse into and choke off the sinkhole should it develop, should not be relied upon. If a sinkhole does form, there is no way to guarantee that the fill material will not just be eroded away and allow the contents of the impoundment to discharge into the mine.
• Basin design may involve dealing with a mine opening or auger holes. In these cases, the design must address strength and piping resistance, including the following: the potential for subsidence above the mining; the strength of the sealing or backfilling material; the anchorage and grouting around bulkheads; the treatment of soft or erodible shale; the control of hydrostatic pressures against the opening areas as the impoundment is raised; the measures to prevent internal erosion of material into the mine openings; and the monitoring provisions.

- An approach that might be considered, for an existing impoundment, is to show that the settled fines have consolidated and gained strength to the point where they will not flow. Whether settled slurry would flow would depend on such factors as its degree of consolidation and cohesive strength, the state of the pore-water pressures, the potential for excess pore-water pressures to be induced, and the size of the opening available to it.

- Just because a soil is shown to be at or below its liquid limit does not, in itself, demonstrate that the material will not flow. The liquid limit is simply the water content corresponding to an arbitrary boundary between the "liquid" and "plastic" states of a soil. It indicates nothing about how the soil will behave when bridging an opening with the pressure of the contents of an impoundment above it, or when excess water pressures are induced. A change in conditions within the impoundment, such as additional water due to a large storm, could help liquefy the fines and result in an unplanned release of slurry and water through underlying or adjacent mine works. Additionally, if a subsidence event occurred under loose fines, the sudden increase in shear stress in the fines would induce increases in pore-water pressure that could trigger "static liquefaction" and cause the fines to flow. For these reasons, this approach would be difficult to justify.

- Finite element programs are available to assist in the analysis of seepage from impoundments into barriers and mine workings. This type of seepage analysis can be a valuable tool in showing the potential for seepage flow, gradients and hydrostatic pressure. However, they should be used with caution because the results can be extremely sensitive to the input parameters, such as permeability. Permeability values are difficult to determine with confidence from small-scale samples and limited field testing. Also, a false sense of confidence may be gained from the simplified or idealized models, whereas in the actual case, even a limited zone of higher permeability could result in a significantly more critical condition. The designer should keep in mind that only one area of weakness in the material separating the mine works from the impoundment basin can lead to a progressive failure and the loss of a substantial amount of the basin contents into the mine workings.

- An approach that may be proposed is the use of additives to stabilize the fine waste in the impoundment (Fiscor [6]). The fine waste would "set up" and have shear strength. To evaluate this alternative, testing would need to be performed to demonstrate the strength and bridging capability of the stabilized fines, and the behavior of the fines when saturated for a long period of time.

- Wherever the potential for a breakthrough exists, critical parameters, such as water discharge, movement, or piezometric level, should be identified and a monitoring program should be put in place that will show whether or not the barrier or overburden is behaving as anticipated. The acceptable range and warning or action levels should be established for all monitored values. Monitoring data should be regularly plotted and evaluated by an engineer familiar with the design requirements of the facility.

### 7.2.3.22 *Emergency action planning*

Standard 77.216-3(e), Title 30, Code of Federal Regulations, requires that each coal-mining related impoundment regulated by MSHA have an approved program for examining the site for structural weakness and other hazardous conditions, monitoring the instrumentation, and evaluating and eliminating hazardous conditions. In the event hazardous conditions develop, the plan includes procedures for notifying the MSHA District Manager, evacuating affected miners, and monitoring the dam and instrumentation on an eight hour, or more frequent, basis. Record keeping is required concerning the examinations made and actions taken.

On June 18, 1994, MSHA issued Program Information Bulletin P94-18 to the coal industry. This bulletin was issued with recognition of MSHA's responsibility, under Section 601 of the Fuel Use Act, to protect public safety as a Federal regulator of dams, and to comply with the Federal guidelines for dam safety (FEMA [7]). The bulletin addressed emergency action plans (EAPs) recommended by the National Dam Safety Program of 1979, as they apply to areas downstream of coal mine waste impoundments. The bulletin encouraged mine operators to develop EAPs which include procedures to delineate the downstream hazard areas, identify and evaluate potential emergencies, notify key personnel and officials, coordinate warning and evacuation activities with State and local officials, prepare contingency plans for preventive action, train involved personnel, and test of the plan. Mine operators may elect to use an EAP to meet the requirements of the mandatory 30 CFR 77.216-3(e) program, provided that it specifically addresses all of the required elements.

The establishment of EAPs for the coal mine impoundments varies from state to state at this time. Currently MSHA is evaluating the EAPs and 30 CFR 77.216-3(e) programs to ensure that, in addition to the downstream areas that would be impacted by a dam failure, the programs also identify areas which could be impacted by a breakthrough of the impoundment into, and out of, underground mine workings.

## 7.2.4   CONCLUSIONS

It's not unusual for impoundments plans to go through several iterations during the review process. Changes may be made as a result of MSHA's review comments, or due to the reviews of other regulatory agencies. Furthermore, once a plan is approved, modifications are common. As a result, especially with facilities that can be in use for decades, it is often difficult to know which specifications or drawings are current, and which parts of the plan have been superceded. To avoid problems which could compromise the safety of the dam, it is recommended that mining companies maintain for their own use, and submit to MSHA, complete, stand-alone, up-to-date plan documents when any significant changes are made.

## REFERENCES

1. U.S. Department of Labor, "Design Guidelines for Coal Refuse Piles and Water, Sediment, or Slurry Impoundments and Impounding Structures (Amendment to IR 1109)," March, MSHA Technical Support Center, Mine Safety and Health Administration, 1983.
2. U.S. Bureau of Reclamation, Design of Small Dams, Government Printing Office, Washington, DC, 1987.
3. Soil Conservation Service, 1985, "Dimensioning of Filter-Drainage Diaphragms for Conduits According to TR-60," Technical Note No. 709, (with supplement dated November 22, 1989), U.S. Department of Agriculture.
4. Babcock, C.O., and Hooker, V.E., 1977, "Results of Research to Develop Guidelines for Mining Near Surface and Underground Bodies of Water," Information Circular 8741, U.S. Bureau of Mines.
5. Sames, G.P., and Moebs, N.N., Hillseam Geology and Roof Instability near Outcrop in Eastern Kentucky Drift Mines, RI 9267, U.S. Bureau of Mines, 1989.
6. Fiscor, S., Design Alternatives for Refuse Disposal, Coal Age, May 2002, pp. 34–36.
7. FEMA, "Federal Guidelines for Dam Safety: Emergency Action Planning for Dam Owners", 1998.

# CHAPTER 8

## Case histories of sustainable mining

Raj Rajaram
*Complete Environmental Solutions, Oak Brook, IL, USA*

### INTRODUCTION

The progress in mining practices in the latter half of the twentieth century has been remarkable when compared to practices of the past. The awareness of the health impacts of many industrial activities and the public demand for a clean environment acted as drivers to foster innovation and led to the development of mining practices that attempt to balance mineral development costs with the cost of maintaining a clean environment. Previous chapters have provided many examples of environmental protection measures, mineland reclamation, tailings management, and other mining practices that contribute to sustainable development. This chapter discusses case histories from the Americas, Asia, and Africa, to provide the range of practices used in recent years. These case histories provide "lessons learned" which may be useful for planning sustainable mining projects in the future.

# 8.1 Americas

William Langer[1], Frederick T. Graybeal[2], Stuart A. Bengson[2]
[1]*U.S. Geological Survey, Denver, Colorado, USA*
[2]*ASARCO LLC, Tucson, Arizona, USA*

## 8.1.1 INTRODUCTION

Sustainable practices should begin during exploration and continue until mine reclamation is completed. The emphasis during the whole process, which could last several decades, is one of communication with the local community and applying the technologies that will minimize environmental impacts. Three case studies from the Americas are presented here to exemplify some unique practices adopted in resource management, exploration, and mine land reclamation. The first case study demonstrates how sustainable resource management was implemented in an aggregate mining operation by Lafarge Company in the absence of existing local government policies or procedures in Jefferson County, Colorado. LaFarge is a large producer of aggregates worldwide, and is operating the quarry in Denver, Colorado. In the second case study the sustainable practices during exploration have been demonstrated by the Camp Caiman gold project in South America, a prospect explored by ASARCO LLC (Asarco). An environmental advisory committee was formed there who periodically advised Asarco and provided oversight with a goal to minimize environmental impact at this site. The third case study involves an innovative reclamation method applied by Asarco for managing copper tailings using livestock and biosolids in the arid southwestern United States. Asarco is a wholly owned subsidiary of the Americas Mining Corporation owned by Grupo Mexico de S.A. de C.V (Grupo Mexico). Asarco is an integrated producer of primary refined copper and associated co-products.

## 8.1.2 A CASE STUDY DEMONSTRATING SUSTAINABLE MANAGEMENT OF NATURAL AGGREGATE RESOURCES (William Langer)

This case study describes how Lafarge, a large multinational construction materials supplier, implemented the principles of sustainability even though there was an absence of existing local government policies or procedures addressing sustainable resource management. Jefferson County, Colorado, USA, is one of three counties in the six-county Denver, Colorado, region that has potentially available sources of crushed stone. Crushed stone comprises 30% of the aggregate produced in the area and plays a major role in regional aggregate resource needs. Jefferson County is home to four of the five crushed stone operations in the Denver region. Lafarge operates one of those four quarries. Lafarge recently proposed to expand its reserves by exchanging company-owned land for existing dedicated open space land adjacent to their quarry but owned by Jefferson County. A similar proposal submitted about 10 years earlier had been denied. In contrast with the earlier proposal, predicated on public relations, the new proposal was predicated on public trust. Although not explicitly managed under the moniker of sustainability, Lafarge used basic management principles that embody the tenets of sustainability.

To achieve the goals of sustainable aggregate management where no governmental policies existed, Lafarge not only assumed their role of being a responsible corporate and environmental member of the community, but also assumed the role of facilitator to encourage and enable other stakeholders to responsibly resolve legitimate concerns regarding the Lafarge quarry proposal. Lafarge successfully presented an enlightened proposal where the county will gain 745 acres of new open space land in exchange for 60 acres of current open space land adjacent to the quarry.

The process involved collaborative efforts by all stakeholders and resulted in an outcome that balances the needs of society, the environment, and business.

Crushed stone and sand and gravel together comprise the top non-energy mineral resource in both the world and in the United States. In the United States, about 2.74 billion tons of aggregate worth about $14.5 billion was produced during 2000 (Tepordei [1], Bolen [2]). Approximately 15 billion tons of aggregate worth about $76 billion are annually produced throughout the world (Regueiro and others [3]).

Natural aggregate can be produced from a broad variety of geologic environments, but even though potential sources of aggregate are widely distributed throughout the world, there are large regions where natural aggregate sources are non-existent (Langer [4]). Furthermore, aggregate resources must meet certain physical and chemical quality parameters that are determined by the final application. Departures from any quality specifications can make potential aggregate unsuitable for some specific uses (Langer and Knepper [5]).

An aggregate operation must consider all costs, including acquisition, operation, compliance with regulations, transportation to market, environmental management, and reclamation, in order to be profitable. These factors make opening a new operation a complicated process that can cost millions of dollars and take many years. Aggregate is a high-bulk, low unit value commodity that derives much of its value from being located near the market. Thus, it is said to have a high place value (Bates [6]). Transporting aggregate long distances can add significantly to the overall price of the product (Leighton [7]). For example, a city of 100,000 can expect to pay an additional $1.3 million for each additional 10 miles that the aggregate it uses must be hauled (Ad Hoc Aggregate Committee [8]). Therefore, aggregate operations frequently are located near population centers and other market areas.

Despite society's dependence on natural aggregate, urban expansion often works to the detriment of the production of those essential raw materials. "Resource sterilization" occurs when the development of a resource is precluded by another existing land use. For example, aggregate resources that exist under a housing development or shopping center commonly will not be extracted.

Before an aggregate resource can be developed, the extraction site must qualify for all necessary permits and the approving officials must be convinced that the operation can take place without adversely affecting the environment. Proposed aggregate operations have become the rallying point for citizens, and citizen involvement in the public hearing process has changed from participatory to absolute opposition (Bauer [9]). Today, we are in a situation where it is extremely difficult to obtain necessary permits to initiate new aggregate operations.

A core issue surrounding the sustainable development of natural aggregate resources is the conflict between regional needs and local opposition to resource extraction. This conflict occurs because the negative impacts of extraction are located near the site of extraction, while the benefits from resource extraction are dispersed throughout an entire region. Dunn [10] termed the conflict, and the consequences arising from it, the "Dispersed Benefit Riddle." The regional benefits commonly are not considered in the local permitting process (Dunn [10]), and if resource extraction is denied because of local opposition, other costs arise, such as those associated with longer haul routes resulting in more traffic, more accidents, more fuel consumption, generation of more greenhouse gases, greater wear and tear on vehicles, and higher vehicle replacement rates. Any gain by the local community that restricts extraction is usually at the expense of the greater public, the greater environment, and some other local area where extraction ultimately takes place.

Sterilization, permits, and regulations restrict development or expansion of aggregate production in established areas more than any actual limitations of suitable resource availability (Poulin and others [11]). Bauer [9] concluded that local units of government were unwilling or seem unprepared to deal objectively with the conflict between regional needs and local opposition, and the failure to plan for the protection and extraction of aggregate resources often results in increased consumer cost, environmental damage, and an adversarial relation between the aggregate industry and the community.

### 8.1.2.1    *History of quarrying in the Denver, Colorado, area*

The area around Denver, Colorado, USA (Figure 8.1), obtains a significant amount of its aggregate resources from quarries that produce crushed stone. Of the six Denver area counties, only three – Boulder, Douglas, and Jefferson – have any source of bedrock suitable for use as crushed stone; Adams, Arapahoe, and Denver Counties have no crushed stone resources (Figure 8.2).

Crushed stone has been produced in the Denver area since the early twentieth century (Schowchow [12]). Four quarries produced crushed stone at South Table Mountain, in Jefferson County, Colorado, with the first operation starting as early as 1905. These operations worked intermittently until the 1950's, and provided concrete and asphalt aggregates. One of these four quarries, the Wunderlich quarry, provided rip-rap for Cherry Creek Dam, in Denver. The Rogers Brothers quarry, started in 1925 and enlarged in 1949, mined crushed stone from North Table Mountain in Jefferson County and provided concrete aggregates for Harlan County Dam near McCook, Nebraska (Argall [13]). The Bertrand quarry, operating at the mouth of Clear Creek Canyon in Golden, started in 1926, but closed in 1975 because of a threatening landslide. Many years ago, two crushed stone quarries operated north of Golden at Ralston Reservoir.

The production of crushed stone has increased greatly from the 1960's, and crushed stone will become an increasingly important source of aggregate in the Denver market. Statewide, starting in the late 1950's, crushed stone has become a significant component of the total aggregate production stream (Figure 8.3). Crushed stone production in the Denver area has increased from just 3% of the State total in the 1960's to 56% in the 1990's. During 1960, 99% of the aggregates in the Denver area were derived from sand and gravel sources. In contrast, during 1997, only 55% of the aggregates demand in the Denver area was being met from sand and gravel sources while 31% was being met from crushed stone. The remaining 14% of the aggregate was derived from recycled concrete or asphalt (Wilburn and Langer [14]).

Today, crushed stone serves an important function in the Denver area beyond just being a replacement for sand and gravel. For example, specifications for aggregates used in concrete for

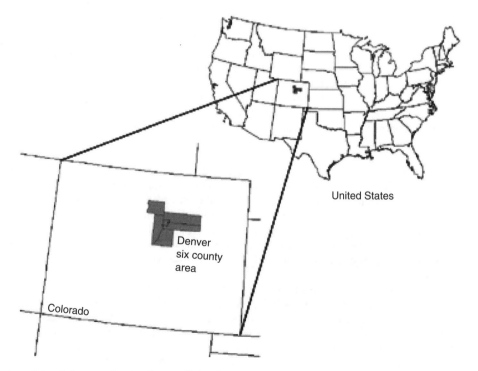

Figure 8.1.    Index map showing Denver, Colorado, USA.

the runways at Denver International Airport required crushed stone. Similarly, asphalt highways typically require crushed stone aggregates in order to achieve the required strength parameters. Highways being constructed with money from the Federal government commonly must meet SUPERPAVE specifications, which in effect require the use of sand manufactured from the crushing of rock and prohibit the use of natural sand. It is clear that crushed stone plays a key role in the sustainable management of aggregate resources in the Denver area.

Many of the potential aggregate resources in the Denver area are not accessible for extraction. The population centers have built out and gradually encroached upon existing deposits, both sand and gravel and crushed stone resources, thus rendering some nearby resources inaccessible. Sheridan [15] predicted that restrictive zoning, lack of general public understanding of sand and gravel occurrence and mining operations, and conflicting land uses would cause a shortage of

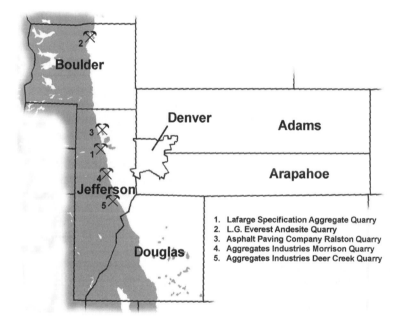

Figure 8.2. Index map showing the six Denver area counties, areas of bedrock, and locations of operating crushed stone quarries.

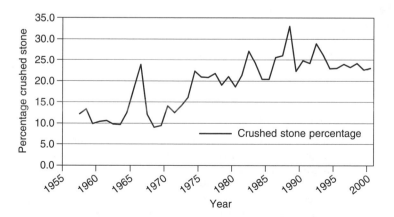

Figure 8.3. Percentage of crushed stone aggregate production in Colorado, USA.

near-by, low-cost aggregates in Denver. James Cooley [16] restated the problem at the 74th National Western Mining Conference. During 1973, the Colorado legislature officially recognized the problem and passed House Bill 1529. The act declared that: (1) the State's commercial mineral deposits were essential to the State's economy, (2) the populous counties of the State faced a critical shortage of such deposits and (3) such deposits should be extracted according to a rational plan that was calculated to avoid waste and would cause the least practical disruption to the ecology and quality of life of the citizens.

By 1974, commercial deposits of significant economic or strategic value to the area were mapped throughout the 10 populous counties on the eastern slope of the Colorado Front Range, including the six Denver area counties (Schwochow and others [17,18]). The mapping efforts focused on sand and gravel deposits; potential sources of crushed stone were only generally described and located. Although counties were required to complete master plans by July 1, 1975, only two counties, neither in the Denver area, met the deadline. Lack of state funding, personnel shortages, and no provision for penalties were reasons given for delays (Stearn [19]). Unfortunately, H.B. 1529 did not succeed at protecting existing aggregate resources in the Denver area. The U.S. Department of Labor [20] pointed out that resource availability in the Front Range had continued to decline. They blamed the decline on adverse zoning, non-compliance with H.B. 1529, increased production, inadequate grain size to meet specifications, environmental concerns, and some notoriously poor operational procedures employed by some aggregate operators.

In 1976, the Jefferson County planning department solicited a task force to prepare an outline and a master plan to manage aggregate resources. The resulting Mineral Extraction Policy Plan (Jefferson County, Colorado [21]) had 17 goals, each of which has numerous associated policies. Those goals and policies overwhelmingly were designed to control the impacts of aggregate mining. Essentially the plan had no provisions to protect aggregate resources from encroaching land uses. In 1986, the Jefferson County Commissioners convened a Roundtable, giving roundtable members the charge to formulate recommendations to guide decision making in the county regarding rezoning and the mining of aggregate. The Aggregate Resources Roundtable Report (Aggregate Resources Mining Roundtable [22]) included a variety of recommendations, including identification of preferred areas of extraction, public involvement, monitoring and enforcement, and mitigation of impacts. One recommendation deserving special mention was the recommendation to conduct informal meetings between the applicant and the citizenry and the appointment of an ombudsperson to facilitate those meetings. The purpose of the meetings was to gain citizen input before the final plans are developed, and the Roundtable report suggested the time between the informal meeting and formal hearing could be as much as 90 days.

In spite of the Mineral Extraction Policy Plan and the Aggregate Resources Roundtable report, the efforts to permit the extraction of aggregate resources remained a contentious issue in Jefferson County. Shortly after the passage of H.B. 1529 by the Colorado legislature, five companies submitted applications to open new quarries or reactivate older sites in Jefferson and Boulder Counties. During 1980–1981, all five of the applications were denied (Schwochow [23]). Although additional new applications to open new crushed stone quarries have been submitted, none has been approved in Jefferson or Boulder Counties since the passage of H.B. 1529, and applications have been denied as recently as 1999.

### 8.1.2.2   *Sustainable management of aggregate resources*

Natural aggregate of suitable quality for an intended use can be in short or non-existent supply on a regional or local scale. In the realm of sustainability, having an accessible local supply of aggregate resources takes on great significance because transporting aggregate long distances not only adds to the overall cost of the product, but also adds to the overall cost to the environment as described in the "Dispersed Benefit Riddle" (Dunn [10]).

Sustainable practices do not always have to be conducted under the title of sustainability. Many countries, provinces, territories, or states in the European Union, Australia, Canada, the United

States and elsewhere are beginning to develop sustainability programs, however, these efforts generally stop short of including aggregate resources (Langer et al. [24]). In spite of the lack of government policy promoting sustainability, some aggregate companies in the United States and elsewhere have already begun implementing some of the concepts of sustainability without waiting for government intervention.

There are a variety of key policies and issues that relate to the sustainable management of aggregate resources (see for example, Department of the Environment, Transport, and the Regions [25]). These policies are applicable for governmental agencies, but have little direct application to the aggregate industry. There is little or no clear guidance to the aggregate industry regarding application of sustainable management principles. Therefore, the industry must design their own sustainable practices by interpreting and drawing from basic sustainability tenets.

The practical application of sustainability of aggregate resources requires that each of the primary stakeholders – government, industry, public, and other non-governmental organizations – assume certain responsibilities (Langer et al. [26]). The government is responsible for developing policies that provide the conditions for success. The industry must work to be recognized as a responsible corporate and environmental member of the community. The public and non-governmental organizations must become informed about aggregate resource management issues and constructively address both their own and a wide range of objectives and interests. All stakeholders – the government, industry, and the public – have the responsibility to resolve legitimate concerns regarding sustainable aggregate extraction.

By working within these broad guidelines, the industry can effectively manage its resources and reserves in a sustainable manner. The company is clearly required to assume a role as a responsible corporate and environmental member of the community. But if sustainable management policies are not in place, the company must also assume the role of the facilitator who both brings the various stakeholders together and encourages each of the stakeholder groups to assume their proper roles.

### 8.1.2.3  *Specification Aggregate Quarry Expansion*

Quarries have operated in Jefferson County, Colorado for many years, but since the early 1980's, new quarry applications have raised serious logistical, environmental, and social concerns. No applications for new quarries have been approved since 1976. For example, a recent attempt to rezone property so that aggregate extraction could take place was unanimously denied in 1999 after almost 17 months of community effort to oppose the proposal.

Lafarge, and its predecessors, have been operating their largest quarry in the Denver area, the Specification Aggregate Quarry, since 1965. The 222 acre quarry is located just west of the metropolitan area and is bordered on the west by Mother Cabrini Shrine, on the North by a small retail and amusement park, on the East by a state road, and on the south by Matthews/Winter Park – land acquired by Jefferson County for use as open space (Figure 8.4).

The quarry is within the City of Golden, Colorado, and operates pursuant to a mine plan approved by the Colorado Mined Land Reclamation Board and according to a Planned Unit Development (PUD) originally approved by the City of Golden in 1977.

Lafarge tendered a proposal that would adjust the southern boundary of its quarry to include 60 acres of land contained in the Matthews/Winter Park. Lafarge proposed to compensate the county by transferring to the county three other parcels of land adjacent to the quarry totaling 60 acres, plus additional land owned by the company elsewhere in the county. A major challenge of this proposal was the fact that the 60-acre parcel in the Matthews/Winter Park had been acquired by the Jefferson County Open Space Program and was owned by the County. The proposal represents a complicated open space transaction with wide ranging political, environmental, and philosophical implications.

Local environmental activists, many of whom were involved in starting the Open Space Program over 30 years ago in 1972, continue to monitor this program. The program is funded by a 0.5% sales tax, which currently generates over $40 million per year, all specifically dedicated for open space

Figure 8.4.    Aerial photograph of the Lafarge Specification Aggregates Quarry.

purposes. The Jefferson County Open Space Program is considered by many to be the premier open space program in the United States, and many other open space programs have been modeled after it.

To provide an historical perspective, in 1992, the company that previously owned the quarry made a similar proposal to trade 400 acres of property on North Table Mountain (located approximately 5 miles north of the quarry) for 100 acres of open space located immediately adjacent to the quarry. This proposal was defeated by a majority vote of the Open Space Advisory Committee, the committee charged with making recommendations to the County Commissioners on all open space matters. The proposal failed for several reasons, including the undesirable precedent of relinquishing open space for a commercial operation. In addition, the previous quarry operator had hired a well respected outside public relations firm with a great reputation to take the lead in interfacing with the elected officials and the community. Lafarge determined that the core problem leading to failure of the proposal was the lack of direct company outreach to the community and the perceived indifference of the company to various issues raised by the community regarding the quarry expansion. It was further determined that the community, including the elected officials, wanted to see a commitment by the company directly through its employees – not through an outside consultant who typically disappears once a project is completed.

This realization embraces one of the basic tenets of sustainability: the industry must work to be recognized as a responsible corporate and environmental member of the community. Top management at Lafarge supported this kind of effort to be recognized and, consequently, took a very different approach in their land exchange proposal. The new proposal was to be built on public trust, not public relations.

Employees chosen for this very important responsibility were those who were willing to work in the local community for a number of years. Building and maintaining grassroots support was "kick started" by an employee who already was connected in the community. The key employee responsible for building and maintaining the grassroots connections was Ms. M.L. Tucker, coauthor of this paper. She had been very involved in the community for a number of years – even before starting work with Lafarge over eight years ago. She had served on various committees appointed by the elected officials, including those dealing with land use, and had worked her way

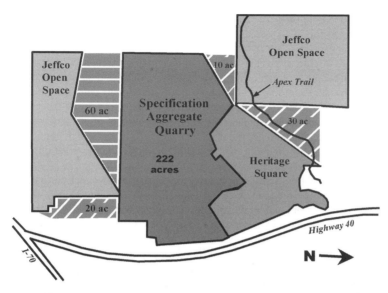

Figure 8.5.   Map of land parcel involved in proposed land exchange.

up to leadership positions in business organizations, ultimately to the Chairperson of both the Chamber of Commerce and the local Economic Council. Each of these positions offered additional opportunities to interface regularly with both elected officials and key business and community leaders, resulting in the establishment and maintenance of strong community relationships. In addition, through these relationships, she was able to help establish contacts between the members of the Lafarge team and the people with whom they needed to interface on the project, and to coach them in developing relationships.

Relationship building has commonly not been a priority of the aggregate industry in the past. Historically, in this business, once ownership of the land or a lease of the mineral rights was secured, you were ready to remove rock. In today's world, you not only need to acquire those rights, you also need to secure the permission of your neighbors to operate – you need a social license to mine. Permission is secured by earning the trust of the neighbors in the community. The way trust is earned is by building and maintaining relationships. Meaningful involvement in the community is critical.

Equally important, this close interaction with the community allowed the company to understand how its proposal fit into the larger community plan. This is a second basic tenet of sustainability – that resource development must address the needs of the community, as well as the needs of the company. Consequently, the new proposal was significantly different from the 1992 proposal and offered compensating, long-term advantages to the community.

Under this innovative new proposal, the boundary of the quarry was to be adjusted by acquiring 60 acres of open space to the south (Figure 8.5). In return, those 60 acres were replaced with the following:

• one prime commercial real estate parcel, consisting of approximately 20 acres
• one 30 acre parcel, currently zoned for 212 residential dwelling units
• one 10 acre parcel of pristine, highly visible mountain property.

In addition, upon obtaining annexation and all necessary permits and zoning, Lafarge would donate all of its land holdings on top of North Table Mountain (approximately 463 acres) to Jefferson County for use as Open Space land (Figure 8.6).

During the negotiations the County asked if Lafarge would transfer the quarry land (222 acres) to the County upon completion of mining. When Lafarge agreed to do so, the County responded that they were not ready to make the decision to accept the quarry at this point in time. Lafarge gave the County an option to acquire the quarry land so that the County could make the decision at a later point in time.

This process was a true application of the mutual gains theory, which seeks to create a "win-win" situation for the community, the governmental entities, and the company involved in the project. As a direct result of this project, the citizens of Jefferson County will gain as much as 745 acres of new open space. Preservation of the three parcels of land surrounding the quarry that were under threat of development and the acquisition and preservation of trailhead access to open space is a worthy outcome. The donation of land on North Table Mountain, which has been on the community's priority list for over 30 years, will substantially complete preservation of that mesa as open space.

Lafarge is given ownership of the 60 acres and the right to remove rock for natural aggregate. In doing so, Lafarge has secured land that extends the life of the quarry by 20 years and secures a location for a ready mix plant and an existing asphalt plant. Securing additional reserves at this location is excellent from both its proximity to market and to the interstate highway, and it exhibits good business planning for all three product lines – aggregate, asphalt, and concrete. Lafarge has eliminated the need for another site at a "greenfield" (undeveloped) location and can utilize existing facilities such as plants, crushers, scale house, roads, and other infrastructure.

A third key tenet of sustainable development is that all stakeholders should be able to participate in the decision process (Shields and others [27]). Lafarge recognized that participation meant more than including stakeholders in the public hearing process. Therefore, Lafarge informally "floated" their proposal to local governmental, citizens, quasi-governmental, and environmental

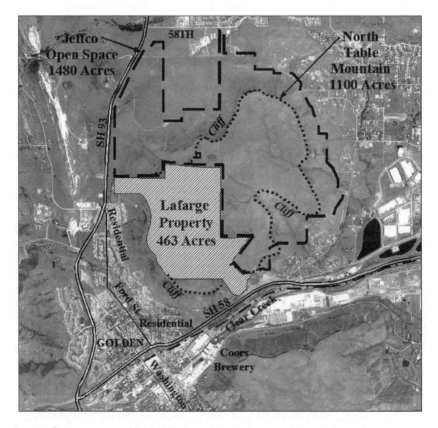

Figure 8.6.    Lafarge property on North Table Mountain.

groups during the Fall of 2000. They listened to stakeholder input, assessed their concerns, and modified the proposal. It was not until over two years later (February 2002) that LaFarge made formal presentations to citizens groups, the Jefferson County Open Space Committee, Jefferson County, and the City of Golden.

Even with this public involvement, the environmental watchdog group that started the Open Space Program was concerned that if this proposal was approved, the land trade would set a dangerous precedent, would discourage future donations of land to the Open Space Program, and would violate the policies and procedures that open space land shall not be disposed of if it is still serving its original purpose. Just prior to Lafarge making its formal proposal to expand the quarry, the watchdog group proposed changes to the open space policies and procedures that added several new requirements regarding disposal of open space land. One of the requirements increased the vote needed to support a disposition from a majority vote to essentially a unanimous vote of the Committee members.

During December 2002, a public hearing was held in Jefferson County to determine if the county would approve the proposed land exchange. Present in the room for this hearing were the County Commissioners and their staff and the applicant and his staff. In addition, there was a State Senator, a previous County Commissioner (from a political party other than the current commissioners), government officials from the City of Golden, a Sister from Mother Cabrini Shrine, the founder of the Jefferson County Open Space Program, the Chair of the Jefferson County Open Space Advisory Committee, the Chair of the Jefferson County Economic Council, the Chair of the Table Mesa Conservation Fund, the President of Save the Mesas, and a number of citizens who described themselves as avid hikers or neighbors of the quarry. All of the people not directly employed by the aggregate company had come to this hearing united in their opinion of the land exchange, and every single one of them spoke out in support of the land exchange. Comments such as "a balance between mineral extraction and open space," "county staff, Lafarge, and citizen groups should be congratulated," and "a great example of merging of environmental interests with business needs" typify the testimony. There was not a single dissenting opinion.

This proposal obtained approval of the County Commissioners as a direct result of an innovative combination of building public trust at the grassroots level, mutual gains approach, and continuous work to involve all stakeholders – three components of sustainable management. The proof of success was in three unanimous votes in favor of the expansion proposal by the independent County Commissioners with jurisdiction over the plan.

Historically, aggregate companies have taken the position that once the fact gathering required for the presentation of a project has been completed and the calculations of reserves, yields, and so forth have been made, the application should be formally submitted for approval (Figure 8.7

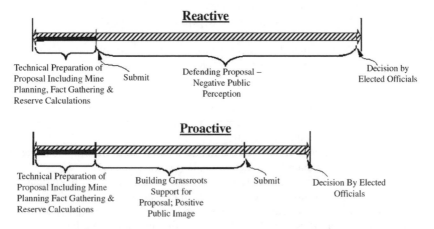

Figure 8.7.   Typical timeline for this type of project in contrast with timeline for this case study.

"reactive" timeline): legally and factually the application is ready for submission. However, at the time the application becomes formal, the project becomes subject to scrutiny, review, and comment – all in the public arena and in the media. Typically, this results in public criticism, and decisions often are based upon emotions and perceptions as opposed to being based upon fact.

Although a company may be technically prepared to formally submit a proposal to the governmental officials, building grassroots support for the proposal prior to formal submission by meeting with the elected officials, community and business leaders, and the citizenry to inform them and solicit input is often advisable (Figure 8.7 "proactive" timeline). This time line often results in positive coverage by the media because the community is informed and in support of the proposal long before the formal submittal, and may even generate additional supportive testimony and letters in support of the proposal – a much more friendly way to achieve community and governmental support.

### 8.1.2.4    *Lessons learned*

National, regional, and local governments are implementing the practice of sustainable management of aggregate resources in only a few parts of the World, and there are few or no guidelines for the industry to follow. Nevertheless, the aggregate industry can independently utilize sustainable management techniques to achieve the goals of meeting their needs for reserves without compromising the ability of future generations to meet their needs.

Lafarge, a major aggregate producer in the area near Denver, Colorado, U.S.A., submitted a proposal to enlarge their quarry holdings through an exchange of land owned by the company for dedicated open space land owned by Jefferson County, Colorado. A similar proposal submitted about 10 years earlier had been denied, and the new proposal represented a complicated open space transaction with wide ranging political, environmental, and philosophical implications.

During the application process, Lafarge employed three key tenets of sustainable management that relate to the aggregate industry:

1. Industry must be responsible corporate members of the community.
2. Resource development must meet the needs of both the company and the community.
3. All stakeholders must have the opportunity to participate in the process.

A number of operational concepts necessary for the sustainable management of aggregate resources grew from these tenets. They are:

- *Establish Community Relationships*: It is never too early for a company to get involved in the community in which they want to do business or do business and hope to continue to do business. In today's world, it is through the permission of neighbors that aggregate companies are able to continue doing business. Company strategies for each of their principal markets need to include continued community involvement by key employees. Those key employees also need to assist other employees in developing and maintaining long-term community relationships.
- *Obtain Executive Support*: The company president, his or her management team, and other appropriate managers must support the proposal. Develop a plan with input from both operations and management. Schedule team-meetings regularly so that team members can exchange information and keep each other up to date.
- *Create Benefits for the Community*: Understand how your proposal fits into the larger community plan and offer the community compensating, long-term advantages.
- *Communicate the Plan*: Identify the tools you will use to communicate the plan, such as videos, computer presentations, pamphlets, and field trips to the site. Make sure that your message to the various community groups, elected officials, and business leaders is consistent. Present the proposal to the elected officials well in advance of the formal presentation so that they are informed and stay informed as the proposal becomes more "public." Identify the real decision

makers; sometimes they are different from the obvious ones. Know who influences the decision makers and why they have the influence.

- *Solicit Feedback on the Plan*: Listen actively, take notes, ask questions, and show true interest in the communication. Make every effort to work in harmony with those of opposite views. Try to view the proposal through their eyes and analyze and address their issues and concerns before the final hearing.
- *Be Visible and Accessible*: Be available beyond just normal business hours, Monday through Friday. Maintaining availability requires commitment of team members' personal time outside normal business hours, on a regular basis, over the term of this proposal.
- *Maintain Your Public Trust*: Remain active in the community after the proposal is approved and operations begin.

The proposal was unanimously approved by the three independent County Commissioners responsible for land use decisions in Jefferson County. The authors are not aware of any other case in the United States, or in the world, where an expansion of a quarry onto pristine, dedicated open space land has been allowed. As a direct result of this project, the citizens of Jefferson County will gain an additional 745 acres of open space including three parcels of land surrounding the quarry that were under threat of development and highly prized land on North Table Mountain, which has been on the community's open space priority list for over 30 years. The County also has the option to acquire the quarry land once quarrying is completed.

Lafarge is given ownership of 60 acres of previously dedicated open space land adjacent to the quarry and the right to remove rock for natural aggregate from that land. In doing so, Lafarge has secured land that doubles the life of the quarry and secures a location for a ready mix plant and an asphalt plant. In addition, Lafarge has eliminated the need to start a new quarry at a "greenfield" (undeveloped) location and can utilize existing facilities, such as plants, crushers, scale house, roads, and other infrastructure.

### 8.1.3   CAMP CAIMAN GOLD PROJECT (Frederick Graybeal)

The Camp Caiman gold project is on the south side of the Kaw Mountains, 50 kilometers southeast of Cayenne, French Guiana, in South America and about 500 kilometers north of the mouth of the Amazon River. The project is accessible by paved road to within 3 kilometers of the exploration camp, and then by forest road requiring four-wheel drive vehicles. The climate is typical of the Amazon rainforest with an average annual rainfall of 4 meters and an average maximum daily temperature of 32°C.

In 1995 Asarco acquired in a public tender two exploration permits covering 50 square kilometers over a gold anomaly originally located by the Bureau de Recherches Géologiques et Minières (BRGM). Three additional permits covering 27 square kilometers were acquired in 1997 and an application for an additional permit was submitted. The exploration permits are adjacent to the Kaw Biotope zone an area of both wetland, which is a principal habitat of the Black Caiman alligator, and adjacent hills, where caves high on the north side of the Kaw Mountains host the elusive bird called the Cock-of-the-Rock. The Biotope zone is roughly coincident with the Kaw Wetland System, is a Ramsar site, i.e., designated for inclusion in the Ramsar List of Wetlands of International Importance. These wetlands were identified for specific protective measures in the Convention on Wetlands in Ramsar, Iran by 90 nations in 1971. Further, a regional park is planned for the area.

A case history of gold exploration at Camp Caiman is reported in detail by Graybeal [28]. By 1996 an exploration camp had been built and a moderate drilling program was underway. Some of the initial findings were as follows:

- Use of bulldozers in land clearing and building of access roads, resulted in removal of forest litter. The rainwater runoff caused silting of the adjacent stream (Figure 8.8).

Figure 8.8.    Downstream silting after driving on initial log bridges.

- In order to preserve the forest canopy drill roads were located with care to avoid felling larger trees and required several re-routings of access roads from the camp to the drill sites. However, these roads were constantly muddy where they crossed the numerous gullies. The rainwater runoff caused silting of small creeks at road crossings, although it was not as serious because the forest floor litter acted as an effective filter.
- Mitigation measures used by Asarco's exploration staff included filling and recontouring of exploration trenches after mapping and recontouring the campsite to divert the runoff to the adjacent forest floor in order to reduce the silting of nearby streams. Care was also taken in locating operations to ensure that streams draining from the camp and drill sites flowed away from the Kaw Biotope. Concrete pads for the containment of small oil spills from the servicing of machinery and equipment and a wastewater treatment plant were built and non-biodegradable trash was hauled to Cayenne.

It however proved to be a constant challenge, despite the measures listed above, to minimize the environmental impacts, control project expenditures when the project viability was unclear while, at the same time advancing the project rapidly. By this time, public awareness on loss of tropical rainforests, loss of biodiversity and sustainable development had risen to high levels. Asarco was very much aware of these issues as well as the often negative perception of the mining industry presented by non-governmental organizations (NGOs) and recognizing that the company lacked expertise to deal with environmental issues unique to rainforest environments decided to seek external advice. This culminated in the formation of an Environmental Advisory Committee (EAC or Committee), consisting of five members to advise Asarco on the environmental and social impacts associated with mineral exploration, mine planning, mining operations, mine closure and post-closure reclamation and monitoring. The members were all biologists with expertise in rainforest environments. The objective was to have an open dialog that while recognizing the potential environmental impacts of an open pit gold mine in the French Guiana rainforest, resulted in improvements in exploration practice and a more environmentally friendly mine, if exploration activities were successful. The Committee members were given full access to all information

Figure 8.9.    Tracked and tired lightweight drill – scout.

being developed by Asarco, but were under no confidentiality restrictions and were not compensated for their advice, other than for travel expenses. The intent was to make the committee part of the planning process, but at the same time have the members feel free to speak about Asarco's work to anyone.

As reported by Graybeal [28], the consultative process with the committee was a success. As a result of advice from the committee:

• Future clearing was accomplished by cutting trees instead of bulldozing; building elevated wooden bridges and reducing the size of drill sites in order to reduce silting in the streams.
• Revegetation of the campsite and re-design of the principal access road was initiated.
• The use of a new rubber-tired, self-propelled combination diamond, reverse circulation and auger drill that could drive on the forest litter was accelerated (Figures 8.9 and 8.10).
• A full-time environmental manager was hired.
• An exploration permit application was withdrawn based the advice against acquisition of exploration permits inside the Biotope zone.

The Committee also took an active role in the design of an environmental baseline study and recommended that two baseline studies be done, one dealing with the natural environment and the other with social environment, including public safety and health. Their advice was instrumental in the final selection of a plant site for a prefeasibility study.

Graybeal also noted that the existence of the committee improved Asarco's access to various NGOs who initially viewed the company and the project with great suspicion.

Asarco since sold the gold deposit to Ariane Gold Corporation, a Canadian company. Ariane has continued and expanded the environmental advisory committee concept as reported by Viens et al. [29]. In 2003, Arianne was merged into Cambior Inc, a mid-size international gold producer that also will continue consultations with the Committee. Asarco has stated that the "consultations with the Committee" approach was one of its most successful environmental initiatives.

Figure 8.10.   Example of drill road using lightweight drill riding on forest litter.

### 8.1.4   INNOVATIVE RECLAMATION PRACTICES IN THE ARID SOUTHWESTERN UNITED STATES (Stuart Bengson)

Asarco has been attempting several innovative techniques for reclamation of copper tailings in Arizona, U.S.A. The use of biosolids (municipal sewage sludge) and cattle in reclamation has been successfully demonstrated in Arizona, and these efforts are briefly described here (Bengson [30,31]).

Much of the early work using biosolids for reclamation involved coal spoils in the Appalachian region. Sopper [32] references 123 sites that are successfully reclaiming mine wastes with biosolids and noted that that many of the concerns over the use of biosolids are unsubstantiated. These include leaching of nitrates and heavy metal contamination from the biosolids. Other references citing biosolids utilization in reclamation include Norland et al. [33] for taconite tailings in Minnesota, Wilson et al. [34] for copper tailings in British Columbia, Canada, and McNearny [35] for copper mine tailings impoundment in Utah, U.S.A.

There are many benefits to the utilization of biosolids for tailings reclamation. Biosolids offer a cost effective source of organics and nutrients necessary for successful reclamation; and the tailings sites offer an economical and environmentally sound solution to the management of biosolids. The primary objectives in using biosolids are:

- meet tailings reclamation objectives
- provide for an environmentally safe method of managing biosolids.

Recently developed rules by Arizona Department of Environmental Quality (ADEQ) allow for the one time use of biosolids for reclamation upto 150 dry tons/acre. The main objective of the use of biosolids for reclamation is to incorporate enough organic matter into the tailings to produce a "growth medium" that can sustain plant growth without the need of "topsoil."

Tests have shown that vegetation can be established and sustained on copper tailings if sufficient organic matter is incorporated into the copper tailings. Biosolids offer a viable and economical source of this organic matter. In addition, the biosolids provide the essential macro- nutrients necessary to establish a sustainable ecosystem. For the municipalities, this use of the biosolids provides

Figure 8.11.   Reclamation of copper tailings with cows (Photo: courtesy Asarco).

a cost effective and environmentally sound disposal alternative. The results of several studies verify that an application rate of from 60 to 100 dry tons per acre have achieved successful reclamation.

The use of cattle/livestock as a tool for reclamation of copper tailings stems from the concept of using livestock as a management tool for enhancing ecosystems. The original idea came from a wildlife biologist working in South Africa. He observed the impacts of large herds of ungulates on the vegetative communities. He noticed that as these large herds migrated across the arid plains they would severely disturb the soil surface, breaking up the soil crust and opening it up for water infiltration and aeration, and trample dead litter & other organics into the soil. Also, the animals would graze old decadent plants, which then opened up the root crown to stimulate vigorous regrowth. His observations evolved into a natural resource management philosophy that became known as "Holistic Resource Management", or "HRM". The concept of using livestock for tailings reclamation is a natural extension of this "Holistic" theme.

Asarco started using "ASARCOws" (also called Four Legged Organic Soil Builders) in 1994 (Figure 8.11). Today more than 600 acres of copper tailings slopes have been stabilized and reclamation begun by using livestock in southern Arizona. The livestock are concentrated on relatively small areas for a very short duration and fed hay. An abundance of organics is incorporated into the tailings by the hoof action of the animals. As the organics build-up in the sterile tailings, a soil-like medium is produced and enhances the reclamation of the tailings site.

The standard methods for reclamation of copper tailings rely almost exclusively on the use of mechanical and other high-tech methods. These include the use of heavy equipment for surface manipulation of slopes and top-soiling; as well as irrigation, fertilizers, and other soil amendments to try and induce plant growth on the tailings reclamation site. A far simpler, and perhaps more ecologically sound, approach is to use livestock. Livestock can trample organics into the tailings to build a soil-like medium that will support plant growth. Livestock also help to stabilize the steep slopes with their hoof action trampling up and down the slopes. As plant communities are developed a self-sustaining ecosystem is established.

In the late 1970's, Asarco had some success revegetating copper tailings by incorporating manure into the top 6-inches of the surface at its Silver Bell mine. Actually, a few years earlier, Ken Ludeke, another researcher, was having success on steep slopes at the Pima mine copper tailings by spraying a slurry of manure and sewage sludge and incorporating the organics by rolling a sheepsfoot roller up and down the slope. Then, in 1989, Noel Gillespie, working with a group of consultants from AZ Ranch Management at the Miami Mine, started to successfully use livestock on copper tailings to stimulate vegetative growth and to stabilize the steep slopes. Since these early beginnings many other mines have started using livestock for copper tailings reclamation. Livestock have been successfully used at Asarco's Mission mine and the Ray Complex Hayden tailings impoundment and at the Pinto Valley Mine, Sierrita Mine, Morenci Mine and Mineral Park Mine in Arizona; as well as several other mines in the western United States.

Electric fences are used to hold 40–70 animals in small padlocks of approximately ¼–½ acre in size for a very short duration. Generally it requires 400–700 animal units of impact per acre. This equates to 100 animals on one acre for 4–7 days. The site is prepared by spreading hay on the surface consisting mostly of sudan grass, Bermuda, oats and barley. The cows are put on the site. Besides a base mulch of hay, the cows are fed alfalfa hay for nutrition and growth, as well as salt and other mineral supplements. The hay and alfalfa is spread over the surface of the tailings from the top of the slope to the bottom. As the hay gets incorporated into the surface more is added. Water troughs are placed at the top of the slope to enhance the movement of animals up and down the slope as well as contouring on the slopes.

The monitoring of the health of the livestock indicates that there are no problems with the health of the animals. Analysis of blood samples collected from livestock before placement on the reclamation site and after removal shows no problems with heavy metals. In fact, the blood levels of copper, molybdenum and zinc show deficiencies of these essential minerals, and the diet of the cattle had to be supplemented with mineral blocks. Many healthy calves have been conceived and born at reclamation sites. Steers that have been taken to sale from the reclamation sites have shown an average weight gain of 0.5 lbs/day.

The future of reclamation using livestock is very promising. Vast improvements in the stability and ecological productivity of reclamation sites impacted by livestock have been noted. The slopes have been stabilized, there was far less wind and water erosion and vegetation started to be established. The Arizona State Mine Inspector honored Asarco's Ray Complex for the innovative use of livestock for copper tailings reclamation.

REFERENCES

1. Tepordei, V.V., *Crushed Stone*. U.S. Geological Survey Minerals Yearbook, 2000, pp. 73.1–73.6.
2. Bolen, W.P.,: *Construction Sand and Gravel*, U.S. Geological Survey Minerals Yearbook, 2000, pp. 66.1–66.4.
3. Regueiro, M., Martins, L., Feraud, J., and Arvidsson, S., Aggregate Extraction in Europe: The Role of the Geological Surveys. In: Geological Survey of North Rhine-Wesphalia (ed.): *Proc. Third European Conference on Mineral Planning*: Raw Materials Planning in Europe – Change of Conditions! New Perspectives? Krefeld, Germany, October 8–10, 2002, pp. 187–198.
4. Langer, W.H., *Natural Aggregates of the Conterminous United States*. U.S. Geological Survey Bulletin 1594, 1988.
5. Langer, W.H., and Knepper, D.H., Jr., Geologic Characterization of Natural Aggregate. In: Bobrowsky, P.T. (ed.): *Aggregate Resources – A Global Perspective*. Balkema A.A., Rotterdam, Netherlands, 1998, pp. 275–293.
6. Bates, R.L., *Geology of the Industrial Rocks and Minerals*: Dover Publications, Inc., New York, 1969.
7. Leighton, M.W, Industrial Minerals Resource Identification and Evaluation. In: Bush, A.L., and Hayes, T.S. (eds.): *Proc. of the Midcontinent Industrial Minerals Workshop*: Industrial Minerals of the Midcontinent – U.S. Geological Survey Bulletin 2111, 1991, pp. 9–20.
8. Ad Hoc Aggregate Committee.: Minnesota's Aggregate Resources – *Road to the 21st* Century. Ad Hoc Aggregate Committee for the Aggregate Resources Task Force, St. Paul, Minn., 1998, accessed on 12 March 2003 at http://www.commissions.leg.state.mn.us/aggregate.resources/.
9. Bauer, A.M., Mineral Resource Management Programs and the Construction Aggregate Industry. *Planning and Zoning News*, (April 1991), Planning and Zoning Center, Inc., Lansing, Mich., 1991 pp. 5–7.

10. Dunn, J.R., Dispersed Benefit Riddle, In: Ault, C.R., and Woodard, G.S. (eds.): *Proc. 18th Forum on Geology of Industrial Minerals*: Indiana Geological Survey Occasional Paper 37, 1983, pp. 1–9.
11. Poulin, R., Pakalnis, R.C., and Sinding, K., Aggregate Resources – Production and Environmental Constraints. *Environmental Geology* 23 (1994), pp. 221–227.
12. Schwochow, S.D., The effects of mineral conservation legislation on Colorado's aggregate industry, In: Schwochow, S.D. (ed.): *Proc. Fifteenth Forum on Geology of Industrial Minerals*: Colorado Geological Survey Resource Series no. 8, 1980, pp. 30–39.
13. Argall, G.O., Jr., *Industrial Minerals of Colorado*. Quarterly of the Colorado School of Mines, v. 44, no. 2, 1949.
14. Wilburn, D.R., and Langer, W.H., *Preliminary Report on Aggregate Use and Permitting Along the Colorado Front Range*. U.S. Geological Survey Open-File Report 00-258, 2000.
15. Sheridan, M.J., *Urbanization and Its Impact on the Mineral Aggregate Industry in the Denver, Colo., Area*. U.S. Bureau of Mines Information Circular 8320, 1967.
16. Cooley, J.B., Our Rapidly Disappearing Sand and Gravel Deposits: *Proc. 74th National Western Mining Conference and Exhibition*. Colorado Mining Association Mining Yearbook, 1971, pp. 17–20.
17. Schwochow, S.D., Shroba, R.R., and Wicklein, P.C., *Sand, Gravel, and Quarry Aggregate Resources – Colorado Front Range Counties*. Colorado Geological Survey Special Publication 5-A, 1974.
18. Schwochow, S.D., Shroba, R.R., and Wicklein, P.C., *Atlas of Sand, Gravel, and Quarry Aggregate Resources – Colorado Front Range Counties*: Colorado Geological Survey Special Publication 5-B, 1974.
19. Stearn, E.W., Master Plans for Minerals Take Effect. *Rock Products*, v. 82, no. 9, 1979, pp. 86–88.
20. U.S. Department of Labor: *Report to the Denver Construction Committee on Sand and Gravel Operations at Chatfield Dam and Recreation Area*. U.S. Department of Labor, Office of Construction Industry Services, 1981.
21. Jefferson County, Colorado: *Mineral Extraction Policy Plan*. Planning Department, Jefferson County, Colorado, 1977.
22. Aggregate Resources Mining Roundtable: *Report of the Aggregate Resources Mining Roundtable*. Planning Department, Jefferson County, Colorado, 1987.
23. Schwochow, S.D., New Quarries for Denver Get "No" Votes. *Rock Products*, v. 84, no. 9, 1981, pp. 38–40, 72, 76.
24. Langer, W.H., Šolar, S.V., Shields, D.J., and Giusti, C., Sustainability Indicators for Aggregates. In: *Proc. Sustainable Development Indicators in the Minerals Industries*: May 21–23, 2003, Milos, Greece., in press.
25. Department of the Environment, Transport, and the Regions.: *Planning for the Supply of Aggregates in England*. Department of the Environment, Transport, and the Regions, Minerals and Waste Planning Division, Draft Consultation Paper, 2000.
26. Langer, W.H., Giusti, C., and Barelli, G., *Sustainable Development of Natural Aggregate with Examples from Modena Province, Italy*. Society for Mining, Metallurgy, and Exploration Preprint 03–45, Littleton, Colorado, 2003.
27. Shields, D.J., Šolar, S.V., and Martin, W.E., The Role of Values and Objectives in Communicating Indicators of Sustainability. *Ecological Indicators*, v. 2 (1–2), 2002, pp. 149–160.
28. Graybeal, F.T., 2001, *Evolution of Environmental Practice During Exploration at the Camp Caiman Gold Project in French Guiana, in Bowles, I.A., and Prickett, G.T., ed., Footprints in the Jungle. Natural Resource Industries, Infrastructure, and Biodiversity Conservation*: New York City, Oxford University Press, pp. 222–232.
29. Viens, F., Gaillou, J., and Jean-Elie, J., 2003, *Exploration and Mining in the rain forest of French Guiana*: Prospectors and Developers Association Annual Convention, Convention Program, pp. 14.
30. Bengson, S.A., 2000, *Reclamation of Copper Tailings in Arizona Utilizing Biosolids*, Presentation at the Mining, Forest and Land Restoration Symposium & Workshop, Golden, CO, July 17–19, 2000.
31. Bengson, S.A., 1999, The *Use of Livestock as a Tool for reclamation in Southern Arizona* in Proceedings of the 16th Annual Meeting of the American Society for Surface Mining & Reclamation, August 13–19, Scottsdale, AZ.
32. Sopper, W.E., 1993, *Municipal Sludge Use in Land Reclamation*, Lewis Publishers.
33. Norland, M.R., Veith, D.L., and Dewar, S.W., 1992, *Vegetation Response to Organic Soil Amendments on Coarse Taconite Tailing* in "Achieving Land Use Potential through Reclamation," Proceedings of the 9th Annual Meeting of the American Society for Surface Mining & Reclamation, June 14–18, Duluth, MN.
34. Wilson, S., Pediie, C., Salahub, D., and Murray, M., 1993, *Environmental Effects of High Rate Biosolids Application for the Reclamation of Copper Tailings*, pp. 11.25–11.36.
35. McNearny, R.L., 1997, Revegetation of a Mine Tailings Impoundment using Municipal Biosolids in a Semi-arid Environment, Utah, U.S.A., v. 6, no. 3, pp. 155–172.

# 8.2   Asia

Ravi Bhargava
*Ecomen Laboratories, Lucknow, India.*

## 8.2.1   INTRODUCTION

There are several countries in Asia with diverse mining practices, varying from the unsophisticated to the most sophisticated state-of-the-art practices. A common thread running through the mines in Asia is their attempt to learn from the technologies developed in Europe and the Americas, and implement sustainable mining practices. A few case histories from India (Bhargava [1]) are presented below.

## 8.2.2   JODA EAST IRON ORE MINE, KEONJHAR, ORISSA, INDIA

The Joda East Iron Ore mine is operated by one of the most sophisticated companies in Asia, the Tata Iron and Steel Company (TISCO). The mine is located in Keonjhar district of Orissa, India. It produces 2 million tons of iron ore per year. Since 1999, the mine has been certified under ISO 14001 for its environmental management system.

The ore is blasted from nine meter high benches, and the ore is loaded with shovels into trucks for transport to the processing plant (see Figures 8.12 and 8.13). Drills are equipped with dust extraction systems and water sprays. Ground vibrations from blasting are monitored to keep the vibrations below the regulatory limits. Haul roads are kept optimally wet to reduce air borne dust. A hearing conservation program has been in place since 1992 and miners are given hearing protection devices.

Training is provided to employees on environmental management and occupational health. With the implementation of the ISO 14001 program, lubricant oil consumption has reduced by

Figure 8.12.   Drilling and mine operation in Konjahar district, Orissa, India (Photo: courtesy TISCO).

32.6% and power consumption has been reduced by 11.7%. Soil erosion through check dams is practiced in all areas of the mine. A tailings dam is used to store the slimes from ore processing, and safety of the dam is monitored on a regular basis.

The mine reclamation program aims at biodiversity and return of the original fauna and the aesthetics. It consists of planting over 400,000 saplings over a 36 hectare (ha) area. A botanical park has been created in a 4-ha area, and a green buffer has been created around the residential areas. Socio-economic development activities are accomplished through the Tata Steel Rural Development Society. The Society helps the villagers in the vicinity of the mining area.

### 8.2.3 VELGUEM/SURLA IRON ORE MINE, GOA, INDIA

The Velguem/Surla iron ore mine, owned by the Salgaocar family, covers an area of 287 ha in North Goa, India. It produces about 3 Mt tons of iron ore annually. Ripping and dozing have replaced conventional drilling and blasting to minimize environmental impacts. Waste rock and overburden is disposed in step dumps, and various check dams and settling ponds are used to minimize erosion.

Continuous monitoring of air and water quality, soil chemistry, and noise levels is conducted to control environmental impacts. The monitoring data is reviewed to make improvements to the mining and processing operations. The tailings dam is designed to recover and recycle 60% of the process water. A tailings reclamation program has recovered about 3.5 Mt of tailings.

A multi-species nursery is maintained to meet the annual plantation program that is a key component of the mine reclamation plan. Mined out areas are backfilled and planted. Distribution of saplings, drinking water, and supporting the local community activities are part of the socio-economic development activities of the company.

### 8.2.4 GEVRA COAL MINE, INDIA

Coal mining in India was started in 1774 in Ranigunj coalfield in the Burdwan district of West Bengal. Till 1971 almost all the coal mining was in the private sector and the only Public Sector companies operating in the country were National Coal Development Corporation (NCDC) and Singareni

Figure 8.13. Truck and shovel mining in Keonjahar, Orissa, India (Photo: courtesy TISCO).

Collieries Company Limited (SCCL). The coking coal mines were nationalized in 1971 and the non-coking coal mines were nationalized in 1973. Presently out of the total coal being produced in the country about 75% is being used to generate thermal power and the demand of coal is expected to become almost double with almost 100% planned increase in the capacity of the thermal power generation in the next 10–12 years.

The nationalization of the coal mines brought the era of opencast mining as a result of which the present trend is that more than 75% of the total coal production is coming from the opencast mines and the production from the underground mines has remained stagnant since more than a decade. The present trend of dominance of opencast mining is likely to continue till such time as the reserves amenable to this method are exhausted. Opencast mines of various capacities ranging from about 0.5 million tonne per year (MTY) to over 20 MTY are found in India. The present trend is to design opencast mines of about 5 to 10 MTY.

In terms of mechanization the country has purely manual to completely mechanized opencast mines. The maximum capacity of the shovels deployed in the opencast mines is around 20 to 25 cubic meters ($m^3$) and that of the trucks is about 180 tons. The coal mines have drag-lines of about 35 $m^3$ bucket size and in the recent years, surface miners have been introduced in some of the mines. Although longwall mining in underground mines has been going on for over 30 years, the system has not gained dominance and still bord and pillar mining is more prominent. The underground mines have various degrees of mechanization.

Gevra coal mine, the largest opencast coal mine in the country, is situated in Korba coalfield in the State of Chattisgarh. The mine was started in 1981, and is mining two seams, Lower Kusmunda (30 to 44 m thick) and Upper Kusmunda (25 to 29 m thick). The leasehold area of the mine is 2,958 ha out of which the active area is about 1,131 ha. The extent of the opencast mine in 2002–2003 is about 5 km along strike and about 2 km in the dip direction.

Overburden consisting mainly of coarse grained sandstones is being removed by drilling and blasting and transported by 85–120 ton trucks. The mine has three overburden benches of 8–15 m thickness and the soils (top soil and subsoil) are not removed separately. The overburden is being stacked outside as well as inside the mine in the mined out area. Coal is being excavated by small capacity loaders of 1.9 $m^3$ capacity after drilling and blasting. It is transported to the feeder breakers by 10 ton tipping trucks. There are more than nine benches in the coal seam. The mining operation is shown in Figure 8.14.

Figure 8.14.   Mining operations at the Gevra coal mine, Korba, India. (Photo: courtesy TISCO).

The run-of-mine coal is fed to feeder breakers of 400 ton/hour capacity and after sizing it is fed to 3 sets of conveyors of 1,400 mm width. The mine has a coal handling plant and an automatic wagon loading system for the dispatch of coal to the thermal power plant. The Project is linked to the 2,130 MW Korba Super Thermal Power Plant of National Thermal Power Corporation.

With the total manpower of about 2,350, the overall output per man-shift of the Project was over 20 tons in 2001–2002. The Project has an Integrated Mine Management System (IMMS) with the sub-systems comprising of Production Planning Control (Truck Dispatch System & Drilling and Blasting System), Equipment Maintenance, Material Management, Personnel Management, Payroll Processing, Survey Data Processing, and Project costing.

Reclamation is by filling the mined out areas and planting trees. Mine closure is getting more attention at present because of several environmental laws passed by the Indian government. Mine planning with reclamation considerations is being emphasized in the Gevra coal mine.

### 8.2.5 WEST BOKARO MINING COMPLEX, INDIA

West Bokaro Mining Complex of Tata Iron & Steel Company (TISCO) in the Hazaribagh district of the State of Jharkhand has three opencast mines and one underground mine. Mining in the complex was started in the underground mine in 1948, which has since been closed. Two of the opencast mines are on the verge of closure and there are plans to open two new opencast mines in near future.

The complex has six workable seams of thickness ranging from 1.3 m to 11.0 m. The seams have medium coking coal, and the overlying strata mainly consist of sandstone and shale. The leasehold area of the complex is about 4,300 ha and has a highly undulating topography with occasional hillocks. The elevations in the area are in the range of 300–385 m above mean sea level. The Bokaro River in the area is fed by numerous tributaries.

Originally the complex was a forest cum agricultural area and more than 65% of the land was used for agriculture. The area was ecologically very rich and was basically a tribal habitat with a few villages. The tribal people were engaged in agriculture and were dependent on agriculture and forest products. Mining has brought about a total change in the area and the tribals in the area have become a small minority. One of the tribes is on the verge of extinction.

The complex has two coal washeries with a total of 3.6 MT Y capacity for preparing coal for steel plant use. The washery rejects are used in two 10 MW fluidized bed thermal power plants. In addition, the complex has a 4 MW thermal power plant to meet the energy needs of the mines and colonies.

The closed underground mine openings are being used for the storage of water for use in the complex. In the underground mine, a limited area has been depillared while the remaining developed pillars are being extracted from the opencast mines. In the opencast mines the overburden is removed by shovel and truck combination after blasting. There is no system for separate handling and management of top soil and sub-soil. Due to shortage of space for dumping overburden outside the opencast mines, a part of the overburden is being dumped inside the mines. Some of the views of the opencast mines are shown in Figures 8.15 and 8.16.

The complex has developed a well planned park on a large overburden dump which is attracting a large number of visitors. The park has a large number of local and exotic plant species. A view of the recreation area at the park is shown on Figure 8.17.

### 8.2.6 JHARIA COALFIELD, INDIA

Coal mining in the coalfield was started in the year 1894. Another unique feature of the coalfield is that it has 40 workable seams, with total proved and indicated reserves of 11,408 and 5,669 million tons, respectively. The coal measure rocks are interbedded sandstone and shale.

Originally the coalfield was predominantly an agricultural area and in 1925, the percentage of land under agriculture was over 65% and the area with village settlement was about 8%. The mining and associated activities have brought about a marked change in the land use of the area with agricultural area reducing to about 39% and the area under settlement increasing to over 33%.

Figure 8.15.   Truck and shovel mining at the West Bokaro Mine (Photo: courtesy TISCO).

Figure 8.16.   View of a deep pit in the West Bokaro Mine (Photo: courtesy TISCO).

The coalfield has mines belonging to Bharat Coking Coal Limited (BCCL), TISCO and Indian Iron & Steel Company Limited (IISCO). There are more than 100 underground and over a dozen opencast mines. Most of the opencast mines are extracting developed coal pillars.

The main method of underground mining has been bord and pillar system, and as a result, large areas are supported by developed pillars. Some of the mines have developed pillars in a number of seams. Longwall stowing and caving methods have also been adopted in a number of mines.

Figure 8.17.   A recreation area at the park, West Bokaro Mine (Photo: courtesy TISCO).

In the opencast mines, various levels of mechanization have been adopted with shovel-truck combination and draglines.

The coalfield has a total area of about 450 sq km, and has witnessed coal mining for about 109 years. The current status of the coalfield is briefly outlined below:

1. The coalfield has more than 70 mine fires spread over an area of about 10 sq km, which are not only polluting the atmosphere but also are not permitting proper mining of coal seams in their vicinity. In the past few decades, some of the fires have been adequately mitigated with encouraging results. According to a rough estimate, the mine fires affect half the number of the collieries in the coalfield and have already consumed about 40 million tons. The fires have isolated about 1,800 Mt tons of coal from possible recovery.
2. The underground mining of coal seams has caused a large number of subsidences and the total area subsided in the coalfield is over 35 sq km. Due to the presence of a large number of settlements on the surface, it is becoming difficult to find areas for extraction with caving. In most of the underground mines, the main source of production is the development of pillars. Some of the subsidence in recent history has not only damaged the surface properties but has also caused social problems. Mine openings underneath some of the settlements may not be stable from a long-term standpoint.
3. There are more than 120 urban and rural settlements in the coalfield belonging to the coal companies and others. The surface construction work in the coalfield area continues even though it is not permitted.
4. The total population of the coalfield area is about 1.1 million and most of the settlements are unplanned. Some of the settlements are quite close to the fire and subsided areas. In some areas, construction is continuing over the subsided areas.
5. The coalfield has a vast network of railway lines and roads, which are blocking a large quantity of coal in the underground mines. It is estimated that about 7,000 Mt tons of coal reserves are blocked by surface development.

Figure 8.18.    Fruit trees growing over an area affected by mine subsidence (Photo: courtesy TISCO).

A study conducted by Norwest Mine Services Limited in 1997 revealed that most of the core and buffer zone of the coalfield has been denuded of forest cover and suffers from excessive soil erosion. The ground water in the coalfield area has been disturbed by the mining activities. Due to the subsidence and other surface activities, the surface contours have changed and in many a situations the rain water finds its way to underground workings.

TISCO has developed a large park over a subsided area. Some views of the park are shown in Figure 8.18.

All those concerned with the coalfield are looking for ways and means to reinvent the coalfield and develop a sustainable mining and reclamation plan. The most prevalent opinion revolves around planning of large scale opencast mining with suitable reclamation measures for the land, and rehabilitation and resettlement measures for the population affected by opencast mining.

REFERENCE

1. Bhargava, R. 2003, Personal communications with Tata Iron & Steel Co., (TISCO) Jamshedpur, India.

# 8.3   Africa

R. Knol[1], Dr. D. Watson-Jones[2] and D. Renner[3]

[1]Former HSE Manager, Geita Gold Mine, Tanzania
[2]Clinical Lecturer, London School of Hygiene & Tropical Medicine, London, England, UK
[3]Managing Director, Iduapriem Gold Mine, Ghana

## 8.3.1   INTRODUCTION

Two case studies, one from Tanzania (East Africa) and another from Ghana (West Africa), are discussed in this section. Two factors affecting sustainability – HIV/AIDS and mine closure – will be the focus of this section.

Inadequate strategic planning by governments and industry for what will follow after mine closures renders economies vulnerable to decline and the communities around former mines vulnerable to dislocation. Similarly, inadequate operational planning and action by governments and industry to address HIV/AIDS in mining communities is likely to result in long term consequences for community health in these regions.

As evidenced in the 2002 World Summit, sustainable development is now seen as a concept that integrates the three pillars – economic, social and environmental. The sustainable development challenge now requires a truly integrated approach, and this is especially true in the mining industry, where the challenges of economically viable extraction, environmental stewardship and social dislocation can no longer be managed in isolation.

Ultimately for closure and HIV/AIDS to be adequately addressed by governments in developing or developed countries, they must become part of government fiscal planning and mining codes, thereby ensuring that these issues are strategically incorporated from the outset for each project. The failure of developed country governments, especially in regard to HIV/AIDS, has no doubt had an impact in developing countries that utilize their expertise and legislation to develop their own frameworks of success.

## 8.3.2   HIV/AIDS

Worldwide, 42 million adults and children are now estimated to be infected with the human immunodeficiency virus (HIV). This infection, transmitted through sexual intercourse, or via infected blood or from mother to child, claimed over 3 million lives in 2002 (UNAIDS/WHO [1]). The impact of the HIV/AIDS epidemic is substantial at both social and economic levels. Left untreated, those infected suffer a gradual deterioration of their immune system, leading to progressive ill-health as a result of opportunistic infections and cancers. Many will die during their most productive years because HIV is often acquired in sexually active, relatively young adults. Deaths of wage-earners or their care givers cause substantial disruption to families because of social and economic losses to the household. The impact of such deaths can be far-reaching since, in resource-poor settings, children may interrupt their schooling to help their relatives, so further threatening their future economic security. Similarly, HIV/AIDS orphans may be left to fend for themselves and, surviving on the streets, will themselves be at increased risk of acquiring the infection.

### 8.3.2.1   Sub-Saharan Africa

The greatest burden of HIV infection is found in sub-Saharan Africa (SSA), where over 29 million people are infected with HIV (Figure 8.19). The overall adult prevalence, the proportion of people living

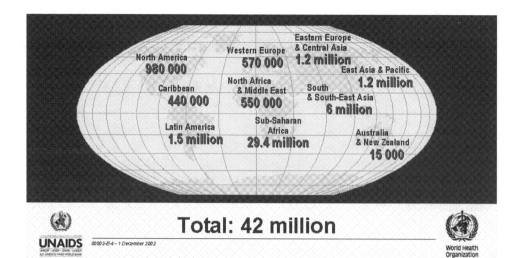

Figure 8.19.   Adults and children estimated to be living with HIV/AIDS as of end 2002. Source: UNAIDS and World Health Organization.

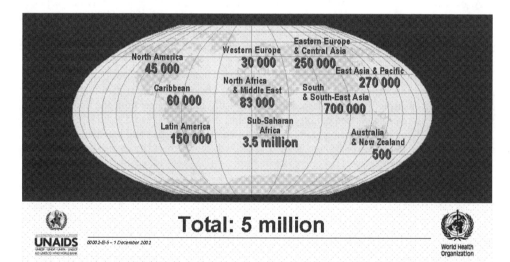

Figure 8.20.   Estimated number of adults and children newly infected with HIV in 2002. Source: UNAIDS and World Health Organization.

with HIV, in SSA is 8.8%. However, this varies widely by country and by various socio-demographic factors. HIV prevalence is generally higher in southern and eastern SSA countries, especially in Botswana, Lesotho, Swaziland, South Africa and Zimbabwe, where up to 30% of people are HIV-infected. Indeed, the impact of the epidemic in southern Africa is so severe that it is changing the population structure of countries such as Botswana (Gottlieb [2]). HIV prevalence is also higher in urban towns and cities compared to many rural areas, and in young women compared to young men of the same age. Although there is evidence of a reduction in the number of new people being infected in several SSA countries, the incidence of new infections is generally high across the region.

Approximately 3.5 million people in SSA are estimated to have been infected with HIV during 2002 (Figure 8.20), primarily through heterosexual transmission. The epidemic has been fueled by poverty, gender inequality, high rates of sexually transmitted infections (STI), stigma, armed conflict and inadequate health infrastructures. In the absence of provision of adequate HIV care, the death rate is predicted to rise further and reach its peak in the next 10 years.

A few countries in SSA are, however, showing signs of an encouraging trend in the number of people living with HIV. In Uganda, between 1995 and 2000/2001, condom use has doubled and the HIV prevalence in pregnant women in Kampala has fallen from 29% in 1992 to 11% in 2000. Similar trends in HIV prevalence have been seen in other sites in the country, along with a significant reduction in high risk sexual behavior (Uganda Ministry of Health, Mbulaiteye et al., Kilian et al. [3–5]). In Senegal, the prevalence of HIV among pregnant women has remained relatively stable at 0.5 to 1.6% over the past decade (Ndoye [6]). There are some signs that the epidemic may be slowing in sub-groups in other countries (Fylkesnes [7]). The HIV prevalence in young South African pregnant women has fallen from 21% in 1998 to 15% in 2001, although HIV prevalence has risen in older age groups, and the overall proportion of those living with HIV has also increased. Reduction in prevalence has been attributed to changes in high risk sexual behavior among other factors. Sustaining HIV prevention messages, and targeting them to all sectors of the population, in order to continue such positive trends, will however pose huge economic and logistic challenges for these resource-constrained countries.

### 8.3.3   HIV AND THE MINING INDUSTRY

The International Labor Organisation estimates that approximately half of those infected with HIV are workers. HIV infection has become a problem for the mining industry, especially in southern Africa where up to 30% of mine workers are HIV-positive. The mining industry employs many migrant laborers (Williams et al. [8]). Migrant labor has been associated with increasing risk of HIV transmission since migrant workers live far from their families (Haour-Knipe et al. [9]) and may then seek other sexual partners. Furthermore, the presence of large numbers of predominantly, male migrant workers attract women who work as sex workers. These will act as a critical mass for the spread of HIV because of their large number of sexual partners and their high prevalence of infection. Most data exploring the risk of HIV in miners come from South Africa. A high percentage (37%) of miners in South Africa believe that they are at risk of acquiring HIV infection (Meekers [10]). In these men, being young, being unmarried and a low educational level are all important factors in the spread of the disease (Ijsselmuiden [11]). Intensive education campaigns established after initial surveys failed to slow the spread of the epidemic in the mining workforces. Studies in South Africa have shown that mine workers have a high threshold of risk and that their working environment, lifestyle, relatively high income and masculine identity makes them especially vulnerable to acquiring HIV(Campbell [12]).

The role of migrant workers in spreading the HIV epidemic goes beyond in-country migration for employment. This can be biologically demonstrated since HIV has a number of circulating genetic subtypes. Subtype B initially predominated in Europe and North America. Increasing evidence has linked returning expatriates, other travelers and immigrants, mainly from South East Asia and sub-Saharan Africa, with the introduction of non-B subtypes to Europe and North America (Thompson et al. [13]). This clearly documents the spread of HIV across continents by people who travel. The implications of genetic mixing of subtypes for HIV therapy and prevention are unclear.

Expatriate migrant workers comprise a potentially high risk population since they report a significant rate of sexual encounters with nationals while overseas and they may underestimate or not comprehend their risk of acquiring HIV or STI in high prevalence countries. One study documented that 41% and 31% of men and women, respectively, had sex with a local partner during an average stay of 26 months, and 23% had unprotected sex (de Graff et al. [14]). Expatriates may sometimes even have a higher prevalence of HIV than their local neighbors. In Senegal, for example, one study found that 27% of expatriate men were infected with HIV compared with less than 1% of local residents from the same rural communities (Kane et al. [15]). Documented risk factors for HIV in expatriates include sex with local women, sex with prostitutes, especially in Africa, having an STI and receiving injections by unqualified staff (Bonneux et al. [16]). Ideally, therefore, companies employing expatriates should provide detailed health information prior to departure and during their term of employment, including reinforcement of HIV prevention messages.

The impact of HIV in the mining industry has fueled a parallel epidemic of tuberculosis (TB) in southern Africa. Goldmine workers have increased risks of acquiring HIV infection and silicosis. Both these factors are associated with an increased risk of infection with TB and nontuberculous mycobacteria (Corbett et al. 2000, Corbett et al. 1999, Sonnenberg et al. [17–19]), and have contributed to the two-fold increase in TB incidence in South African miners from 1990 to 1996 (from 1,174 to 2,476 cases per 100,000 per year) (Churchyard et al. [20]).

Multinational companies have begun using their resources and capacity to implement programs for HIV prevention and care for their workforces. These have varied widely in scale and focus. In the gold mining industry, prevention programs have been implemented by a number of companies (Clift et al., Talgaard et al. [21,22]). However, a study of the gold mining industry in the countries of the Southern African Development Community (SADC) suggest that many programs implemented by mines in the region initially had little impact on the epidemic. It is recommended that an integrated and sequential approach that addresses cultural and community factors that can facilitate HIV transmission, as well as biomedical interventions to reduce STI, be implemented (Campbell, 1999 [23]).

A number of studies, again mainly in South Africa, have identified social factors that might aid planners of sexual health interventions such as peer-education and condom promotion (Campbell, 2000 [24]). Recently, there is evidence that these initiatives may have had some impact on high risk behavior and STI in South Africa. There has been a reduction in the proportion of South African miners with four or more sexual partners in the past year, 25% in 1995 compared with 13% in 1997, and a reduction in the prevalence of STI (Steen et al. [25]). It is too early for the new large-scale mines in Tanzania to show any impact of their HIV prevention programs. However, repeated HIV/STI prevalence and risk behavior surveys will give some indication of how well control efforts are being performed and such surveillance is recommended where data are scarce. Unfortunately, this is not done systematically in many mining areas. It is worth noting that this type of survey is feasible, uptake is usually high and the cost is not prohibitive. For example, integrated STI prevalence and behavior surveys in Mali have been estimated to cost approximately $30 per participant for biological testing of four STI including HIV, with participation rates of 84–100% (Maclachlan et al. [26]).

### 8.3.3.1   *HIV care*

To try and combat the spread of the infection, many countries, including 40 in sub-Saharan Africa, now have their own national HIV control strategies. The main facets of these programs are to prevent sexual transmission of HIV, to reduce mother-to-child transmission and to prevent blood-borne transmission through safe blood services. Global initiatives have also begun, such as the Global Fund to fight AIDS, Tuberculosis and Malaria, with a budget of more than $2 billion (Brugha [27]).

Despite extensive efforts, the HIV epidemic has unfortunately continued to spread in many parts of the world. For those individuals who are already infected and living with HIV/AIDS, few will receive adequate care and treatment. Effective prophylactic treatments have been used for many years in high-income countries to protect those who have HIV from developing other opportunistic infections. The picture of HIV care is very different in many parts of the developing world. In resource-poor settings, only 10% of HIV-infected patients will ever receive these prophylactic treatments. More recently, antiretroviral drugs to control the immune damage caused by HIV infection have become available. However, less than 4% of those living outside the developed world who require antiretroviral therapy (ART) will receive it. Until recently antiretroviral drugs were costly, but recent campaigns have led to a fall in their price and guidelines have been developed for sourcing and scaling-up the provision of ART in resource-constrained settings [28,29]. Nevertheless, at the time of writing, access to ART is limited in SSA since few resource-poor countries are able to afford and distribute these drugs at the scale to which they would be required in this region. In addition, few health services currently have the skills and resources to adequately assess, manage and monitor HIV patients who may require treatment. Monitoring of treatment effectiveness in the developed country setting requires expensive assays and machines to routinely measure immune markers and HIV viral load. Efforts are now being made to develop affordable laboratory monitoring of HIV patients for use in resource-poor countries but until such technology becomes available, countries in SSA implementing ART often rely on sending samples to a few well-equipped laboratories, that are usually only located in large cities [30]. Furthermore, the cost of these assays is prohibitive for many patients.

The economic loss that HIV infection results in for workforces is immense. The reasons for this are highlighted in a study of South African HIV-infected miners who were followed over 12 months. The HIV prevalence in this workforce was 24% at the time of the study. The rate of those leaving employment was 18.4 per 100 person years of follow-up and there was a nearly three-fold higher rate of hospital admissions in HIV-positive men compared to infected men [31]. The reasons for admission differed by HIV status; 70% of HIV-positive men were admitted for treatment of infectious diseases, mainly tuberculosis, compared to 30% of HIV-negative men who were predominantly admitted for trauma. Men with HIV infections also had a much higher rate of re-admission to hospital. In terms of mortality of the workforce, HIV-positive men had nine-fold greater risk of death over the follow-up period compared to HIV-negative men (5.5 and 0.6 deaths per 100 person years, respectively).

It is easy to see with this level of morbidity and mortality that HIV infection results in significant impacts on productivity, with additional costs for treatment, worker absence, and costs and time to recruit and train replacement workers. As a consequence, several companies in Africa have started schemes to make ART available to workers and their dependents. These include De Beers Consolidated Mines, Debswana, [32] Heineken and AngloGold in South Africa. Businesses are often able to provide high standards of care for their employees. With a planned HIV care program, they can send medical personnel for training in HIV management, ensure regular and efficient supply of antiretroviral drugs and can purchase the equipment and assays currently in use to monitor and guide the provision of ART. However, little is available to help those who cannot afford private health care or who are not employed by multinational companies. In 2002 Botswana, a wealthy country by SSA standards, took the first step in SSA to introduce free HIV care and provision of anti-retroviral drugs to individuals who had reached the stage of infection which required such treatment. Scaling up this initiative has proved a major challenge, even in a country of only 1.5 million, and currently only a small number of people are being treated. Clearly, introducing similar interventions in other countries that have much larger populations than Botswana will prove logistically difficult, and may be economically impossible, given the low per capita expenditure on health in SSA.

Apart from the examples above, provision of ART to mine workers and dependents has been limited and has been contentious, with different companies initiating different treatment strategies. Some companies limit ART to their employees only and do not extend provision of treatment to dependents. Others require workers to contribute to the cost of treatment. Data on follow-up and compliance with therapy are scarce to date. However, workplace programs in the mining setting can take significant steps to improve HIV care in the workforce. Implementation of workplace HIV care programs by a company should follow clear clinical practice guidelines and should include strategies for auditing and evaluating the quality of the services. These interventions may need to be adapted for specific settings. They can be implemented alongside prevention programs, in a step-wise process, initially with HIV information, health and dietary advice, TB screening, STI treatment, provision of prophylactic treatment to prevent opportunistic infections, and provision of ART to pregnant women and ultimately to the HIV-infected individuals in the workforce.

### 8.3.4   MINE CLOSURE

Globally there is now a significant body of information on mine closure planning processes. Increasingly, in many parts of the world, regulators are insisting on, and industry is adopting, mine closure strategies. However, there is still comparatively little information on government agendas for "industry" closures, few mining companies are considering closure planning at the pre-feasibility stage of projects and few mining companies are preparing meaningful closure plans for external discussion.

There is limited published information on the cost of closure. However, examples include the Summitville mine in Colorado where cleanup costs post closure are likely to reach $225 million and the Yerrington copper mine in Nevada to be about $200 million. The US based Mineral Policy Center, a non-governmental organization, suggests that it will cost $50–60 billion to clean up abandoned mine sites in the US alone (MMSD, 2002). Unpublished research suggests that a closure not involving significant long term issues, such as acid drainage, is likely to cost $50 million. This is a significant expenditure at the end of an operation's life when reserves are typically diminishing, and one that company's and regulators need to be well prepared for, to ensure that the closure does not become a community and environmental liability.

### 8.3.5   TANZANIA CASE STUDY

Tanzania is part of the East Africa Community whose other members are Uganda and Kenya. The member countries cooperate in areas of trade, immigration, transport and telecommunications, among others. These three countries border the shores of Lake Victoria, the world's second largest freshwater lake, and share that largely untapped non-mineral resource.

Tanzania is rich in mineral resources (Figure 8.21) and has supported an important gold and diamond industry since the 1930s. Other minerals of economic significance include gemstones, phosphates, salt, coal, kaolin and tin. Tanganyika, formerly German East Africa from 1883 to 1918 and a British Colony from 1918 to 1961 achieved independence in 1961. Tanganyika and Zanzibar united to become the United Republic of Tanzania in 1964.

In the 1970s the unification strategies of the Tanzanian government led to the formation of a State Mining Corporation, commonly known as STAMICO. This company operated many successful ventures such as Buck Reef and exploration at Bulyanhulu, now the site of the Barrick Gold Corporation owned by Kahama Mining Corporation Limited (KMCL). By the mid 1980s, declining production resulting from poor management and a low gold price was making most STAMICO-owned operations non-viable. It was at this time that privatization of the mining industry in Tanzania occurred. Privatization caused the demise of STAMICO, since there were no longer operational profits to support exploration ventures. The only mines to survive this period were Williamson Diamonds, which was 49% privately owned by the De Beers, and Kiwira Coal Mines, which continued to receive strong support from China.

Large and small-scale mining in Tanzania, before the recent modern mining boom, has left a legacy of contamination. Many of the orebodies contained sulfides and the acid drainage that followed the extraction of the ore has impacted water quality in many areas. In addition, tailings from processing operations were not contained and have, over successive wet seasons, migrated into waterways, taking with them heavy metals and acidic contamination. Infrastructure from these former large-scale mining operations is another legacy of the earlier mining era. While much equipment has been salvaged or used for other purposes, open adits and mine workings remain a safety risk in many places.

Tanzania also has a large artisinal mining population, currently estimated to consist of half a million miners. Artisinal mining is small-scale mining and in most situations continues illegally. Mercury is used as the primary means of extraction and the waste from these small-scale processes are poorly managed. To a large extent, artisinal mining has centered on previously mined orebodies or has opportunistically exploited exploration sampling.

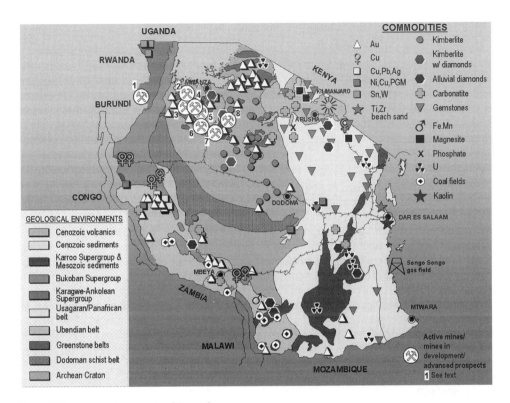

Figure 8.21. Mineral prospects of Tanzania.

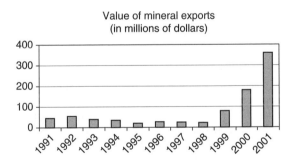

Figure 8.22. Value of mineral exports – Tanzania. (Source: World Bank.)

#### 8.3.5.1 *Value of mineral exports*

In 1994 the government sought assistance from the World Bank for its privatization initiatives. The World Bank, through the Mining Sector Development Technical Assistance Project (MSDTA), provided a strategy for private sector development-aimed at encouraging foreign investment by international mining houses-and regulatory reform through the revision of the Mining Act. This legal, regulatory and fiscal reform was the first phase of a program to address issues of foreign debt and dependency on foreign nation aid. Subsequent phases of World Bank assistance have focused on institutional strengthening, investment promotion and environmental management.

The success of this foreign investment strategy is apparent in Figure 8.22.

Williamson Diamond Mine commenced production in 1940 and is the only large-scale mine to have continued operating through independence and into the present day. Williamson is currently jointly owned by De Beers (75%) and the Tanzanian Government (25%). The mine received a capital injection in 1994 of $23 million and is managed by De Beers.

### 8.3.5.2    *Mine closure in the region*

Since 1994, three major mines have opened:

- Geita Gold Mine (managed by Geita Gold Mining Limited, a 50/50 joint venture (JV) between Ashanti Goldfields of Ghana & AngloGold of South Africa).
- Bulyanhulu Gold Project (managed by Kahama Mining Corporation Limited (KMCL), a wholly owned indirect subsidiary of Barrick Gold Corporation).
- Golden Pride Project (an unincorporated 50/50 JV between Resolute Limited (operator) and Samax (Mabangu) Mining Limited).
- North Mara Gold Mine (managed by North Mara Gold Mining Limited, a wholly owned subsidiary of Afrika Mashariki Gold Limited of Australia) North Mara.

Other deposits have been identified but have not yet been developed into operating mines. These include:

- Kabanga Nickel Project (Barrick Resources) – 19.7 million tons (currently in feasibility stage).
- AngloGold/Pangea/Madaba – 2 million ounces of gold.
- Randgold/Pangea – 1.67 million ounces of gold.
- Anglo/El-Hillal – 2.25 million ounces of gold.

In addition, Meremeta Limited (a 50/50 JV between the Government of the United Republic of Tanzania and a South African Company) is a gold beneficiation/collection project that manages small to medium scale mining operations and small scale mining assistance programs.

The Tanzanian Mining Act (1998) and Regulations (1999) are comprehensive regulations compiled on behalf of the Tanzanian government by World Bank-sponsored consultants. The Act and Regulations are based on policies, strategies and legislation from around the world, customized to suit Tanzanian conditions. Of particular relevance is the inclusion of mine closure requirements in Parts IV and V of the Act and Regulations, which make specific provisions for reclamation.

The Act and Regulations include provision for closure planning. Unfortunately most government officers lack the experience and expertise to enforce these regulations. While the government has invested time and money to train its personnel, its capacity to enforce closure planning and other environmental regulations in an immature industry depends to a large degree on the good corporate citizenship of mining operators. The government remains exposed to the risks of avoidance by unscrupulous operators and inadequate supervision and enforcement by its own inexperienced regulators.

This problem becomes acute in closure planning. Although the projected life of the mine for each of the operations is comparatively short, the regulatory authorities have as yet made no moves to address closure planning issues. The Act and Regulations have made bold expectations, such as those presented in Table 8.1, that do not take into account the fact that local infrastructure, such as drinking water provided now to townships by the mines, will, in many cases, disappear when mines close. The regulations, drafted with an emphasis on site restoration, allow mining operations to limit their long-term liabilities to local communities by strict compliance with the law (Table 8.1).

The Regulations make provision for the posting of bonds (Table 8.2), but to date none of the four major companies operating in Tanzania have been requested to post a bond. This further demonstrates the absence of regulatory enforcement.

Table 8.1.   Extracts from Mining (Environmental Management and Protection) Regulations, 1999.

Part IV – Reclamation requirements

| Reference | Provisions |
|---|---|
| Regulation 25 | Prior to abandonment and unless the Chief Inspector has made a ruling with respect to National heritage consideration:<br>(a)  All machinery, equipment and building superstructures shall be removed; |
| Regulation 29 | Prior to mine closure:<br>(d)  All roads shall be reclaimed in accordance with land use objectives unless permanent access is required to be maintained |

Table 8.2.   Extracts from Mining (Environmental Management and Protection) Regulations, 1999.

Part V – Rehabilitation bond

| Regulation 31 (Part 1) | The Minister may require the holder of a Special Mining Licence to provide for posting of a rehabilitation bond, which may be in various forms |
|---|---|
| Regulation 31 (Part 2) | The bond and financial guarantee will form a separate agreement between the Government and the licensee |

Table 8.3.   Projected closure dates for major Tanzanian mines.

| Mine | Ownership | Reserve (MT) | Expected LOM |
|---|---|---|---|
| Williamson Diamond Mine | De Beers/Government of Tanzania (75/25) | 12.5 | 2006 |
| Geita Gold Mine | Ashanti/AngloGold (50/50) | 50 | 2014 |
| Bulyanhulu Gold Project | Barrick Gold Corporation (100) | 30.8 | 2021 |
| Golden Pride Project | Resolute/Samax (50/50) | NA | 2009* |
| North Mara Gold Mine | Afrika Mashariki Gold Limited of Australia (100) | 30 | 2012 |

* Inclusive of a possible 2 year extension.
Sources: Various literature cited under references.

While the development of four major modern mining operations, as well as the Williamson Diamond Mine, is bringing immediate growth to Tanzania in terms of employment, revenue, exports and infrastructure, the sustainability of this growth may be difficult. With the exception of the Kabanga Nickel Project there are no other mines scheduled to open in Tanzania in the next five years. Further, as detailed above, only Kabanga Nickel promises a longer term resource.

Out of the existing operating mines in Tanzania, the projected life of mine is summarized in Table 8.3.

While exploration is continuing at each of these locations and there is hope that additional reserves will be found, without further major discoveries, Tanzania may face shrinkage or even closure of the modern mining industry in little over ten years. With the exception of the Bulyanhulu Gold Project, delineated resources of sufficient size to sustain the industry beyond 20 years have not been identified. It is likely that Bulyanhulu Gold Project will replace Williamson Diamonds as the mainstay of the industry. If the gold industry contracts in the way it seems destined to, the country's fiscal position will come under strain; and if it does not attend to closure planning that perpetuates community infrastructure in areas where mines have operated, it will have to face up to social dislocation and other related issues that are sure to follow.

### 8.3.5.3    *HIV/AIDS in the region*

Recent estimates suggest Tanzania has an urban adult HIV prevalence of 16–22%, with a lower rural prevalence of 3–4% [33]. Surveillance data estimates that 16–17% of pregnant women were HIV-infected in 2000. As in other countries in SSA, the HIV prevalence is higher in specific core groups such as female sex workers. In the early 1990's, 50% of sex workers in Dar es Salaam and 40% in Mwanza town were HIV-positive. The country currently has an annual life expectancy at birth of only 51 years.

Tanzania currently faces an unusual situation where new gold mining communities have recently been established in the Lake Victoria Goldfields, an area endemic for HIV infection. In Mwanza region, where the Geita Gold Mine is located, there has only been intermittent HIV sentinel surveillance. Data from 1995 documented a 9% seroprevalence in pregnant women in urban Mwanza and 5% in rural areas [33]. Several research studies have more recent HIV prevalence estimates and these suggest that the epidemic is still on the increase in parts of the region. In Kisesa ward, HIV seroprevalence has increased in adults from 6% in 1994–5 to 8% in 2000 [34]. Prevalence figures appear stable for some other rural areas in Mwanza region, at approximately 4% in 1992 and 2001 [35].

Mining communities have been associated with explosive epidemics of HIV/STI in southern Africa [36]. Prompted by the concern that these mines may increase HIV transmission in the region, a private-public partnership project has been initiated to develop and implement HIV and STI interventions in communities neighboring goldmines and in the mines themselves. The partnership (known as the AMREF Mine Health Project) is a collaboration of the African Medical and Research Foundation (AMREF), a non-governmental organization, the regional and district health authorities, the London School of Hygiene and Tropical Medicine (LSHTM) and the National Institute for Medical Research Tanzania with financial support and input from the mining and related companies. Baseline surveys, conducted as part of this project, around the Geita Gold Mine found that 39% of female bar, guest-house and food-hall workers, 19% of men and 16% of other women from Geita town were HIV-infected compared to 4% of miners [21]. With in-migration and expansion of these mining communities, the HIV incidence in high risk women, such as bar workers, in these new sites is likely to be high. Interventions against STIs for HIV prevention are most effective at this growth stage of an HIV epidemic [37] and the project combines targeted interventions for sex workers and mine staff with STI treatment services, condom promotion and peer-education programs.

### 8.3.6    GHANA CASE STUDY

Ghana, like Tanzania, provides a useful example of imminent industry closures. Ghana has also profited from the removal of barriers to foreign investment in the mining industry, with the opening of resource deposits to modern mining techniques in the late 1980's. Bogosu and Konongo were the first developments opened on old sites at this time.

In 1983 the Ghanaian Government launched an Economic Recovery Program (ERP) in an effort to reverse the steep deterioration in the national economy that had occurred since 1960. The mining sector was one of the areas that received the greatest attention under the program. Foreign exchange earnings from mineral production increased from $108 million in 1985 to $710 million in 1999, accounting for about 40% of all foreign exchange revenue in the country (Barning, 2001). The value of gold exports in Ghana is presented in Figure 8.22.

Ghana has now reached the stage of a mature industry (in economic terms) with an initial leveling off and now a decline in export earnings. This decline is a result of depleting reserves and much reduced development activity coupled with, as would be expected, some imminent mine closures. Unfortunately, during the mining boom there was little emphasis on strategic industry diversification, resulting in declining exports and a lack of industry sustainability. The impact of

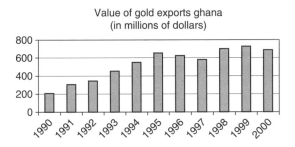

Figure 8.23.   Value of gold exports – Ghana. (Source: Minerals Commission, Ghana.)

Table 8.4.   Projected closure dates for major Ghanaian gold mines.

| Mine | Ownership | Reserve (MT) | Expected LOM |
|---|---|---|---|
| Abosso (Damang) | Goldfields SA (90%), Government of Ghana (GoG) 10% | 280,000 | 2005 |
| Obuasi | AGC/AGC (100%) | 580,000 | 2025 |
| Bibiani | AGC/AGC (100%) | 250,000 | 2004* |
| Bogosu | Biliton Bogosu Gold/Golden Star (90%), GoG (10%) | 100,000 | 2006 |
| Iduapriem | AGC (80%), IFC (20%) | Approx. 220,000 | 2012 |
| Obotan | Resolute Amansie Resources Ltd/ Resolute (90%), GoG (10%) | 120,000 | Closed 2002 |
| Tarkwa | Goldfields Guernsey Ltd (71.1%), Repadre Capital Corp. (18.9%), GoG (10%) | Approx. 500,000 from 2001 to 2005, thereafter 219,000 until end of mine life | 2015 |
| Wassa | Golden Star Resources (90%) GoG (10%) | Start year 2004 Approx. 100,000 | 2009 |
| Yamfo-Sefwi | Newmont Mining Ltd (90%), GoG (10%) | Start year 2005, 148,000 oz in start year, 340,000 oz for rest of mine life | 2014 |

\* Possible extension underground.
Source: (Renner, 2001).

mine closures is starting to be felt and is likely to have far-reaching economic ramifications if no new developments are brought on stream.

Figure 8.23 reflects the successful results of the economic recovery program in the mid 1980's as well as the challenge ahead in sustaining such growth in a maturing market.

The projected mine closure situation for Ghana is presented in Table 8.4. It demonstrates that Ghana is also facing substantial closure of the modern mining industry inside ten years. While Newmont's Yamfo-Sefwi Mine is scheduled for production in 2005, the mine has a life of 8–10 years, leaving Obuasi, like the Bulyanhulu Gold Project in Tanzania, as a lone survivor in the modern mining field.

A few new potential deposits have been identified recently that are mostly situated in forest reserve areas such as Ashantis Kubi, Redback's and Chirano and Newmont's Yamfo and Akim properties. The government of Ghana is still contemplating legislation that will enable these projects to move beyond the feasibility stage. If these projects commence however, it is not expected that they will exceed a maximum 10 year lifespan.

The World Bank and International Monetary Foundation (IMF) today classify Ghana as being a Heavily Indebted Poor Country (HIPC). This is largely a result of the absence of industry diversification and the downturn in the modern mining industry.

### 8.3.6.1   *HIV/AIDS in the region*

Detailed data on HIV trends in Ghana are limited although some information is available through HIV sentinel surveillance in pregnant women conducted at selected sites in the country since 1990. There has been a slow rise in reported HIV prevalence in antenatal attendees from 0.7–1.0% in 1991/92 to a relatively low overall adult HIV prevalence of 3.0% in 2001 [38]. Approximately 3.8% of pregnant women in urban areas and 1.0–7.8% in rural areas are HIV-positive and an estimated 360,000 adults and children are currently living with HIV/AIDS [38]. Data on sex workers are also scarce but suggest that there has been a more rapid rise in incidence in this group, from 2% in 1986 to 40% in 1991. Currently, up to 76% and 82% of sex workers in Accra and Kumasi respectively are HIV-positive. Ghana currently has a life expectancy of 56 years.

Although West Africa previously had a lower population prevalence of HIV/AIDS compared with many other regions of SSA, many countries in West Africa are now showing signs of an impending generalized HIV epidemic. The HIV prevalence has risen to over 5% in a number of countries in this region that previously had a similar HIV statistics to Ghana, including populous Nigeria. The infection has spread particularly quickly in some countries. Nearly 10% of people in Cameroon, for example, are now estimated to be infected and the proportion of pregnant women aged 20–24 years with HIV infection has doubled over a two-year period. This rapid spread should serve as a warning to other countries in the region, including Ghana, that they should not be complacent about HIV prevention, especially since there is already a high core group HIV prevalence in sex workers and condom use is low. In 1998, only 30% of men and 14% of women reported using condoms with the last reported sex with a non-regular partner [38]. There are little data on knowledge of HIV prevention messages or myths about HIV/AIDS. No HIV seroprevalence surveys have been conducted in mining communities in Ghana.

### 8.3.7   CONCLUSION

### 8.3.7.1   *HIV/AIDS*

While the draw of migrating to a mine to seek work may decrease on news of its imminent closure, this may lead to early resignation of staff and migration of local community members, including high risk groups such as sex workers, to other areas. However, many people in the local population are unlikely to move purely because a mine has closed. Some individuals will have settled in the area, investing in houses, land and business. They may be unwilling or economically unable, to travel to another region. High HIV transmission rates, if present during the mine life, are therefore likely to continue. This was seen in the Lake Zone of Tanzania in areas that had previously had artisanal mining activities. The HIV and STI prevalence was higher in these communities than neighboring rural areas, [21,35] presumably because high risk groups were still residing in these settlements despite the decline in artisanal mining activity and the migration of these small scale miners to another site. It is therefore important that HIV prevention efforts are sustained beyond the lifetime of the mine. To date, there have been no examples of program sustainability in the face of large-scale mine closures in HIV-endemic regions in SSA. The potential impact of mine closure on HIV rates and challenges facing sustainability of programs can only be surmised.

Closure of long-established mines will not affect the impact that expatriates and national migrant workers returning to their homes may have already had on the HIV epidemic prior to the mine closing. However, further transmission may be reduced by providing reinforcement of HIV prevention messages before employment ceases. The impact of migration on HIV transmission will be minimized to its greatest extent by providing permanent housing for mine workers for their families in local communities during the mine's working life.

## 8.3.7.2   *Mine closure and prevention programs*

Mine HIV/AIDS/STI prevention programs are likely to comprise strategies to reduce sexual transmission of HIV in the mine workforce and possibly also the surrounding communities. Sustainability of an HIV prevention program that has been solely funded by a mining company will be impossible once a mine closes unless early plans have been made for suitable donor agency funding to take over this program and/or integration of the program into the government health sector. However, it is worth noting that national services may be reluctant to take on such programs unless they have been involved in the design and implementation of these programs from the start.

HIV prevention activities are likely to be heavily dependent on funding from mining companies, especially in the early phases of the prevention program. Services may also depend on the company for equipment, drugs and consumables e.g. for implementation of HIV voluntary counseling and testing centers or services for STI treatment. Expectations in the community will be high once such programs are underway. The danger is that normal routine requisitioning of supplies may have been neglected when health care providers realized that they could obtain sufficient quantities of drugs and consumables quickly and more efficiently through the company. When the mine plans to close, it may be difficult for these personnel to access routine government supplies from the district to continue the service. The community may then be left in a position worse than before the mine-supported programs were initiated. Potentially this could lead to an escalation in HIV transmission at the end of the mine life.

It is essential, therefore, that capacity-building of staff and plans for integration and alternative sources of the program into governmental or other existing services, begin at the start or during the early phases of the project. In Tanzania, the HIV prevention project for Geita Gold Mine, funded by the mine and a number of supporting companies, is part of the district management plan. As part of this project, the mine has funded an HIV Information and Voluntary Counseling and Testing (VCT) centre that also functions as a reproductive health service. It will be important for this service, currently the only one of its kind in the district currently, to be sustained when the mine does eventually close.

## 8.3.7.3   *Mine closure and HIV care and treatment programs*

Mining companies in SSA have already started, or may be planning to implement, HIV care and treatment programs for their workforce and dependents. If steps have nor been taken to ensure continuity of care at the onset of the program, a major impact of any mine closure will therefore be the collapse of these interventions. This is particularly crucial for the provision of ART. Current regimens are designed to be given daily for life once ART has been initiated in an HIV-infected patient. HIV is able to rapidly develop resistance to antiretroviral drugs if treatment is not taken correctly. Evidence suggests that antiretroviral drug resistance is already emerging in SSA, despite limited access to this treatment [39].

Ideally, ARV regimens used in a workplace treatment program should be selected following national guidelines for the provision of ART, if these are available. This will help to ensure some continuity of drugs when the mine supply has ceased, assuming these drugs are available through a national drug distribution program. Where guidelines are not available, company guidelines should use standardized and simple regimens as recommended by the World Health Organization [29]. Although access to free ART is widening in sub-Saharan Africa, access to these services in many regions is likely to be limited in the first few years owing to capacity of health services in the region. Clearly, long-term adherence to therapy will be poor if mine staff and dependents are used to receiving free treatment and then, at the end of the mine-life, have no provision to access ART. The drugs are relatively expensive and patients may also have to find resources to pay for medical checks. In this situation, it is conceivable that patients will buy only as many drugs as they can afford and may take only one or two drugs out of the regimen or may take their treatment intermittently. Both these scenarios may potentially lead to emergence of antiretroviral drug-resistant strains that could then spread into the local population. This would undermine the impact on HIV care of any national or other local programs beginning provision of ART. This is particularly important since neighboring communities may potentially contain the highest proportion of infected individuals requiring ART in the country since HIV prevalence is likely to be higher around long-established mines.

Arguments concerning problems with long-term compliance should not delay the introduction of HIV care in these high risk populations. Preserving the health of the workforce is essential for both ethical and economic reasons. Mines that have started HIV care programs, including ART, for HIV-positive workers, should put in place policies that allow such workers to continue receiving treatment, at least for a period, after the mine has closed, for example by providing health insurance or healthcare compensation to HIV-infected workers.

### 8.3.7.4   *Mine closure*

While the need to attend to site-specific closure planning remains a high priority for regulators, community and industry, a more urgent priority is emerging for governments to evaluate the impact of industry closures on the fiscal health of their countries.

The World Bank has contributed significantly to the fiscal initiatives and strategic directions of developing countries. Tanzania and Ghana provide useful demonstrations of the closure predicament in Africa. This is not the first time these countries have faced industry closure and yet the preparation or lack thereof, is the same. There is a complacency that appears to result from the relative prosperity of the mining industry, and Africa needs to ensure that its future is not further undermined by this complacency.

It is extremely difficult for governments of developing countries to successfully manage the opening, operating and closure of mining operations in such a short period of time. This is evidenced by the current situation in Ghana and is very likely to be seen again in Tanzania. It is important to note however, that despite the longevity of the mining industry in developed countries such as North America and Australia, these countries are still struggling with the concept and implementation of mine closure. For countries that have provided so much expertise and training in emerging markets, their own failure in effective closure planning seems to have been transferred to developing regions such as Africa.

### 8.3.7.5   *Mine closure preparation*

If developing countries are to be prepared for industry closure, there are a number of specific issues that need to be addressed:

- improved mine closure awareness of regulatory agencies
- improved long term community infrastructure planning
- improved long term industry diversification strategies
- improved self management (without assistance from the World Bank or other financial institutions).

1. Today regulatory agencies are struggling to keep abreast of operational regulations and have given little attention to mine closure. Closure issues should be a significant topic at a ministerial level to ensure the country is well prepared for imminent closures.
2. There is an absence of long-term community infrastructure planning. Many examples worldwide demonstrate the need for sound long-term community infrastructure planning if the transition from operation to closure is to be successful. Regions that have focused on infrastructure have demonstrated how the area can then attract other industries when mining is gone. The town of Port Hardy, British Columbia (formerly BHP Island Copper's operation) is an excellent example of where sound cooperative planning and appropriate allocation of mining revenue funds have provided a future for the town's inhabitants.
3. Developing countries need to put in place strategies for industry diversification. While the mining industry can provide a considerable injection of capital into a country, the Government should look to use this capital to develop new industries that are sustainable in

the longer term – for example, in Tanzania the development of fisheries, hatcheries and processing factories for the fish export market to take advantage of the Lake Victoria resource. While the World Bank is taking steps to encourage diversification, there appears to be a reluctance to use the "mining pilot" (term used by the World Bank to describe the process of reinvigorating an economy through the use of modern mining development) as a model for developing other industries.

4. There is a need for African countries to develop strategies to manage without the intervention of organizations such as the World Bank. Strategic mine closure planning is a means of taking responsibility for its own future, but in the absence of strategic long-term planning there is the very real threat of financial collapse. Such a collapse occurred in the 1960s when there were a number of mine closures in Tanzania. Tanzania has won a second chance with the renaissance of the mining industry, but it is not clear that the country is any better prepared for closure than it was in the sixties.

On the one hand, developing countries in Africa are particularly vulnerable to financial collapse as a result of the mining sector, and on the other, their exposure is limited, because mining activity remains comparatively small. For example, Tanzania has just four mines on which it hopes to build a future. It will be hard to develop alternative industries and support services from these four mines alone. In other regions of the world, where there has been a long history of profitable mining, there are only a few examples of sound community infrastructure planning. Most regions have not planned well for closure and have not positioned themselves for industry diversification or transition.

Ultimately, for closure to be adequately addressed by governments in developing or developed countries it must become part of government fiscal planning and mining codes, thereby ensuring that closure is strategically incorporated from the outset for each project. The failure of developed country governments in this regard has no doubt had an impact in developing countries that utilize developed country expertise and legislation to develop their own frameworks for success.

A continued reluctance to address mine closure planning and HIV/AIDS from a national perspective is likely to have serious sustainability ramifications. The need to take time out from the day-to-day activities to plan well for closure is paramount. The industry has provided guidelines to help with this process and there is enough information on closure planning to provide guidance to the industry. Our readiness to contribute to higher level planning, however, will be the key to the economic survival of the industry and the sustainability of the economies in which it operates.

# REFERENCES

1. UNAIDS/WHO. AIDS Epidemic Update, December 2002. UNAIDS/02.58E. 2002. Geneva, Switzerland.
2. Gottlieb, S., UN says up to half the teenagers in Africa will die of AIDS. British Medical Journal 2000;321: 67.
3. Uganda Ministry of Health. 2001 HIV/AIDS Surveillance Report. 2001. Kampala, Uganda, STD/AIDS Control Programme, Ministry of Health.
4. Mbulaiteye, S.M., Mahe, C., Whitworth, J., Ruberantwari, A., Nakiyingi, J.S. and Ojwiya A et al., Declining HIV-1 incidence and associated prevalence over 10 years in a rural population in south-west Uganda: a cohort study. Lancet 2002;360: 41–6.
5. Kilian, A.H., Gregson, S. and Ndyanabangi, B., Reductions in risk behaviour provide the most consistent explanation for declining HIV-1 prevalence in Uganda. AIDS 1999;6: 391–8.
6. Ndoye, P.M., Gueye-Gaye, A., Gueye-Ndiaye, A., Toure-Kane, N.C., Gaye-Diallo, A. and Ndoye, I et al., Reversal trends of HIV1 and HIV2 sero-prevalence among pregnant women, in four urban major areas in Senegal (1989–2001). Abstract (WePeC6097) presented at XIV International AIDS Conference, Barcelona, 7–12 July 2002.
7. Fylkesnes, K., Musonda, R.M., Sichone, M., Ndhlovu, Z., Tembo, F. and Monze, M., Declining HIV prevalence and risk behaviours in Zambia: evidence from surveillance and population-based surveys. AIDS 2001;15: 907–16.
8. Williams, B., Gilgen, D., Campbell, C., Taljaard, D. and MacPhail, C., The natural history of HIV/AIDS in South Africa: a biomedical and social survey in Carltonville. Johannesburg: Council for Scientific and Industrial Research; 2000.

9.  Haour-Knipe, M., Leshabari, M. and Lwihula, G., Interventions for workers away from their families. In: Gibney et al., editor. Preventing HIV in Developing Countries: Biomedical and Behavioural Approaches. New York: Plenum Press; 1999.
10. Meekers, D., Going underground and going after women: trends in sexual risk behaviour among gold miners in South Africa. International Journal of STD & AIDS 2000; 11: 21–6.
11. Ijsselmuiden, C.B., Padayachee, G.N., Mashaba, W., Martiny, O. and Van Staden, H.P., Knowledge, beliefs and practices among black goldminers relating to the transmission of human immunodeficiency virus and other sexually transmitted diseases. South African Medical Journal 1990; 78: 520–3.
12. Campbell, C., Migrancy, masculine identities and AIDS: The psychosocial context of HIV transmission on the South African gold mines. Social Science and Medicine 1995; 45: 273–81.
13. Thomson, M.M. and Najera, R., Travel and the introduction of human immunodeficiency virus type 1 non-B subtype genetic forms into Western countries. Clinical Infectious Diseases 2001; 15: 1732–7.
14. de Graaf, R., Van Zessen, G., Houweling, H., Ligthelm, R.J. and van der Akker, R., Sexual risk of HIV infection among expatriates posted in AIDS endemic areas. AIDS 1997; 11: 1173–81.
15. Kane, F., Alary, M., Ndoye, I., Coll, A.M., M'boup, S. and Gueye, A. et al. Temporary expatriation is related to HIV-1 infection in rural Senegal. AIDS 1993; 7: 1261–5.
16. Bonneux, L., Van der Stuyft, P., Taelman, H., Cornet, P., Goilav, C. and van der Groen, G et al. Risk factors for infection with human immmunodeficiency virus among European expatriates in Africa. British Medical Journal 1988; 297: 581–4.
17. Corbett, E.L., Churchyard, G.J., Clayton, T.C., Williams, B.G., Mulder, D. and Hayes, R.J. et al., HIV infection and silicosis: the impact of two potent risk factors on the incidence of mycobacterial disease in South African miners. AIDS 2000; 14: 2759–68.
18. Corbett, E.L., Churchyard, G.J., Clayton, T., Herselman, P., Williams, B. and Hayes, R. et al., Risk factors for pulmonary mycobacterial disease in South African gold miners. A case-control study. American Journal of Respiratory & Critical Care Medicine 1999; 159: 94–9.
19. Sonnenberg, P., Murray, J., Glynn, J.R., Thomas, R.J., Godfrey-Faussett, P. and Shearer, S., Risk factors for pulmonary disease due to culture-positive M. tuberculosis or nontuberculous mycobacteria in South African gold miners. European Respiratory Journal 2000; 15: 291–6.
20. Churchyard, G.J., Kleinschmidt, I., Corbett, E.L., Mulder, D. and De Kock, K., Mycobacterial disease in South African gold miners in the era of HIV infection. International Journal of Tuberculosis and Lung Disease 1999; 3: 791–8.
21. Clift S, Kanga Z., Ndeki, L., Changalucha, J., Gavyole, A. and Watson-Jones, D et al. Baseline prevalence of HIV infection & other STIs and their associated risk factors in 2 gold mining communities of the Lake Zone of Tanzania – a cross-sectional survey. Abstract presented at XII International Conference on AIDS/STD in Africa, Ouagadougou, 9–13 December 2001.
22. Taljaard, D., Williams, B., Campbell, C., MacPhail, C., Gouws, E. and Van Dam, J. et al., The response of a South African mining community to the epidemic of HIV. Abstract (ThPeD7715) presented at XIV International AIDS conference, Barcelona, 7–12 July 2002.
23. Campbell, C. and Williams, B., Beyond the biomedical and behavioural: towards an integrated approach to HIV prevention in the southern African mining industry. Social Science and Medicine 1999; 48: 1625–39.
24. Campbell, C., Selling sex in the time of AIDS: the psycho-social context of condom use by sex workers on a Southern African mine. Social Science and Medicine 2000; 50: 479–94.
25. Steen, R., Vuylsteke, B., DeCoito, T., Ralepeli, S., Fehler, G. and Conley, J et al. Evidence of declining STD prevalence in a South African mining community following a core-group intervention. Sexually Transmitted Diseases 2000; 27: 1–8.
26. MacLachlan, E.W., Baganizi, E., Bougoudogo, F., Castle, S., Mint-Youbba, Z. and Gorbach, P et al. The feasibility of integrated STI prevalence and behaviour surveys in developing countries. Sexually Transmitted Infections 2002; 78: 187–9.
27. Brugha, R. and Walt, G., A global health fund: a leap of faith. British Medical Journal 2001; 323: 152–4.
28. World Health Organisation. Sources and prices of selected drugs and diagnostics for people living with HIV/AIDS. Geneva, Switzerland. WHO/EDM/PAR/2002.2. 2002. World Health Organisation.
29. World Health Organisation. Scaling up antiretroviral therapy in resource-limited settings: guidelines for a public health approach. 2002. Geneva, Switzerland, World Health Organisation.
30. Jani, I., Tugume, S., Bradley, N., Pitfield, T., Glencross, D. and Barnett, D. et al. New concepts in affordable CD4+ T cell enumeration for resource-poor settings. Abstract presented at XIV International AIDS conference, Barcelona, 7–12 July 2002.

31. Corbett, E.L., Churchyard, G.J., Charalambos, S., Samb, B., Clayton, T. and Grant, A.D., et al. Morbidity and mortality in South African gold miners: impact of untreated disease due to human immunodeficiency virus. Clinical Infectious Diseases 2002; 34: 1251–8.
32. Fantan, T. and Villafana, T., Implementing an HIV and AIDS treatment programme at Debswana diamond company. Abstract (LbPeF9054) presented at XIV International AIDS conference, Barcelona, 7–12 July 2002.
33. UNAIDS/WHO. Epidemiological fact sheets on HIV/AIDS and sexually transmitted infections: United Republic of Tanzania. 2002. Geneva, Switzerland, World Health Organisation.
34. Urassa, M., Isingo, R., Kaatano, G., Mwaluko, G., Ngalula, J. and Zaba, B. et al., Continuing spread of HIV in rural Tanzania: prevalence and incidence trends in Kisesa during 1994–2000. Abstract (TuPpC2058) presented at XIV International AIDS conference, Barcelona, 7–12 July 2002.
35. Mwita, W., White, R., Chilongani, J., Zaba, B., Anemona, A. and Ross, D et al. Population Mobility and the Spread of HIV in Rural Villages in Mwanza Region, Tanzania. Abstract presented at XII International Conference on AIDS/STD in Africa, Ouagadougou, 9–13 December 2001.
36. Auvert, B., Ballard, R., Campbell, C., Carael, M., Carton, M. and Fehler, G et al. HIV infection among youth in a South African mining town is associated with herpes simplex virus-2 seropositivity and sexual behaviour. AIDS 2001; 15: 885–98.
37. Orroth, K.K., Gavyole, A., Todd, J., Mosha, F., Ross, D. and Mwijarubi, E et al. Syndromic treatment of sexually transmitted diseases reduces the proportion of incident HIV infections attributable to these diseases in rural Tanzania. AIDS 2000; 14: 1429–37.
38. UNAIDS/WHO. Epidemiological fact sheets on HIV/AIDS and sexually transmitted infections: Ghana. 2002. Geneva, Switzerland, World Health Organisation.
39. Adje, C., Cheingsong, R., Roels, T.H., Maurice, C., Djomand, G. and Verbiest, W. et al., High prevalence of genotypic and phenotypic HIV-1 drug-resistant among patients receiving antiretroviral therapy in Abidjan, Cote d'Ivoire. Journal of Acquired Immune Deficiency Syndrome 2001; 26: 501–6.

# Subject index

Page numbers in *italics* refers to figure caption, <u>underline</u> refers to lists and **bold** refers to title.